Theory and Practice in Microbial Enhanced Oil Recovery

Theory and Practice in Microbial Enhanced Oil Recovery

KUN SANG LEE
Professor
Hanyang University
South Korea

TAE-HYUK KWON
Associate Professor
Korea Advanced Institute of Science of Technology
South Korea

TAEHYUNG PARK
Postdoctoral Researcher
Korea Advanced Institute of Science and Technology
South Korea

MOON SIK JEONG
Postdoctoral Researcher
Hanyang University
South Korea

Gulf Professional Publishing
An imprint of Elsevier

Gulf Professional Publishing is an imprint of Elsevier
50 Hampshire Street, 5th Floor, Cambridge, MA 02139, United States
The Boulevard, Langford Lane, Kidlington, Oxford, OX5 1GB, United Kingdom

Notices
Knowledge and best practice in this field are constantly changing. As new research and experience broaden
our understanding, changes in research methods, professional practices, or medical treatment may become
necessary.

Practitioners and researchers must always rely on their own experience and knowledge in evaluating and using
any information, methods, compounds, or experiments described herein. In using such information or
methods they should be mindful of their own safety and the safety of others, including parties for whom they
have a professional responsibility.

To the fullest extent of the law, neither the Publisher nor the authors, contributors, or editors, assume any
liability for any injury and/or damage to persons or property as a matter of products liability, negligence
or otherwise, or from any use or operation of any methods, products, instructions, or ideas contained in the
material herein.

Library of Congress Cataloging-in-Publication Data
A catalog record for this book is available from the Library of Congress

British Library Cataloguing-in-Publication Data
A catalogue record for this book is available from the British Library

ISBN: 978-0-12-819983-1

For information on all Gulf Professional Publishing publications visit our website at
https://www.elsevier.com/books-and-journals

Publisher: Brian Romer
Acquisitions Editor: Katie Hammon
Editorial Project Manager: Chris Hockaday
Project Manager: Kiruthika Govindaraju
Cover designer: Matthew Limbert

Typeset by TNQ Technologies

Preface

As oil and gas price fluctuates with global economy, development of cost-effective technologies, which yields the maximum oil recovery, is of main interest in today's petroleum researches. Microbial enhanced oil recovery (MEOR) has a strong potential as low-cost techniques with less impact to environment, in which different microorganisms and their metabolic products are implemented to increase production rate and efficiency in hydrocarbon reservoirs. Despite drastic advantages of MEOR technology, the technique still remains poorly supported due to lack of knowledge on microbial activities and complexity of the associated processes. Some of the MEOR-related strategies have demonstrated their feasibility on a mass scale through both lab and field trials; however, much work still remains to implement MEOR into oil industry practices.

The authors review and summarize engineering fundamentals of MEOR with emphasis on microbial mechanisms and reservoir-scale modeling. This book provides comprehensive description on fundamental and critical aspects of MEOR and establishes the credibility of field applications. Newest experimental measurements and observations on MEOR-related mechanisms as well as recent development in numerical assessment of MEOR applications can be of interest for the main audience. The main audience would be the enhanced oil recovery (EOR) (R&D, reservoir and operational) community, potentially geologists and bio-/microbial geologists, and reservoir modelers. This book also updates the current progress in research and practical applications related to MEOR, complementary to several literature and books published in the past decades.

Chapter 1 serves as an introduction to the overall strategy of the MEOR method and the subsurface environment affecting the process efficiency. It also presents the screening criteria of the reservoir for the successful applications of MEOR. Chapter 2 reviews the microbial communities in deep subsurface associated with oil reservoirs and addresses various MEOR-related microbial products and their characteristics. Chapter 3 provides fundamental pore-scale mechanisms of microbial activities and their effect on oil production in porous media at a core scale. This chapter also complies extensive laboratory experiment data that are relevant to MEOR processes. Chapter 4 describes numerical simulation of MEOR processes including selective plugging, microbial surfactant generation, and other mechanisms of bacteria. Chapter 5 presents considerations and practical examples on the field applications of MEOR in terms of lithology, type of application, recovery mechanisms, and used microorganisms.

This book would never have been published without the able assistance of Elsevier staffs for their patience and excellent editing job. We shall appreciate any comments and suggestions.

Kun Sang Lee
Seoul, Korea

Nomenclatures

A

A	constant in Eqs 4.65 and 4.85
A_0	initial surface area
A_{cs}	cross-sectional area
A_H	empirical parameter in Eq 4.93
A_{ave}	average cross-sectional area along a breakthrough channel
A_{cum}	cumulative wall surface area along a breakthrough channel
A_j	constant in Eqs 4.23 and 4.24
A_{j+1}	constant in Eq 4.24
a	activity of the active enzyme
a_i	inner radius
a_{ini}	initial radius
a_r	radius of the water-filled pore space (reduced radius)
a_w	pore radius with biopolymer saturation S_{BP}
a_{wo}	pore radius with no biopolymer
ad	adsorbed moles of bioproducts
ad_{max}	maximum adsorption capacity

B

B	constant in Eqs 4.16, 4.65, and 4.84
B_H	empirical parameter in Eq 4.93
b	endogenous decay coefficient
b_1	constant in Eq 4.31

C

C	constant in Eq 4.65
C_g	empirical geometrical constant
C_H	empirical constant
C_{ij}	volume fraction of component i in phase j
C_p	parameter used to fit the laboratory measurements
CK	Carman-Kozeny constant
c_1	constant in Eq 4.31
c_2	regression coefficient
c_3	regression coefficient
c_{att}	attached cell concentration
c_{BAP}	BAP concentrations
c_{NaCl}	NaCl concentration
$c_{NaCl_{max}}$	maximum NaCl concentration for bacterial metabolism

$c_{NaCl_{opt}}$	optimum NaCl concentration for bacterial metabolism
c_{poly}	biopolymer concentration
c_{UAP}	UAP concentrations
c_s	concentration of the insoluble products

D

d_p	plate thickness
D	decimal reduction time
D_0	initial diameter of pore throat
D_1	changed diameter of pore throat
D_{bond}	diameter of network bond
D_i	decimal time at ith pressure
D_E	equatorial diameter in the drop

E

E	enzyme
E_a	activation energy
E_{tot}	total number of enzymes

F

F	constant in Eq 4.21
F_0	formation factor in a fully water-saturated rock
F_1	constant in Eq 4.22
F_2	constant in Eq 4.22
F_c	capillary force
F_e	electrical formation factor in Eq 3.25
F_v	viscous force
F_{max}	force required to raise the ring from the liquid's surface
f	flow-efficiency coefficient
f_d	fraction of the active biomass which is biodegradable
f_c	correction factor

G

G	free energy
g_G	geothermal gradient (°F/100 ft)
g	gravitational force

H

H	fitting parameter
H_c	shape-dependent constant
h	height of raised water

I

I_S	salinity inhibition constant
I_S^*	constant depending on the cultures

K

K	concentration giving one-half the maximum rate
K_0	baseline permeability
K_a	absolute permeability
K_A	half-maximum rate concentration of the second substrate
K_{BAP}	half-maximum rate concentrations for BAP
K_M	Michaelis-Menten constant
K_{UAP}	half-maximum rate concentrations for UAP
K_i	substrate inhibition constant
$K(S_{BP})$	permeability with biopolymer saturation
$K_{p/s}$	saturation constant for metabolite to consumption of substrate S
K_r	relative permeability reduction ratio
K_N	normalized permeability
k	first-order rate constant
k_1	dissociation constants for E
k_2	dissociation constants for E^-
k_3	rate constant for association ($E + S \rightarrow ES$)
k_3^θ	rate constant when $[Na^+] = 1$ M
k_4	rate constant for dissociation ($E + S \leftarrow ES$)
k_4^θ	rate constant when $[Na^+] = 1$ M
k_5	rate constant for chemistry step ($ES \rightarrow EP$)
k_e	effective permeability
k_f	changed permeability
k_{mul}	permeability multiplier factor
k_o	original permeability
k_{rp}	relative permeability of phase p
k_{rp}^e	end-point relative permeability of phase p at its maximum saturation

L

L	ratio of the specific growth rate at the beginning of the deceleration state to previous exponential stage
L_a	length of actual flow
L_c	length of core
L_s	length of sample
l	length of pore throat
$l_{network}$	length of network

M

m	microemulsion phase or component
m_g	empirical tortuosity factor
m_{cell}	mass of bacterial cell

N

N	number of surviving microbes after pressure treatment
N_a	number of capillaries per unit cross-sectional area
N_c	capillary number
$(N_c)_c$	critical capillary number
$(N_c)_{max}$	maximum desaturation capillary number
N_o	initial number of microbes
N_t	number of capillary tubes
n	constant in Eqs 3.34, 4.42, 4.43, and 4.85
n_A	Archie saturation exponent
n'	constant in Eq 4.47
n_p	exponent of phase p

O

o	oil phase or component

P

P	product
P_a	pressure (absolute)
pH_{max}	maximum pH for microbial growth
pH_{min}	minimum pH for microbial growth
p_c	capillary pressure
p_d	displacement pressure
p_i	pressure at ith step
Δp_{max}	maximum pressure drop

Q

Q	flow rates for influent and effluent
$\hat{q}$	maximum specific rate of substrate utilization
$\hat{q}_{BAP}$	maximum specific rates of BAP degradations
$\hat{q}_{UAP}$	maximum specific rates of UAP degradations

R

R	gas constant
R_{os}	solubilization ratio $\left(\frac{c_{osurf}}{c_{surf}}\right)$
R_{ws}	solubilization ratio $\left(\frac{c_{usurf}}{c_{surf}}\right)$
r	growth rate
r_c	radius of capillary tube
$rrft$	constant for residual resistance factor

r_{BAP}	production rate of BAP
r_{UAP}	production rate of UAP
$r_{\deg-BAP}$	degradation rates of BAP
r_{BAP}	degradation rate of UAP
r_{lim}	value of r at this physical limit
r_{max}	maximum specific rate of metabolite production
r_{net}	net growth rate for bacteria
r_p	production rate
r_r	average of the inner and outer radii of the ring
r_{ut}	rate of substrate utilization

S

S	concentration of the rate-limiting substrate
S_f	fluid saturation
$S_{hydrate}$	hydrate saturation
$\overline{S}_p$	normalized saturation
S^*	critical concentration of the substrate for metabolic production
S'	concentration of the second substrate
S_g	gas saturation
S_o	oil saturation
S_p	saturation of phase p
S_{sv}	specific surface area per unit volume
S_w	water saturation
S_{pr}	irreducible or residual saturation of phase p
$S_{p'r}$	residual saturation of the other phase
S_{BP}	biopolymer saturation
$\overline{S}_{pr}$	normalized residual saturation
$surf$	surfactant

T

T	temperature
T_0	conceptual temperature of no metabolic significance
T_f	formation temperature (°F)
T_{max}	maximum temperature for microbial growth
T_{min}	minimum temperature for microbial growth
T_{opt}	optimum temperature for bacterial growth
T_s	mean surface temperature (°F)
t	time
t_a	time at the transition from exponential to the deceleration stage
t_d	hydraulic detention time

U

u	Darcy velocity of the displacing fluid
$\frac{u}{\varphi}$	interstitial velocity

V

V	volume of chemostat
V_0	initial volume
V_{bulk}	volume of bulk volume of the reservoir rock
V_{fluid}	volume of fluid in the reservoir rock
V_{pore}	volume of entire pores
V_{solid}	volume of the solid grains
$V_{biofilm}$	volume of biofilm
$(V_b)_{p_c}$	fractional bulk volume occupied by the displacing fluid at any capillary pressure
$(V_b)_{p_\infty}$	fractional bulk volume occupied by the displacing fluid at infinite pressure
V_{sm}	surfactant volume in the microemulsion phase
v	pore flow velocity of the displacing fluid

W

w	water phase or component
w_p	width of plate

X

X_0	initial biomass
X_A	bacterial concentration at time t_a
X_a	concentration of active biomass
X_b	concentration of bioproducts
X_i	inert biomass concentration
X_{max}	maximum possible microbial biomass

Y

Y	yield coefficient

Z

z_p	negative reciprocal slope of the log D vs. p

Greek Symbols

α	linear growth rate
α_{BAP}	BAP formation coefficient
α_{UAP}	UAP formation coefficient
α_d	destruction rate constant
β	constant of proportionality between thermodynamic changes and kinetic changes due to [Na$^+$]

δ	constant in Eq 4.84
ε	constant in Eq 4.85
ϕ	porosity
ϕ_f	changed porosity
ϕ_o	original porosity
φ_o	$\frac{V_{om}}{(V_{wm}+V_{om})}$
φ_w	$\frac{V_{wm}}{(V_{wm}+V_{om})}$
γ	constant in Eq 4.18
γ_{wa}	surface tension between water and air
$\widehat{\mu}$	maximum specific growth rate
$\widehat{\mu}_p$	maximum specific production rate
μ	net specific growth rate
μ_f	visocisty of fluid
$\mu_{displacing}$	displacing fluid viscosity
μ_{dec}	specific growth rate due to decay
μ_{opt}	specific growth rate at the optimum conditions
μ_{syn}	specific growth rate for cell synthesis

ρ	density of fluid
ρ_a	density of air
ρ_w	density of water
$\rho_{biofilm}$	biofilm density
ρ_s	density of insoluble products
σ	interfacial tension
σ_{max}	maximum IFT from experimental measurements
σ_{min}	minimum IFT from experimental measurements
$\sigma(c_{surf})$	IFT at a given surfactant concentration
τ	tortuosity
τ_B	Bingham yield stress
τ_s	shear stress
ω	constant in Eq 4.25
ω_v	angular velocity
θ	contact angle
θ_{sf}	shape factor in Eq 3.23

Contents

1 **Introduction**, *1*

2 **Microbiology and Microbial Products for Enhanced Oil Recovery**, *27*

3 **Theory and Experiments**, *67*

4 **Modeling and Simulation**, *109*

5 **Field Applications**, *169*

INDEX, *195*

CHAPTER 1

Introduction

1.1 MICROBIAL PROCESSES FOR OIL RECOVERY

1.1.1 Strategy Overview

The microbial enhanced oil recovery (MEOR) is not a completely new concept. In 1926, Beckman introduced that microorganisms can be used to release oil from porous media (Lazar et al., 2007). Since then, Zobell (1947) has utilized sulfate-reducing bacteria in enhanced oil recovery. Recent MEOR researches have predominantly focused on ex situ and in situ methods to transport the metabolites into oil wells as well as on the fundamental challenges of oil production, which include the immiscibility of oil in water, the high viscosity of the oil, and the size of oil components (Patel et al., 2015).

The ex situ method is similar to the chemical enhanced oil recovery (CEOR) approach. In this method, the desired bioproducts are produced externally and then injected into the wellhead to improve oil recovery. Because the specific composition, compounds, and products can be selected and injected, such a method is attractive in that direct control is possible by reservoir operators. The microbes used in the ex situ MEOR processes are either grown or engineered in the laboratories to improve sweep and/or displacement efficiency. Target bioproducts such as biosurfactants can be extracted from these microorganisms and mixed with water before injection, sometimes in combination with synthetic chemicals. In other approaches, the isolated bacteria may be injected into the well, with the hope that they will generate the desired metabolites within the reservoir (Patel et al., 2015).

Although the ex situ method seems to be quite feasible, numerous concerns exist. First of all, the cost of producing ex situ bioproducts is significantly high. While using the crude forms of bioproducts can greatly reduce the price, the high cost of ex situ process still remains a large concern for the advance of petroleum industry (Pornsunthorntawee et al., 2008(b); Zheng et al., 2012). Furthermore, microbes, modified in laboratory and directly injected, are expected to outcompete the reservoir indigenous microbes already adapted to the harsh environments. However, this expectation is not the general case. Since the ex situ MEOR process faces many problems, it must overcome these hurdles to establish itself as a widespread industry practice (Patel et al., 2015).

In contrast with the ex situ process, the in situ process stimulates the indigenous microorganisms in the reservoir to generate the desired metabolites. While the ex situ process yields predictable results with controlled laboratory settings, the results of the in situ process have considerable uncertainty depending on the field application. Indigenous bacteria of interest are stimulated with injected substrates to generate and release bioproducts such as biopolymers, biosurfactants, bioacids, and biosolvents (Fig. 1.1). The biofilm production to decrease the pore volume is also applied in MEOR process. Although both ex situ and in situ methods are potential and applicable simultaneously, the available literatures indicate that in situ method is more important technology in the oil industry (Sen, 2008; Bao et al., 2009; Gudiña et al., 2012; Youssef et al., 2013).

In addition to the reservoir environments that may affect bacterial growth such as pH, temperature, and pressure, many other challenges remain research topics for MEOR applications. For example, a unique characteristic of microorganisms that must be considered in MEOR applications is that they can be grown in an anaerobic environment. Most reservoir conditions are oxygen deficient, and injecting oxygen to grow microbes can cause metal corrosion and equipment damage. Injecting oxygen, as an electron acceptor, can also cause imbalances in the microbial environment and lead to target microbes being outcompeted by other indigenous bacteria (Bryant, 1990; Lazar et al., 2007). These reservoir environments and bacterial growth characteristics constitute the basic parameters of MEOR applications. Employed microorganisms and bioproducts must be resistant to the reservoir environments and be activation in those conditions. As each well has its own unique environment, a variety of microbial consortiums and mixed bioproducts must be used for the success of MEOR application (Lal et al., 2005; Wang et al., 2007; Sen, 2008; Darvishi et al., 2011).

Theory and Practice in Microbial Enhanced Oil Recovery. https://doi.org/10.1016/B978-0-12-819983-1.00001-6

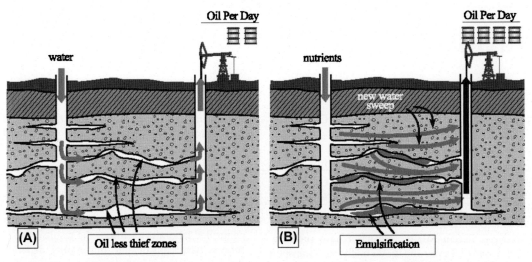

FIG. 1.1 In situ MEOR processes: **(A)** the injected water breakthrough in thief zones, **(B)** injection of nutrients to stimulate indigenous bacteria for producing bioproducts. *MEOR*, microbial enhanced oil recovery. (Credit: from Sen, R., 2008. Biotechnology in petroleum recovery: the microbial EOR. Progress in Energy and Combustion Science 34 (6), 714–724.)

The primary challenges associated with tertiary oil recovery are related to interrelationships between oil and reservoir circumstances. For example, the relatively highly permeable regions can make the thief zones that reduce the sweep efficiency. They make it impossible to recover the oil that has not been in contact with the injected water via traditional waterflooding (Sen, 2008; Ohms et al., 2010; Okeke and Lane, 2012). Water and oil have different viscosities and also have surface tension due to the immiscibility. The combination of them complicates the production mechanism of waterflooding. Most oil reservoirs around the world have very complex biological systems, making laboratory experiments of microbial activity difficult. The microbes injected into the oil reservoir must compete with the indigenous microorganisms (Sen, 2008). Analyzing the data collected from the 322 projects that performed the same MEOR processes, Portwood (1995) provides useful information to analyze the technical and economic effectiveness of the processes and to predict the treatment responses in the given reservoirs. These MEOR applications resulted in a substantial and sustained increase in oil production compared with other operating results in the same reservoir.

In the MEOR processes, microorganisms produce a variety of metabolites that contribute to increasing oil recovery (Sen, 2008). These metabolites affect not only the petrophysical properties including porosity, permeability, and wettability but also chemical properties such as viscosity, interfacial tension (IFT), and so on (Guo et al., 2015). In general, the metabolites related to enhanced oil recovery can be classified into seven major groups as biomass, biopolymers, biosurfactants, biogases, bioacids, biosolvents, and emulsifiers (Patel et al., 2015; Safdel et al., 2017). Biomass can significantly improve oil recovery by bypassing the injected water to residual oil as a result of selectively plugging the porous media. Biopolymers can increase the oil recovery by decreasing reservoir permeability and increasing water viscosity, which improve the mobility ratio. Biosurfactants have a significant effect on wettability alteration by lowering surface and interfacial tensions. Biogases are produced by certain microbial species and contribute to the repressurization of the reservoir to increase oil recovery. Bioacids and solvents dissolve some parts of the reservoir rock, increasing porosity and permeability and consequently reducing entrapped oil. Oil emulsification can be achieved under conditions where emulsifiers produced by a variety of microbes form a stable emulsion with hydrocarbon (commonly oil-in-water) (Patel et al., 2015). Fig. 1.2 shows the state of the oil droplets in a porous rock before and after the emulsification process (Sen, 2008). Table 1.1 shows the list of microbial metabolites and microorganisms along with the production problems, major effects, and best reservoir candidates for the MEOR process.

1.1.2 Selective Plugging

One of the major factors in reducing oil recovery is the high permeability zones of the reservoirs. These zones make it difficult to extract the oil remaining in relatively

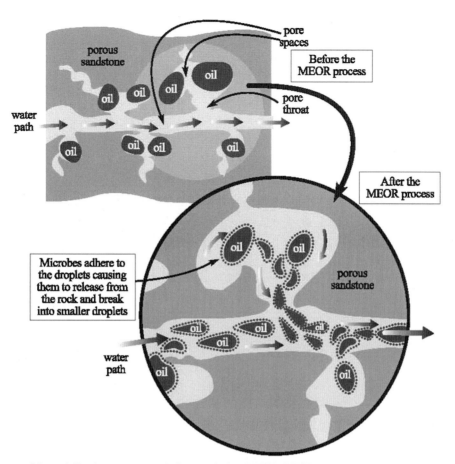

FIG. 1.2 Oil emulsification processes before and after MEOR. *MEOR*, microbial enhanced oil recovery. (Credit: from Sen, R., 2008. Biotechnology in petroleum recovery: the microbial EOR. Progress in Energy and Combustion Science 34 (6), 714–724.)

low permeability zones. In MEOR process, selective plugging diverts the injected water into low permeability areas to produce additional oil (Fig. 1.3). This technique is mainly performed using biomass and biopolymer.

When the indigenous microbes in oil reservoir grow, they occupy the space of porous media, and their surface molecules often allow them to attach to the injected nutrients. As a result, the microorganisms grow themselves in the porous media to biofilms that inhibit the flow of reservoir fluids (Karimi et al., 2012). These microorganisms form colonies and cluster together as groups of biomass, and such clustering has an evolutionary advantage (Xavier and Foster, 2007). Some MEOR studies are interested in reducing oil flow paths using biomass plugging. This method is usually done by stimulating indigenous microbes or injecting selected microbes. The accessible regions of the injected water are increased to

improve the sweep efficiency, which in turn increases the oil recovery. Such a biomass production is not the only way to reduce reservoir permeability.

The targeted growth of specific microorganisms, which cause selective plugging of porous media, is a very important research topic in MEOR. As mentioned earlier, biomass can accumulate in the highly permeable zones and divert injected water toward the remaining oil in the low permeable zones (Satyanarayana et al., 2012). In addition, the biomass can have favorable surface properties for oil production, causing wettability alteration by adsorbing on the rock surface (Karimi et al., 2012). For successful biomass plugging, four criteria were suggested (Jenneman, 1989). The criteria included that the size of the cells must be able to pass through the porous media, suitable nutrients for bacterial growth must be provided, microbes must grow and/or generate

TABLE 1.1
Microbial Metabolites.

Microbial Product Class	Sample Products	Microorganisms	Production Problem	Effect	Type of Formation/Reservoir
Biomass	· Microbial Cells and EPS (mainly Exopolysaccharides)	· Bacillus · Licheniformis · Leuconostoc mesenteroides · Xanthomonas · Campestris	· Poor microscopic displacement efficiency	· Permeability reduction · Selective and nonselective plugging · Emulsification · Wettability alteration of mineral surfaces · Oil viscosity reduction · Oil pour point desulfurization reduction · Oil degradation	· Stratified reservoirs with different permeable zones
Biopolymers	· Xanthan gum · Pullulan · Levan · Curdlan · Dextran · Scleroglucan	· Xanthomonas sp. · Aureobasidium sp. · Bacillus sp. · Alcaligenes sp. · Leuconostoc sp. · Sclerotium sp. · Brevibacterium	· Poor volumetric sweep efficiency	· Permeability reduction in water-swept regions · Injectivity profile and viscosity modification · Mobility control · Selective or nonselective plugging	· Stratified reservoirs with different permeable zones
Biosurfactants	· Emulsan · Alasan · Surfactin · Rhamnolipid · Lichenysin · Glycolipids · Viscosin · Trehaloselipids	· Bacillus sp. · Pseudomonas · Rhodococcus sp. · Acinetobacter · Arthrobacter	· Poor microscopic displacement efficiency	· IFT reduction · Emulsification · Pore-scale displacement improvement · Wettability alteration	· Sandstone or carbonate reservoirs with moderate temperatures ($<50°C$) and relatively light oil (API>25)
Gases	· CO_2 · CH_4 · H_2 · N_2	· Fermentative bacteria · Methanogens · Clostridium · Enterobacter	· Heavy oil	· Reservoir repressurization · Oil swelling · Permeability improvement · IFT and viscosity reduction · Flow characteristics improvement	· Heavy oil-bearing formations (API<15)

Acids	• Propionic acid • Butyric acid	• Fermentative bacteria • *Clostridium* • *Enterobacter* • Mixed acidogens	• Low porosity • Poor drainage • Formation damage	• Dissolve carbonaceous minerals or deposits • Permeability and porosity improvement • Emulsification • CO_2 production due to reaction between acids and carbonate minerals • Oil viscosity reduction	• Carbonate or carbonaceous reservoirs
Solvents	• Alcohols and ketones that are typical cosurfactants • Acetone • Butanol • Propan-2-diol	• Fermentative bacteria • *Clostridium* • *Zymomonas* • *Klebsiella*	• Heavy oil • Poor microscopic displacement efficiency	• Oil viscosity reduction • Wettability alteration • Permeability improvement due to rock dissolution • Heavy, long chain hydrocarbons removal from pore throats • Emulsification • IFT reduction	• Heavy oil-bearing formations (API<15) • Strongly oil-wet, waterflooded reservoirs
Emulsifiers	• Some kinds	• *Acinetobacter* sp. • *Candida* • *Pseudomonas* sp. • *Bacillus* sp.	• Paraffin and oil sludge deposition • Poor microscopic displacement efficiency	• Oil emulsification	• Waxy oil (>C22 alkanes); paraffinic oil and asphaltene-bearing formations
Hydrocarbon metabolism	• Some kinds	• Aerobichydro carbon degraders	• Paraffin deposition • Poor microscopic displacement efficiency	• Paraffin deposits removal • Oil mobility improvement	• Wells with paraffin deposition • Mature waterflooded reservoirs

IFT, interfacial tension.

Safdel, M., Anbaz, M.A., Daryasafar, A., Jamialahmadi, M., 2017. Microbial enhanced oil recovery, a critical review on worldwide implemented field trials in different countries. Renewable and Sustainable Energy Reviews 74, 159—172.

FIG. 1.3 Selective plugging process performed by biomass or biopolymer. (Credit: from Patel, J., Borgohain, S., Kumar, M., Rangarajan, V., Somasundaran, P., Sen, R., 2015. Recent developments in microbial enhanced oil recovery. Renewable and Sustainable Energy Reviews 52, 1539–1558.)

proper metabolites for the selective plugging, and microbial growth rate must not be fast enough to clog the well bore. When a strain known as *Bacillus licheniformis* BNP29 was injected into the low permeability cores, it satisfied all of the criteria and produced the appropriate metabolites for MEOR purposes (Yakimov et al., 1997). This strain has a better selective plugging effect than sulfur-reducing bacteria, which release harmful products to MEOR operations. Therefore, it is important to selectively stimulate microbes causing the selective plugging rather than to stimulate all kinds of microorganisms in the reservoir (Patel et al., 2015).

Biopolymers released by microorganisms are another way of causing selective plugging. Microorganisms growing in the reservoir often generate surface molecules that form biopolymers. Many biopolymers are exopolysaccharides that function to increase cell adhesion and protect the cells from predation and desiccation (Poli et al., 2011). Other polymers, such as xanthan gum, are usually implemented in MEOR process as thickening agents for waterflooding (Sandvik and Maerker, 1977; Yakimov et al., 1997; Sen, 2008). These polymers generated by naturally plugging microbes can be formed ex situ and injected into the well directly. Other methods stimulate the injected or indigenous microbes to generate the desired biopolymers to increase oil recovery.

A variety of microorganisms produce the biopolymers that help improve oil recovery. Representative microbes that produce the polymers are *Xanthomonous,*

Aureobasidium, and *Bacillus* (Sen, 2008). Xanthan gum and curdlan are especially important biopolymers. Xanthan gum is one of the most versatile biopolymers applicable to food, cosmetics, chemicals, and many other industries (Palaniraj and Jayaraman, 2011). Because xanthan gum is resistant to high temperatures and salinities, it can also be used in drilling operations (Palaniraj and Jayaraman, 2011). Therefore, many researches have been carried out to find high producing mutant strains and strains that can grow with cheap substrates (Kurbanoglu and Kurbanoglu, 2007; Rottava et al., 2009; Palaniraj and Jayaraman, 2011). Increasing the viscosity of the injected water, xanthan gum helps to achieve higher oil recovery than traditional waterflooding. Generally, it is added directly to injected water. It is also produced by stimulating the indigenous microbes in the reservoir.

Although not as effective as xanthan gum, curdlan is also a biopolymer used in the MEOR process to increase oil recovery (Sen, 2008). Curdlan can change the rock permeability. In the experimental study, several types of biopolymers and acid-producing bacteria mixtures were injected into a Berea sandstone cores. The mixtures reduced the core permeability from 850 to 2.99 and from 904 to 4.86, respectively, and the residual resistance factors were 334 and 186 (Fink, 2011).

The general application of biopolymers in MEOR is to mix them directly with the injected water rather than to stimulate the injected or indigenous microbes to produce them (Fox et al., 2003). In addition to biopolymers such as xanthan gum and curdlan, other bioproducts can also cause selective plugging in porous media. Bioproducts such as emulsions have a viscosity high enough to aid the plugging effect and form better flow channels for oil recovery (Zheng et al., 2012). Other bioproducts also have various advantages applicable to the MEOR process.

1.1.3 Wettability Alteration

The wettability of reservoir rocks is an important problem of lowering oil recovery. Large amounts of oil are present in fractured carbonate reservoirs (Salehi et al., 2008), and the matrix blocks are mixed or oil wet. Under these circumstances, adsorption of water to the reservoir rocks becomes difficult and lowers oil productivity by lowering sweep and displacement efficiencies (Sen, 2008). The oil recovery can be improved by increasing water wetness of the reservoir rocks (Lazar et al., 2007; Sen, 2008; Armstrong and Wildenschild, 2010). Methods of improving wettability that increase oil recovery include the introduction of surfactants or the stimulation of the microbes that adhere directly to reservoir rocks to increase mixing between water and oil (Fig. 1.4). These techniques increase oil recovery by improving sweep and displacement efficiencies.

Altering the wettability of reservoir rock in MEOR process involves the formation of biofilm and the application of biosurfactants. They can increase the resistance to antimicrobial agents over suspended cells, and they can also adsorb directly to the reservoir rock to alter the surface properties (Karimi et al., 2012). One particular experiment was performed to observe the change in wettability of the glass surface due to the *Enterobacter cloacae* strain. The results showed that many bioproducts, including biosurfactants, affected wettability changes, but biofilm formation had the greatest effect (Karimi et al., 2012). In addition, as the surface roughness and the aging of microbes increased, the wettability of water increased (Karimi et al., 2012).

Among the bioproducts affecting wettability alterations, the biosurfactants are expected to have the greatest effect. According to one laboratory study, when the ion charge and polarity compounds were mixed with the surfactants, the wettability alteration effect was improved (Salehi et al., 2008). In the same study, a potent biosurfactant, surfactin, was found to be more effective than synthetic surfactants (Salehi et al., 2008). Thus, the generation of surface-altering products by microbes contributes to enhanced oil recovery by increasing the wetness of reservoir rock to water. Whether the production of such bioproducts occurs in situ or ex situ, improving wettability of the rock surface is a practical way to increase oil recovery in the MEOR process (Patel et al., 2015).

1.1.4 Surface Tension Alteration

Biosurfactants are of the greatest interest among all materials used in the MEOR process. Biosurfactants,

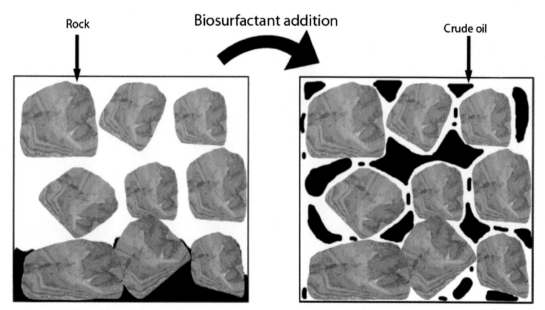

FIG. 1.4 Wettability alteration of reservoir rock by biosurfactant. (Credit: from Patel, J., Borgohain, S., Kumar, M., Rangarajan, V., Somasundaran, P., Sen, R., 2015. Recent developments in microbial enhanced oil recovery. Renewable and Sustainable Energy Reviews 52, 1539–1558.)

amphipathic molecules, are biodegradable, temperature tolerant, and pH hardy and can be an alternative to conventional surfactants used for oil recovery. Biosurfactants, such as conventional surfactants, reduce surface, and interfacial tensions to increase oil recovery (Pornsunthorntawee et al., 2008(a); Sen, 2008; Gudiña et al., 2012). Although the cost of biosurfactants is a problem, the use of crude biosurfactants instead of expensive biosurfactants derived for medical purposes makes it possible to compete with conventional surfactants (Pornsunthorntawee et al., 2008(b); Mukherjee and Das, 2010; Zheng et al., 2012). Biosurfactants released by microbes alter the surface tension, adsorb onto immiscible interfaces, emulsify crude oil, and increase the mobility of bacterial cells (Lazar et al., 2007; Sen, 2008; Yan et al., 2012). In fact, biosurfactants are known to help the mass motility of microbial colonies across different local environmental regions (Déziel et al., 2003). Biosurfactants have similar performance at low concentrations compared with conventional chemical surfactants (Sen, 2008; Gudiña et al., 2012). When the threshold concentration, also known as critical micelle concentration (CMC), of a given biosurfactant is reached, the emulsified ring structure is transformed into spherical micelles. This is the basic mechanism of the MEOR process using biosurfactants (Fig. 1.5).

For MEOR application, biosurfactants should first be screened, developed, and economically scaled up. The optimal biosurfactants should have strong interfacial activity, a low CMC, wide tolerance to temperature and pH, high solubility, and high emulsion capacities (Walter et al., 2010). The target bioproducts can be formed by ex situ process or by stimulating the injected or indigenous microbes. Sometimes, a various metal ions added to the biosurfactants increase the efficacy of the biosurfactants due to the polar interaction of the ions with the biosurfactants (Thimon et al., 1992).

FIG. 1.5 Oil-washing experiments in which the oil sands were treated with surfactants **(A)** surfactin (LB), **(B)** surfactin (G), **(C)** sodium dodecyl sulfate (SDS), **(D)** myristyltrimethylammonium bromide (TTAB), **(E)** polyethylene glycol monododecyl ether (PGME), **(F)** water. (Credit: from Liu, Q., Lin, J., Wang, W., Huang, H., Li, S., 2015. Production of surfactin isoforms by Bacillus subtilis BS-37 and its applicability to enhanced oil recovery under laboratory conditions. Biochemical Engineering Journal 93, 31–37.)

The biosurfactant process is a promising MEOR technique with oil recovery of higher than 95% in sandpack column experiments (Banat, 1993).

However, IFT reduction by biosurfactants in MEOR process is not the only factor that improves the oil recovery, and further research must be done (Zheng et al., 2012). Biosurfactants can also increase oil recovery by altering wettability, aiding in the degrading of long alkyl chains, and cleaning up contaminated soil (Pornsunthorntawee et al., 2008(a); Sen, 2008; Gudiña et al., 2012; Yan et al., 2012). This shows that many MEOR strategies are interconnected due to the multifaceted nature of bioproducts properties. Because many factors can influence each other, the cascade effect caused by applying biosurfactants should always be considered in the field application of MEOR.

The effectiveness and hardness of biosurfactants have been well documented through experimental studies. For example, Gudiña et al. (2012) isolated various *Bacillus* strains and utilized the bacteria to generate biosurfactants. In this study, biosurfactants with bacteria lowered the surface tension of water from 72 to 30 mN/m at the temperatures of 40°C and 121°C when the biosurfactants were used alone (Gudiña et al., 2012). These results show that biosurfactants have high potential for single use and can be mixed directly with injected water, after considering the limitations of temperature and salinity. Moreover, biosurfactants often show better performance than synthetic surfactants (Pornsunthorntawee et al., 2008(a)). Depending on the type of biosurfactants, they have an advantage in production time or surface activity (Fig. 1.6). Therefore, it is important for MEOR success to understand the individual characteristics and apply the proper biosurfactants.

There are a variety of biosurfactants with their own characteristics. The major biosurfactants groups are classified as glycolipids, fatty acid biosurfactants, lipopeptides, emulsifying protein, and particulate biosurfactants (Mukherjee and Das, 2010). Among these, rhamnolipids (which is a type of glycolipid) and lipopeptides are the most frequently used biosurfactants in MEOR process because they can lower the interfacial tension between hydrocarbon and aqueous phases to below 0.1 mN/m (Youssef et al., 2009).

1.1.5 Bioacids/Solvents/Gases

Microorganisms can produce a variety of acids, solvents, and gases applicable to the MEOR. The extracellular acids produced by microbes dissolve some parts of the carbonate rock and help to increase oil recovery (Lazar et al., 2007; Sen, 2008). The process using

microbial acid dissolves the rocks to make it easier to extract the trapped oil. In particular, acetic and propionic acids are most commonly used in this respect (Sen, 2008). Bacteria such as *Clostridium* release acetate and butyrate that lead to bioacids, and *Bacillus* strains, known to produce biosurfactants, can also generate bioacids (Harner et al., 2011).

Solvents such as acetone and ethanol can also dissolve the carbonate rocks (Bryant, 1987). Thus, microbial acids and solvents change the porosity and permeability of the reservoir to allow the production of trapped oil. Biosolvents are generated by microbes such as *Zymomonas mobilis*, *Clostridium acetobutylicum*, and *Clostridium pasteurianum* and generally produced by stimulating the indigenous or injected microbes (Bryant, 1990). Although widespread production of biosolvents such as ethanol is an emerging technology (Pinilla et al., 2011), the amount does not enough to be directly injected into the reservoir.

By applying additional pressure to the reservoir, the remaining oil can be forced out through a similar mechanism with primary recovery. To induce such a phenomenon, gas-producing reservoir microbes are stimulated. Microbes ferment carbohydrates to produce gases such as methane, carbon dioxide, and hydrogen (Sen, 2008). It is also possible to generate microbial acids and solvents through the same fermentation process. In addition to the pressure increasing effect, the gases can dissolve into the reservoir oil and reduces its viscosity, which has a positive effect on oil production (Lazar et al., 2007; Sen, 2008).

Field applications for bioacids, biosolvents, and biogases are still lacking. Compared with selective plugging and biosurfacting methods, the MEOR processes using these three microbial products do not seem to be effective. While it is possible to stimulate microorganisms that produce these bioproducts to increase the efficiency of the MEOR, it is difficult to apply this technique just for the purpose of increasing oil recovery. Other methods remain more effective, controllable, and efficient (Patel et al., 2015).

1.1.6 Degradation/CleanUp

Many microorganisms in the oil reservoirs consume hydrocarbons and use them as nutrients. To do this, they have to break the long alkyl chains in heavy oil (Fig. 1.7). By degrading these chains, microbes increase the amount of light oil in the reservoirs (Sen, 2008; Gudiña et al., 2012). The light oils are more advantageous for production due to their low viscosity. The oil degrading ability of microorganisms is noted as one of the most attractive and practical techniques in

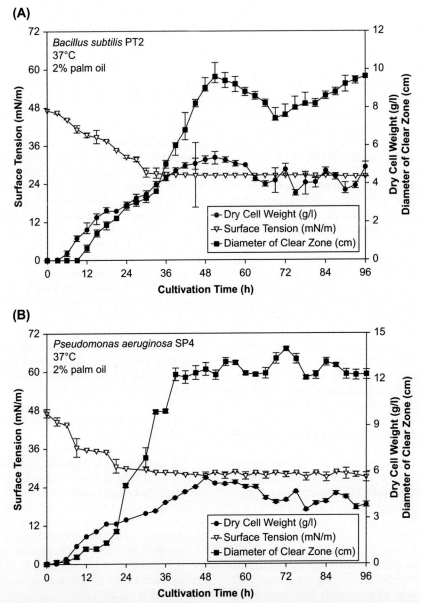

FIG. 1.6 Surface tension and cell concentrations depending on time: **(A)** *Bacillus subtilis* PT2 and **(B)** *Pseudomonas aeruginosa* SP4. (Credit: from Pornsunthorntawee, O., Arttaweeporn, N., Paisanjit, S., Somboonthanate, P., Abe, M., Rujiravanit, R., Chavadej, S., 2008a. Isolation and comparison of biosurfactants produced by Bacillus subtilis PT2 and Pseudomonas aeruginosa SP4 for microbial surfactant-enhanced oil recovery. Biochemical Engineering Journal 42 (2), 172–179.)

MEOR processes. The lighter oil is obtained by stimulating in situ or injecting ex situ strain of hydrocarbon-degrading bacteria. These bacteria can get closer to rock surface and change the wettability (Kowalewski et al., 2006). Furthermore, the experiments confirmed that certain strains of *Bacillus subtilis* can degrade long alkyl chains in 40°C environment, which lowered the viscosity of the oil and increased the oil recovery (Gudiña et al., 2012). Therefore, microbes capable of degrading long alkyl chains

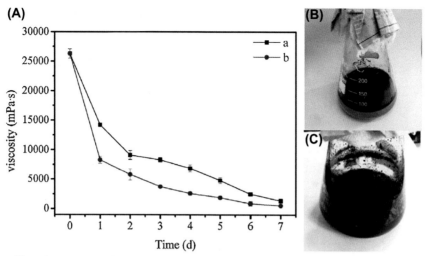

FIG. 1.7 Viscosity reduction of the crude oil after degradation by *Pseudomonas* sp.: (**A**) Time course profiles of (a) cultivation without addition of waste cooking oil (WCO) and (b) cultivation with addition of WCO. (**B**) Surface view of crude oil before degradation. (**C**) Surface view of crude oil after degradation. (Credit: from Lan, G., Fan, Q., Liu, Y., Liu, Y., Liu, Y., Yin, X., Luo, M., 2015. Effects of the addition of waste cooking oil on heavy crude oil biodegradation and microbial enhanced oil recovery using Pseudomonas sp. SWP-4. Biochemical Engineering Journal 103, 219–226.)

and producing biosurfactants at the same time have a very large potential in the MEOR (Gudiña et al., 2012).

The use of these bacteria is a promising technology that is attracting attention in terms of not only oil productivity but also many other applications. Oily sludge remaining after drilling operations has a risk that can cause soil, air, and groundwater contamination (Yan et al., 2012). The oil-degrading bacteria degrade the oil in these contaminated areas and remediate the environment problems (Yan et al., 2012). While other existing strategies may generate their toxic by-products, the use of suitable microbes is much more environmentally friendly (Yan et al., 2012). The use of rhamnolipid is one successful case for bioremediation (Yan et al., 2012). According to another study, suitable biosurfactants for using in oil-contaminated soil remediation have the characteristics such as low CMCs, high miscibility with oil and soil (Fig. 1.8), and low molecular weights (Urum and Pekdemir, 2004). Furthermore, these microbes can help keep industrial equipment available. The microbes are used to keep tanks and storage containers clean, and they can also be injected into oil wells to keep them clean and eliminate plugging (Yakimov et al., 1997; Yan et al., 2012; Zheng et al., 2012). While some solutions use harsh chemicals, the use of microbes, which can easily break down these alkyl chains (Fig. 1.9), can be a practical alternative. In one study, microbial consortiums designed to degrade

paraffin precipitates were used successfully (Lazar et al., 1999). Therefore, bacteria degrading long alkyl chains provide numerous opportunities for the petroleum industry and increase the chances of MEOR success.

1.2 SUBSURFACE ENVIRONMENT AND SCREENING CRITERIA

1.2.1 Lithology

The various rocks and minerals compose the reservoir formation. The sedimentary rocks are known as common reservoir rocks because they have higher porosity than most igneous and metamorphic rocks, which may also store the oil (Amyx et al., 1960). The major sedimentary rocks in reservoir systems are classified as sandstone and carbonate (limestone and dolomite).

Although silicates and carbonates have little effect on microbial activity, the adsorption capacity of clays and other minerals within porous media can interfere in bacterial metabolism. The surface charge of rocks adsorbs the bacteria and disturbs the bacterial migration under the proper pH and ionic strength. Montmorillonites show the largest surface charge, whereas kaolinites are least surface charge. Illites exhibit intermediate surface charge. Clay swelling can also arise with water adsorption, which restricts the bacterial transport through the porous media. The presence of salts

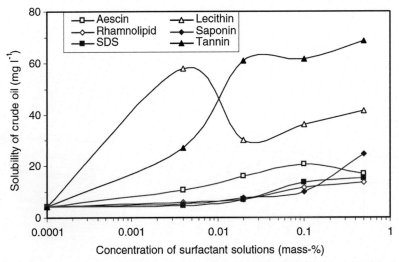

FIG. 1.8 Solubility of crude oil to various surfactant solutions. (Credit: from Urum, K., Pekdemir, T., 2004. Evaluation of biosurfactants for crude oil contaminated soil washing. Chemosphere 57 (9), 1139–1150.)

(NaCl, KCl, CaCl$_2$) in reservoir brines alleviates the clay swelling. Clays also can interfere with the gas diffusion and nutrient transport, which are essential for microbial activity (Jenneman, 1989).

The massive surface area of porous media within reservoir (a cubic foot of rock can possess a surface area same as several football fields) can also affect bacterial growth and metabolism. The surface can enrich the nutrients, which allows the bacteria to grow at low nutrient concentration (Zobell and Grant, 1943). The attachment and colonization of microbes at rock surfaces are able to generate biofilms, which can prevent the liquid flow through the reservoir. Increasing the surface area improves the sporulation ability and manganese oxidization ability of a marine *Bacillus* (Kepkay and Nealson, 1982). Speier and Malek (1982) have shown that the cytotoxic effects occur at rock surfaces. The silica surface can attach cationic chemicals, which would render the surface toxic to contact bacteria (Fig. 1.10). However, it is not clear if the cytotoxicity is due to an increase in effective concentration of the chemical or due to an activation imposed by the surface.

In MEOR operations, the transport of bacteria cells and nutrients into the reservoir is necessary. The retention of cells and nutrients by the effect of rocks and clays can occur. Jang et al. (1983) have shown that a *Clostridium* spores and synthetic microspheres (1.0 μm in diameter) are less adsorbed when they flow into oil-saturated sandstone. When the vegetative cells of several bacterial types are injected into brine-saturated Cleveland sandstone, they are strongly attached and subsequently generated a filter cake at the inlet surface of the core. Jenneman et al. (1983, 1984) also observe the phenomena when a *Pseudomonas* isolate is injected into Berea sandstone. The nutrients such as proteins and phosphates are strongly adsorbed in Berea sandstone, whereas glucose and ammonia nitrogen are easily transported through the sandstone cores within the first several pore volume injections (Jenneman et al., 1984).

1.2.2 Porosity and Permeability

The pore size of rock matrix is very important consideration in bacterial transportation because various morphologies (e.g., rods, cocci, curved rods, tetrads, chains, etc.) of bacteria have dimensions of 0.5–10.0 μm length and 0.5–2.0 μm width. It means that the bacteria transportation in porous media is severely restricted when the pore size is less than 0.5 μm. The pore size must be larger than twice cell diameter of cocci or short bacilli to transport through rock matrix (Updegraff, 1983). Median pore diameter distributions of various Berea sandstones (high permeability: 278–400 md; medium permeability: 130–162 md; and low permeability: 17.7–48.3 md) are 5.5–6.0 μm for the high-permeability cores, 4.5–5.0 μm for the medium-permeability cores, and 3.5–4.0 μm for the low-permeability cores. However, the greatest frequency diameter of each permeability class is smaller than 0.5 μm (Kalish et al., 1964). According to Forbes (1980), average pore throat diameters for reservoir permeability of 100 and 400 md are 7 and

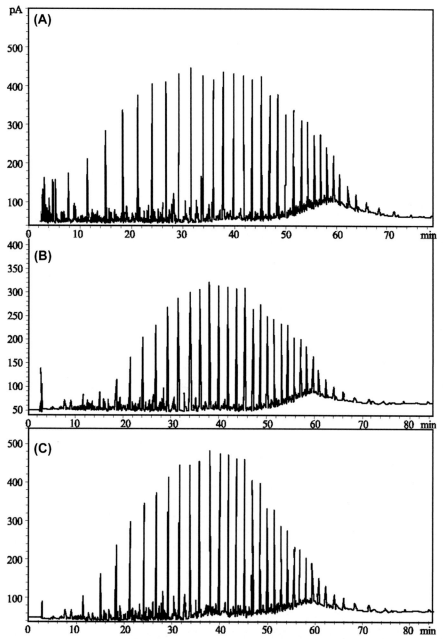

FIG. 1.9 Analysis of alkane compositions after aerobic and anaerobic treatments of *Rhodococcus ruber* Z25:
(A) negative control, **(B)** aerobic biodegradation, **(C)** anaerobic biodegradation. (Credit: from Zheng, C., Yu, L.,
Huang, L., Xiu, J., Huang, Z., 2012. Investigation of a hydrocarbon-degrading strain, Rhodococcus ruber Z25,
for the potential of microbial enhanced oil recovery. Journal of Petroleum Science and Engineering 81, 49–56.)

15 μm, respectively. Although bacterial transport occurs
in low permeability (less than 75 md) core experiments,
permeability of 75–100 md is considered as practical
limitation for effective bacterial transport (Fekete,
1959; Kalish et al., 1964). Conversely, there are some
opinions that microbial penetration and plugging are
independent of permeability (Fekete, 1959; Raleigh,
1962). Raleigh (1962) explained that microbial

(A)

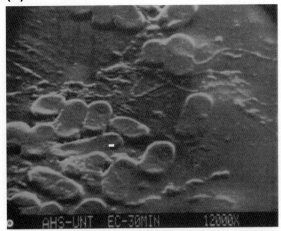

(B)

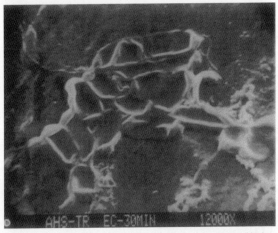

FIG. 1.10 Microscopic images of **(A)** *Escherichia coli* being placed on untreated cellulosic fabric without cell wall destruction and **(B)** cell wall destructions on cellulosic fabric treated with $(MeO)_3Si(CH_2)_3NMe_2C_{18}H_{37}Cl$. (Credit: from Speier, J.L., Malek, J.R., 1982. Destruction of microorganisms by contact with solid surfaces. Journal of Colloid and Interface Science 89 (1), 68–76.)

transport under nongrowth circumstances could be directly related to the pore geometry factor G. The pore geometry factor represents interconnected pore volume and pore distribution, which is calculated from capillary pressure and fractional porosity as follows (Raleigh and Flock, 1965):

$$\frac{(V_b)_{p_c}}{(V_b)_{p_\infty}} = e^{\frac{-G}{\log(p_c/p_d)}} \qquad (1.1)$$

where $(V_b)_{p_c}$ is fractional bulk volume occupied by the displacing fluid at any capillary pressure, $(V_b)_{p_\infty}$ is fractional bulk volume occupied by the displacing fluid at infinite pressure, p_c is capillary pressure, and p_d is displacement pressure. Therefore, the correlation between microbial penetration ability and permeability is still not clear. However, permeability and porosity are the easiest properties to measure and the most important petrophysical parameters to explain the microbial penetration through the reservoir.

The permeability, porosity, and pore size can affect microbial growth and metabolism as well as bacterial transport. Zvyagintsev (1970) has shown that the pore size can contribute to growth rate and cell size of bacteria using rectangular cross-section capillaries. The populations of *Staphylococcus aureus* and *Saccharomyces vini* grown in the 400×150 μm (thickness × width, ID) capillaries increased from 350 to 3560 and from 400 to 2750 cells in 24 hours, respectively. There were no increases in cell population when replaced with 5×3 μm capillaries. Furthermore, the average size of *Saccharomyces vini* was reduced from 6.11 μm in large capillaries to 4.25 μm in smaller capillaries. Nazarenko et al. (1974) showed the correlation between bacterial growth rate and capillary thickness. The growth rates of *Methylosinus trichosporium*, the methane-oxidizing bacteria, did not significantly decrease as the capillary thickness decreased from 500 to 20 μm. The final number of cells in each capillary, however, decreased significantly as capillary thickness decreased. It was also observed that the lag phase of the bacteria was 48 hours in 500~ μm capillary thickness but increased over 72 hours in 20~ μm capillary thickness. Furthermore, the cell generation rate of several bacterial isolates (*Serratia marcescens*, *Staphylococcus aureus*, B. subtilis, and *Rhodotorula glutinis*) significantly decreased as decreasing capillary size from 500×250 μm to 8×4 μm (Zvyagintsev and Pitryuk, 1973). These results were observed in capillaries where a continuous flow of nutrients was applied. The population of B. subtilis did not increase in 8×4 μm capillaries both flow and nonflow conditions. Therefore, it is possible to assume that bacterial growth could be affected by the physical state of the growth medium.

1.2.3 Depth/Temperature/Pressure
The depth distribution of the reservoirs is very diverse. Although depth does not affect microbial growth by itself, pressure and temperature, which are influenced by depth, can affect microbial growth (Fig. 1.11) and metabolic activity.

(A)

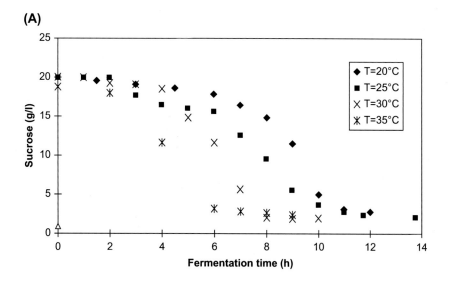

(B)

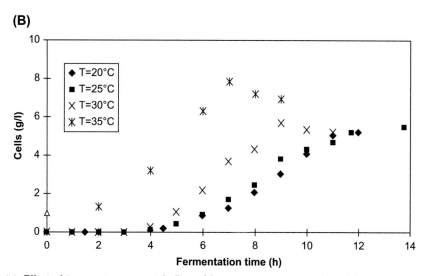

FIG. 1.11 Effect of temperature on metabolites of *Leuconostoc mesenteroides*: **(A)** sucrose consumption and **(B)** cell growth. (Credit: from Santos, M., Teiseira, J., Rodrigues, A., 2000. Production of dextransucrase, dextran and fructose from sucrose using Leuconostoc mesenteroides NRRL B512(f). Biochemical Engineering Journal 4 (3), 177−188.)

Since the temperature will generally increase with depth, the reservoir depth can be used to obtain the reservoir temperature. The downhole temperature profile of reservoir can be calculated by temperature-depth relationship as follows (Hilchie, 1982):

$$T_f = T_s + \frac{g_G D_f}{100} \qquad (1.2)$$

where T_f is formation temperature (°F), T_s is mean surface temperature (°F), g_G is geothermal gradient (°F/ 100 ft), and D_f is the formation depth (ft). Geothermal gradients are commonly 1−2°F/100 ft, and surface temperatures are various according to the geographical location (Jenneman, 1989).

From the studies on thermal springs and fumaroles found in Yellowstone National Park, Iceland, New Zealand, and elsewhere, there is an interesting research relating to bacterial survival at high temperatures (Brock, 1978). Stanier et al. (1970) found that the thermophilic bacteria have their optimal growth temperature between

55°C and 75°C. Thermophilic bacteria and fungi can survive within this temperature range, but the maximal growth for fungi is generally observed below 60°C.

The temperature of oil reservoir is often above 100°C. Due to the high overburden pressure, however, water can exist in the liquid state even under these extreme temperature conditions. Bacteria growth can be found in water at or near boiling point temperature. Some of these bacteria can withstand very low pH (less than 2.0) at these high temperatures (Brock et al., 1972). Ecosystems with high temperature and pressure conditions, such as oil reservoirs, are found in deep-sea hydrothermal springs on the Galapagos Rift (Corliss et al., 1979; Karl et al., 1980; Jannasch and Wirsen, 1981). Baross et al. (1982) confirmed that the temperature of the vents was 350°C at 261.5 atm. Baross and Deming (1983) found that the bacteria presented in these hydrothermal vents were capable of chemolithotrophic growth with a doubling time of 40 minutes at 250°C. The research studies for this ecosystem can be the basis for microbial growth at high temperatures and pressures. However, the optimal growth temperature for most of the bacteria isolated from these hydrothermal vents is below 45°C (Yayanos and Dietz, 1982). Leigh and Jones (1983) reported that the optimal growth temperature of methanogen (anaerobic methane-producing bacteria) isolated from these vents is 86°C. Baross et al. (1982) isolated the mixed cultures of thermophilic bacteria that grow only above 70–75°C from Galapagos vents. Under the assumption that microbial growth at high temperatures is limited by the availability of liquid water rather than by the temperature itself, the upper limit of microbial growth temperature in the oil reservoir may be possible to extend.

The thermophiles that can grow anaerobically are the main concern of the MEOR processes. Clostridia are spore-forming anaerobes that ferment organic matter (sugar, cellulose, etc.) to produce alcohols, organic acids, and gases (CO_2, H_2), and some of them are thermophilic (McClung, 1935; Fontaine et al., 1942; Mercer and Vaughn, 1951). Grula et al. (1983) isolated *Clostridia* species that were able to grow near 45°C and found that they were significantly less able to produce solvents and gases at high concentrations of NaCl (5% w/v). Rozanova and Khudyakova (1974) isolated the anaerobic sulfate-reducing bacteria (*Desulfovibrio thermophilus*) that could grow above 50°C. The bacteria, however, are not suitable for use in MEOR because they can cause the well corrosion and nonselective plugging in formation with generation of water-insoluble sulfides. Methane-producing bacteria such as *Methanobacterium thermoautotrophicum* were also isolated above 50°C (Zeikus and Wolfe, 1972). The bacteria convert carbon dioxide and hydrogen into methane to get the energy needed for metabolism. Zilling et al. (1981) isolated the extremely thermoacidophilic anaerobe, *Thermoproteus tenax*, from solfataras in Iceland. The bacteria can grow at temperatures above 90°C on carbohydrate substrates, and some of them are known to generate hydrogen sulfide.

Pressure, like temperature, also increases as the depth increases. In general, pressure gradients are known to range from 0.43 to 1.0 psi/ft depending on the geographical characteristics. In some areas, however, the gradients of pressure change with depth (Amyx et al., 1960). Extreme pressures are known to affect microbial growth (Fig. 1.12) and metabolic activity, and research studies have been conducted on the effects (Zobell, 1970; Morita, 1972; Marquis and Matsumura, 1978). Marquis (1983) discussed the pressure effects on bacterial metabolism in deep oil formations. Pressures below 100–200 atm have little effect on microbial metabolism, whereas pressures above 500–600 atm significantly inhibit microbial growth for most known bacteria. The ocean floor provides a good research environment for bacterial growth in high pressures. According to Kinne (1972), the average pressure on the ocean floor is 380 atm with a maximum of 1160 atm. There are only a few studies on truly barophilic organisms that require high pressure for growth (Zobell and Morita, 1957; Yayanos et al., 1981). Yayanos et al. (1981) isolated a barophilic organism that could grow at pressures above 1000 atm from Mariana Trench. The growth rate, however, was very slow, taking 33 hours of generation time at 2°C environment. Barophilic bacteria are not essential for most MEOR processes. Barotolerant bacteria, microbes that can grow at high pressure not being the optimal condition, are sufficient.

Marquis and Matsumura (1978) showed that the pressure tolerance of bacteria is dependent on the biophysical conditions such as the energy source present, inorganic salts present, Eh, pH, and temperature. Since these parameters affect the bacterial growth in combination with pressure, it is difficult to accurately determine the impact of individual parameter in most cases. It is particular interest in MEOR processes that the salts (such as NaCl) and divalent cations (such as Mg^{2+} and Ca^{2+}) commonly present in the oil reservoir increase the pressure tolerance of marine organisms (Albright and Henigman, 1971; Marquis and Zobell, 1971; Landau and Pope, 1980). Marquis (1976) found that some mesophiles and a thermophile (*Bacillus stearothermophilus*) have greatest pressure tolerance at a temperature slightly

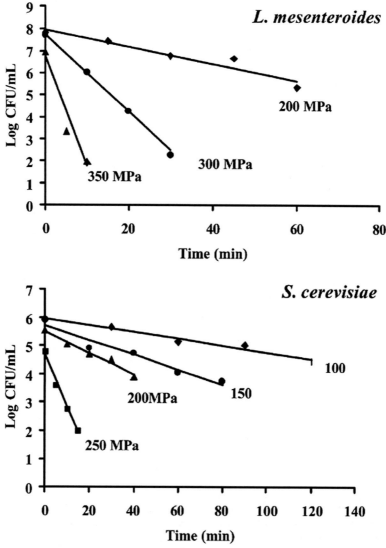

FIG. 1.12 Effect of pressure on destruction of *Leuconostoc mesenteroides* and *Saccharomyces cerevisiae* in single strength orange juice. (Credit: from Basak, S., Ramaswamy, H.S., Piette, J.P.G., 2002. High pressure destruction kinetics of Leuconostoc mesenteroides and Saccharomyces cerevisiae in single strength and concentrated orange juice. Innovative Food Science & Emerging Technologies 3 (3), 223–231.)

above the optimum growth temperature. Bubela (1983) reported that increasing the pressure from atmospheric to 20,000 kPa increases the maximum temperature for bacterial growth. It was also shown that the mean generation time of a rod-shaped anaerobes decreased from 17 hours at 50°C to 12 hours at 65°C with increasing pressure. Furthermore, the shape of the bacteria changed from rod to coccoidal form with increasing pressure. Marquis and Bender (1979) found a variant of *Streptococcus faecalis* that could grow at pressures above 1000 atm by gradually increasing the pressure of the growth medium (normal variant of *S. faecalis* tolerates a pressure of up to 750 atm). The results showed the possibility of adapting bacteria to higher-pressure environments.

1.2.4 Salinity/pH
Sodium chloride accounts for most of the dissolved solids in the reservoir. Salinity is a factor in microbial activity (Fig. 1.13), and tolerance to sodium chloride is one of the most important characteristics in selecting

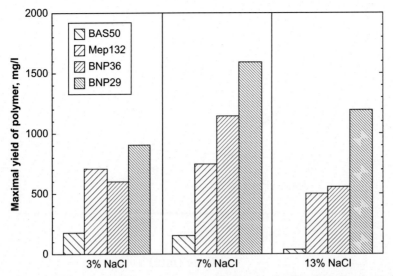

FIG. 1.13 Effect of NaCl concentration on polymer production by *Bacillus licheniformis* strains BAS50, Mep132, BNP36, and BNP29 isolated from Northern German oil reservoirs. (Credit: from Yakimov, M.M., Amro, M.M., Bock, M., Boseker, K., Fredrickson, H.L., Kessel, D.G., Timmis, K.N., 1997. The potential of Bacillus licheniformis strains for in situ enhanced oil recovery. Journal of Petroleum Science and Engineering 18 (1–2), 147–160.)

bacteria for MEOR. Bacteria are known to grow in salty brine (concentration of NaCl is approximately 32% w/v), but little is known about the physiology of halophilic bacteria. It is only known that the minimum NaCl concentration is 10%–12% and complex nutrients are required for the bacterial growth. The role of anaerobic halophiles in MEOR processes is unclear, due to the lack of information about them and need for specific growth conditions. However, several studies found that one of halophilic methanogens (anaerobic methane-producing bacteria) can grow optimally at 15% w/v NaCl brine (Belyaev et al., 1983; Paterek and Smith, 1983; Yu and Hungate, 1983).

Bacteria applicable to MEOR may be the salt-tolerant forms that can grow over a wide range of salt concentrations, and sometimes the moderate halophilic forms are preferred. Kushner (1978) has identified bacteria growing in the range of 0–30% w/v NaCl. Moderate halophiles, which can grow in anaerobic environments above 50°C, are particularly attractive for MEOR applications. Boyer et al. (1973) have found that the *Bacillus* sp. (*Bacillus alcalophilus* subsp. *halodurans*) can grow in brine with a concentration of 7–15% NaCl at a high temperature of 55°C and a high pH of 10.0. When the concentration of NaCl is reduced to 1–5%, this organism can grow at a neutral pH range. It is also possible to isolate *Bacillus* species that can anaerobically grow at high NaCl concentration (>5% w/v), high temperature

(50°C), and the presence of nitrates conditions. According to a number of researchers, the ability to grow bacterial at high salinity has a positive relationship with the ability to grow at high temperatures, which are similar to those of oil reservoirs (Stanley and Morita, 1968; Bilsky and Armstrong, 1973; Novitsky and Kushner, 1975; Keradjopoulos and Halldorf, 1977).

In continuous cultures mixed with moderate and extreme halophiles, moderate halophiles were superior to extreme halophiles at high dilution rates of limiting nutrients and 20–30% w/v NaCl concentration (Rodriguez-Valera et al., 1980). Since the moderate halophiles can survive high salinity and limiting nutrient conditions, application to MEOR processes may be considered. Although these microorganisms can grow in high salinity conditions, it should always be noted that their specific metabolic functions such as gas generation, solvent generation, expolymer generation, and so on may be degraded or eliminated.

Among the biochemical parameters affecting microbial growth and metabolism, pH appears to be the most moderate in the oil reservoirs. In general, the optimum pH for bacterial growth is 4.0–9.0. However, bacteria, which can grow at the pH values as low as 1.0 and as high as 12.0, have been isolated.

The pH can affect bacterial growth and metabolism (Fig. 1.14) not only directly but also indirectly by affecting solubility of toxic materials. Among these

(A)

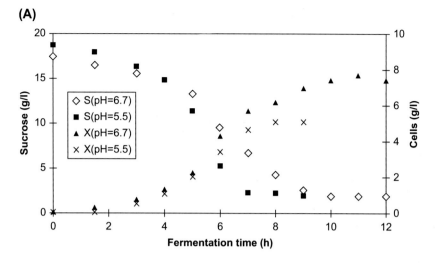

(B)

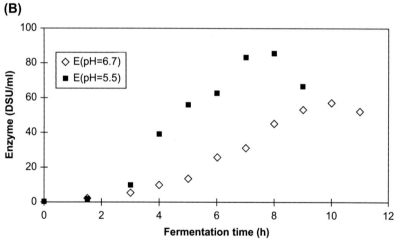

FIG. 1.14 Effect of pH on metabolites of *Leuconostoc Mesenteroides*: **(A)** sucrose consumption and cell growth and **(B)** enzyme production. (Credit: from Santos, M., Teiseira, J., Rodrigues, A., 2000. Production of dextransucrase, dextran and fructose from sucrose using Leuconostoc mesenteroides NRRL B512(f). Biochemical Engineering Journal 4 (3), 177–188.)

phenomena, the solubilization of heavy metals is most important. Heavy metals can be very toxic to microorganisms if they are present over the range required for microbial growth, generally in the range of 10^{-3}–10^{-4} M. Heavy metals in reservoir brine, seawater, and freshwater such as copper, ferric iron, zinc, and so on can exist beyond this concentration range. However, microorganisms are affected in different ways (not as toxicity) and are known to be resistant to most heavy metals at high concentrations (Daniels, 1972).

Although it is not impossible to overcome the problems caused by the effect of pH on microbial growth in oil reservoirs, the control of specific metabolisms required for MEOR processes can be complex by pH

gradients or pH changes in the reservoir. Gottschalk and Bahl (1981) showed the effect of pH on the metabolism of *Clostridium acetobutylicum* to produce butanol. At a pH above 5.0, this anaerobic microorganism lowers the pH below 5.0 by fermenting the nutrient (glucose) to butyrate. Below pH 5.0, butyrate production is lowered, and butanol begins to form.

1.2.5 Screening Criteria

There are several factors that must be considered, which can significantly affect the final results of MEOR such as temperature, pressure, salinity, pH (Fig. 1.15), and economical aspects (Guo et al., 2015). Temperature, which has a significant impact on the survival, growth,

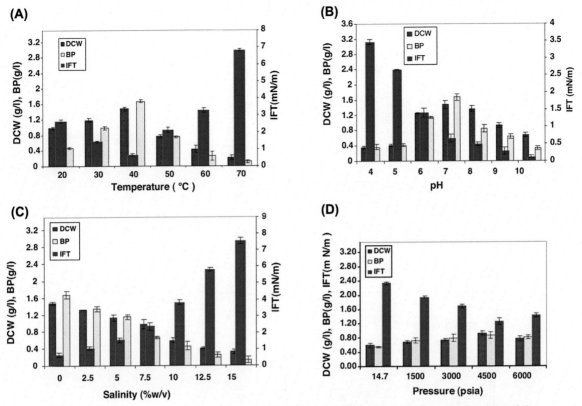

FIG. 1.15 Effects of **(A)** temperature, **(B)** pH, **(C)** salinity, and **(D)** pressure on microbial activities of the consortium ERCPPI-2. (Credit: from Darvishi, P., Ayatollahi, S., Mowla, D., Niazi, A., 2011. Biosurfactant production under extreme environmental conditions by an efficient microbial consortium, ERCPPI-2. Colloids and Surfaces B: Biointerfaces 84 (2), 292−300.)

and metabolic activity of microorganisms, is the most important factor in the MEOR processes. For example, at too low temperatures, the microbial growth and transport processes may be slowed down. On the other hand, at too high temperatures, enzymes and proteins for metabolic activity are possible to be modified or destroyed (Jimoh, 2012).

Pressure is also a key parameter that plays an essential role in the MEOR process. At higher pressure conditions, microbial growth is adversely affected. However, there are also claims that this pressure can be negligible in terms of microbial survival (Bartlett, 2002). The increase in pressure affects the redox potential of gases such as carbon dioxide. In addition, the increasing pressure also affects gas solubility (Saikia et al., 2013).

Salinity, which is affected by both temperature and pressure, is another important parameter that has a significant effect on viscosity reduction (Nmegbu, 2014). Salinity of reservoir brines is a range of 100 mg/L to over 300 g/L (Gran et al., 1992). This salinity depends

on the corresponded depth which dramatically affects the MEOR process (Jimoh, 2012; Nmegbu, 2014). The microorganisms have a reduced metabolic production rate of biosurfactants, gases, alcohols, and acids in high salinity conditions (Bubela, 1989). On the other hand, less ex situ biosurfactant is needed to reduce surface tension because of the CMC reduction as a positive effect of salinity (Safdel et al., 2017).

Permeability, pore size, and pH are also key parameters affecting the success and failure of MEOR processes. Permeability and pore size are major factors affecting bacterial transport. pH can affect surface charge and enzymatic activities (Safdel et al., 2017).

Table 1.2 shows the reservoir screening criteria parameters that are significantly important for bacterial growth, survival, and metabolic activities. The ranges of screening criteria parameters are gathered from different research studies, which are based on various sources including Institute of Reservoir Studies (IRS), US Department of Energy (DOE), and China National Petroleum Company (CNPC).

TABLE 1.2
MEOR Screening Criteria Parameters.

Parameter	IRS[a]	US DOE[a]	CNPC[b]	Reviewed projects[c]	Lazar (1990)[c]	24 Norwegian fields[c]	Bryant and Burchfield[d]	Lake and Walsh[e]
Porosity (%)	–	–	17–25	8–32	≥20	11–35	–	–
Permeability (md)	>50	>100	≥150	0.1–5770	≥150	1–20,000	>100	>150
Depth (ft)	<8,000	<10,000	-	122–2103	-	1300~4208	<3048	<8000
Temperature (°C)	<90	<71	30–60	19–82	≤70	61–155	<71	<140
Salinity (g/L)	<10	<10	≥100	1.4–104	≤150	14–273	<100,000	100,000
pH	6–9	–	–	–	–	–	–	–
Oil gravity (°API)	>20	18–40	–	–	–	–	–	>15
Oil viscosity (cp)	<20	–	30–150	3–50	5–50	0.1–4.83	–	–
Residual oil saturation (%)	>25	>25	–	–	–	–	>25–30	Not critical
Water cut (%)	30–90	–	60–85	–	–	–	–	–

CNPC, China National Petroleum Company; *IRS*, Institute of Reservoir Studies; *MEOR*, microbial enhanced oil recovery; *US DOE*, US Department of Energy.
[a] Patel et al. (2015).
[b] Guo et al. (2015).
[c] Awan et al. (2008).
[d] Bryant and Burchfield (1989).
[e] Lake and Walsh (2008).
Safdel, M., Anbaz, M.A., Daryasafar, A., Jamialahmadi, M., 2017. Microbial enhanced oil recovery, a critical review on worldwide implemented field trials in different countries. Renewable and Sustainable Energy Reviews 74, 159–172.

REFERENCES

Albright, L.J., Henigman, J.F., 1971. Seawater salts—hydrostatic pressure effect upon cell division of several bacteria. Canadian Journal of Microbiology 17, 1246—1248.

Amyx, J.W., Bass, D.M., Whiting, R.L., 1960. Petroleum Reservoir Engineering. McGraw-Hil, New York, USA.

Armstrong, R.T., Wildenschild, D., 2010. Designer-wet micromodels for studying potential changes in wettability during microbial enhanced oil recovery. In: Proceedings of the AGU Fall Meeting, San Francisco, California, USA, 13—17 December, 2010.

Awan, A.R., Teigland, R., Kleppe, J., 2008. A survey of North Sea enhanced-oil-recovery projects initiated during the years 1975 to 2005. SPE Reservoir Evaluation and Engineering 11 (3), 497—512.

Banat, I.M., 1993. The isolation of a thermophilic biosurfactant producing *Bacillus* sp. Biotechnology Letters 15 (6), 591—594.

Bao, M., Kong, X., Jiang, G., Wang, X., Li, X., 2009. Laboratory study on activating indigenous microorganisms to enhance oil recovery in Shengli Oilfield. Journal of Petroleum Science and Engineering 66 (1—2), 42—46.

Baross, J.A., Liley, M.D., Gordon, L.I., 1982. Is the CH_4, H_2 and CO venting from submarine hydrothermal systems produced by thermophilic bacteria? Nature 298, 366—368.

Baross, J.A., Deming, J.W., 1983. Growth of 'black smoker' bacteria at temperatures of at least 250°C. Nature 303, 423—426.

Bartlett, D.H., 2002. Pressure effects on in vivo microbial processes. Biochimica et Biophysica Acta (BBA) - Protein Structure and Molecular Enzymology 1595 (1—2), 367—381.

Basak, S., Ramaswamy, H.S., Piette, J.P.G., 2002. High pressure destruction kinetics of *Leuconostoc mesenteroides* and *Saccharomyces cerevisiae* in single strength and concentrated orange juice. Innovative Food Science & Emerging Technologies 3 (3), 223—231.

Belyaev, S.S., Wolkin, R., Kenealy, W.R., DeNiro, M.J., Epstein, S., Zeikus, J.G., 1983. Methanogenic bacteria from the Bondyuzhskoe Oil Field: general characterization and analysis of stable-carbon isotopic fractionation. Applied and Environmental Microbiology 45 (2), 691—697.

Bilsky, A.Z., Armstrong, J.B., 1973. Osmotic reversal of temperature sensitivity in *Escherichia coli*. Journal of Bacteriology 113 (1), 76—81.

Boyer, E.W., Ingle, M.B., Mercer, G.D., 1973. *Bacillus alcalophilus* subsp. *halodurans* subsp. nov.: an alkaline-amylase-producing, alkalophilic organism. International Journal of Systematic Bacteriology 23 (3), 238—242.

Brock, T.D., 1978. Thermophilic Microorganisms and Life at High Temperatures. Springer-Verlag, New York, USA.

Brock, T.D., Brock, K.M., Belly, R.T., Weiss, R.L., 1972. *Sulfolobus*: a new genus of sulfur-oxidizing bacteria living at low pH and high temperature. Archives of Microbiology 84 (1), 54—68.

Bryant, R.S., 1987. Potential uses of microorganisms in petroleum recovery technology. In: Proceedings of the Oklahoma Academy of Science, Oklahoma, USA.

Bryant, R.S., 1990. Microbial Enhanced Oil Recovery and Compositions Therefore. Patent.

Bryant, R.S., Burchfield, T.E., 1989. Review of microbial technology for improving oil recovery. SPE Reservoir Engineers 4 (2), 151—154.

Bubela, B., 1983. Combined effects of temperature and other environmental stresses on microbiologically enhanced oil recovery. In: Proceedings of the International Conference on Microbial Enhancement of Oil Recovery, Afton, Oklahoma, USA, 16 May, 1982.

Bubela, B., 1989. Geobiology and microbiologically enhanced oil recovery. In: Donaldson, E.C., Chilingarian, G.V., Yen, T.F. (Eds.), Microbial Enhanced Oil Recovery. In: Chilingarian, G.V. (Ed.), Developments in Petroleum Science, vol. 22. Elsevier, Amsterdam, pp. 75—97.

Corliss, J.B., Dymond, J., Gordon, L.I., Edmond, J.M., von Herzen, R.P., Ballard, R.D., Green, K., Williams, D., Bainbridge, A., Crane, K., van Andel, T.H., 1979. Submarine thermal springs on the Galápagos Rift. Science 203 (4385), 1073—1083.

Daniels, S.L., 1972. The adsorption of microorganisms onto surfaces: a review. Developments in Industrial Microbiology 13, 211—252.

Darvishi, P., Ayatollahi, S., Mowla, D., Niazi, A., 2011. Biosurfactant production under extreme environmental conditions by an efficient microbial consortium, ERCPPI-2. Colloids and Surfaces B: Biointerfaces 84 (2), 292—300.

Déziel, E., Lépine, F., Milot, S., Villemur, R., 2003. *rhlA* is required for the production of a novel biosurfactant promoting swarming motility in *Pseudomonas aeruginosa*: 3-(3-hydroxyalkanoyloxy)alkanoic acids (HAAs), the precursors of rhamnolipids. Microbiology 149 (8), 2005—2013.

Fekete, T., 1959. The Plugging Effect of Bacteria in Sandstone Systems (M.S. thesis). University of Alberta, Edmonton, Alberta, Canada.

Fink, J., 2011. Petroleum Engineer's Guide to Oil Field Chemicals and Fluids. Gulf Professional Publishing, Houston, Texas, USA.

Fontaine, F.E., Peterson, W.H., McCoy, E., Johnson, M.J., Ritter, G.J., 1942. A new type of glucose fermentation by *Clostridium thermoaceticum* N. Journal of Bacteriology 43 (6), 701—715.

Forbes, A.D., 1980. Micro-organisms in oil recovery. In: Harrison, D.E.F., Higgins, I.J., Watkinson, R. (Eds.), Hydrocarbons in Biotechnology. Heyden, Lodon, pp. 169—180.

Fox, S.L., Xie, X., Schaller, K.D., Robertson, E.P., Bala, G.A., 2003. Permeability Modification Using a Reactive Alkaline Soluble Biopolymer. Idaho National Engineering and Environmental Laboratory (Technical Report).

Gottschalk, G., Bahl, H., 1981. Feasible improvements of the butanol production by *Clostridium acetobutylicum*. In: Hollaender, A. (Ed.), Trends in the Biology of Fermentations for Fuels and Chemicals. Plenum Press, New York, pp. 463—471.

Gran, K., Bjørlykke, K., Aagaard, P., 1992. Fluid salinity and dynamics in the North Sea and Haltenbanken basins derived from well log data. In: Hurst, A., Griffiths, C.M., Worthington, P.F. (Eds.), Geological Applications of Wireline Logs II, vol. 65. Geological Society, London, pp. 327–338.

Grula, E.A., Russell, H.H., Bryant, D., Kenaga, M., Hart, M., 1983. Isolation and screening of *Clostridia* for possible use in microbially enhanced oil recovery. In: Proceedings of the International Conference on Microbial Enhancement of Oil Recovery, Afton, Oklahoma, USA, 16 May, 1982.

Gudiña, E.J., Pereira, J.F.B., Rodrigues, L.R., Coutinho, J.A.P., Texeira, J.A., 2012. Isolation and study of microorganisms from oil samples for application in microbial enhanced oil recovery. International Biodeterioration & Biodegradation 68, 56–64.

Guo, H., Li, Y., Yiran, Z., Wang, F., Wang, Y., Yu, Z., Haicheng, S., Yuanyuan, G., Chuyi, J., Xian, G., 2015. Progress of microbial enhanced oil recovery in China. In: Proceedings of the SPE Asia Pacific Enhanced Oil Recovery Conference, Kuala Lumpur, Malaysia, 11–13 August, 2015.

Harner, N.K., Richardson, T.L., Thompson, K.A., Best, R.J., Best, A.S., Trevors, J.T., 2011. Microbial processes in the Athabasca Oil Sands and their potential applications in microbial enhanced oil recovery. Journal of Industrial Microbiology & Biotechnology 38, 1761–1775.

Hilchie, D.W., 1982. Applied Openhole Log Interpretation. Douglas W. Hilchie, Colorado, USA.

Jang, L.K., Sharma, M.M., Findley, J.E., Chang, P.W., Yen, T.F., 1983. An investigation of the transport of bacteria through porous media. In: Proceedings of the International Conference on Microbial Enhancement of Oil Recovery, Afton, Oklahoma, USA, 16–21 May, 1982.

Jannash, H.W., Wirsen, C.O., 1981. Morphological survey of microbial mats near deep-sea thermal vents. Applied and Environmental Microbiology 41 (2), 528–538.

Jenneman, G.E., Donaldson, E.C., Chilingarian, G.V., Yen, T.F., 1989. The potential for *in-situ* microbial applications. In: Chilingarian, G.V. (Ed.), Microbial Enhanced Oil Recovery, Developments in Petroleum Science, vol. 22. Elsevier, Amsterdam, pp. 37–74.

Jenneman, G.E., Knapp, R.M., Menzie, D.E., McInerney, M.J., Revus, D.E., Clark, J.B., Munnecke, D.M., 1983. Transport phenomena and plugging in Berea sandstone using microorganisms. In: Proceedings of the International Conference on Microbial Enhancement of Oil Recovery, Afton, Oklahoma, USA, 16–21 May, 1982.

Jenneman, G.E., Knapp, R.M., McInerney, M.J., Menzie, D.E., Revus, D.E., 1984. Experimental studies of *in-situ* microbial enhanced oil recovery. Society of Petroleum Engineers Journal 24 (1), 33–37.

Jimoh, I.A., 2012. Microbial Enhanced Oil Recovery (Ph.D. thesis). Aalborg University, Denmark.

Kalish, P.J., Stewart, J.A., Rogers, W.F., Bennett, E.O., 1964. The effect of bacteria on sandstone permeability. Journal of Petroleum Technology 16 (7), 805–814.

Karimi, M., Mahmoodi, M., Niazi, A., Al-Wahaibi, Y., Ayatollahi, S., 2012. Investigating wettability alteration during MEOR process, a micro/macro scale analysis. Colloids and Surfaces B: Biointerfaces 95, 129–136.

Karl, D.M., Wirsen, C.O., Jannasch, H.W., 1980. Deep-sea primary production at the Galápagos hydrothermal vents. Science 207 (4437), 1345–1347.

Kepkay, P.E., Nealson, K.H., 1982. Surface enhancement of sporulation and manganese oxidation by a marine *Bacillus*. Journal of Bacteriology 151 (2), 1022–1026.

Keradjopoulos, D., Holldorf, A.W., 1977. Thermophilic character of enzymes from extreme halophilic bacteria. FEMS Microbiology Letters 1 (3), 179–182.

Kinne, O., 1972. Pressure, general introduction. In: Kinne (Ed.), Marine Ecology. Wiley, New York, pp. 1323–1360.

Kowalewski, E., Rueslatten, I., Steen, K.H., Bødtker, G., Torsæter, O., 2006. Microbial improved oil recovery—bacterial induced wettability and interfacial tension effects on oil production. Journal of Petroleum Science and Engineering 52 (1–4), 275–286.

Kurbanoglu, E.B., Kurbanoglu, N.I., 2007. Ram horn hydrolysate as enhancer of xanthan production in batch culture of *Xanthomonas campestris* EBK-4 isolate. Process Biochemistry 42 (7), 1146–1149.

Kushner, D.J., 1978. Life in high salt and solute concentrations: halophilic bacteria. In: Kushner, D.J. (Ed.), Microbial Life in Extreme Environments. Academic Press, London, pp. 317–368.

Lake, L.W., Walsh, M.P., 2008. Enhanced Oil Recovery (Eor) Field Data Literature Search (Technical Report).

Lal, B., Reddy, M.R.V., Agnihotri, A., Kumar, A., Sarbhai, M.R., Singh, N., Khurana, R.K., Khazanchi, S.K., Misra, T.R., 2005. A Process for Enhanced Recovery of Crude Oil From Oil Wells Using Novel Microbial Consortium. Patent.

Lan, G., Fan, Q., Liu, Y., Liu, Y., Liu, Y., Yin, X., Luo, M., 2015. Effects of the addition of waste cooking oil on heavy crude oil biodegradation and microbial enhanced oil recovery using *Pseudomonas* sp. SWP-4. Biochemical Engineering Journal 103, 219–226.

Landau, J.V., Pope, D.H., 1980. Recent advances in the area of barotolerant protein synthesis in bacteria and implications concerning barotolerant and barophilic growth. Advances in Aquatic Microbiology 2, 49–76.

Lazar, I., 1990. An overview on MEOR field trials. In: Proceedings of the Biotechnology for Energy, Faisalabad, Pakistan, 16–21 December, 1989.

Lazar, I., Voicu, A., Nicolescu, C., Mucenica, D., Dobrota, S., Petrisor, I.G., Stefanescu, M., Sandulescu, L., 1999. The use of naturally occurring selectively isolated bacteria for inhibiting paraffin deposition. Journal of Petroleum Science and Engineering 22 (1–3), 161–169.

Lazar, I., Petrisor, I.G., Yen, T.F., 2007. Microbial enhanced oil recovery (MEOR). Petroleum Science and Technology 25 (11), 1353–1366.

Leigh, J.A., Jones, J.W., 1983. A new extremely thermophilic methanogen from a submarine hydrothermal vent. In: Proceedings of the 83rd Annual Meeting of the American Society for Micobiology, New Orleans, Louisiana, USA, 6–11 March, 1983.

Liu, Q., Lin, J., Wang, W., Huang, H., Li, S., 2015. Production of surfactin isoforms by *Bacillus subtilis* BS-37 and its applicability to enhanced oil recovery under laboratory conditions. Biochemical Engineering Journal 93, 31–37.

Marquis, R.E., 1976. High-pressure microbial physiology. Advances in Microbial Physiology 14, 159–241.

Marquis, R.E., 1983. Barobiology of deep oil formations. In: Proceedings of the International Conference on Microbial Enhancement of Oil Recovery, Afton, Oklahoma, USA, 16 May, 1982.

Marquis, R.E., Bender, G.R., 1979. Isolation of a variant of *Streptococcus faecalis* with enhanced barotolerance. Canadian Journal of Microbiology 26 (3), 371–376.

Marquis, R.E., Matsumura, P., 1978. Microbial life under pressure. In: Kushner, D.J. (Ed.), Microbial Life in Extreme Environments. Academic Press, London, pp. 105–158.

Marquis, R.E., Zobell, C.E., 1971. Magnesium and calcium ions enhance barotolerance of *Streptococci*. Archives of Microbiology 79 (1), 80–92.

McClung, L.S., 1935. Studies on anaerobic bacteria: IV. Taxonomy of cultures of a thermophilic species causing "swells" of canned foods. Journal of Bacteriology 29 (2), 189–203.

Mercer, W.A., Vaughn, R.H., 1951. The characteristics of some thermophilic, tartrate-fermenting anaerobes. Journal of Bacteriology 62 (1), 27–37.

Morita, R.Y., 1972. Pressure: bacteria, fungi and blue-green algae. In: Kinne (Ed.), Marine Ecology. Wiley, New York, pp. 1361–1388.

Mukherjee, A.K., Das, K., 2010. Microbial surfactants and their potential applications: an overview. In: Sen, R. (Ed.), Biosurfactnats, Advances in Experimental Medicine and Biology, vol. 672. Springer, Berlin, pp. 54–64.

Nazarenko, A.V., Nesterov, A.I., Pitryuk, A.P., Nazarenko, V.M., 1974. Growth of methane-oxidizing bacteria in glass capillaries. Mikrobiologiya 43, 146–151.

Nmegbu, C.G.J., 2014. The effect of salt concentration on microbes during microbial enhanced oil recovery. International Journal of Engineering Research in Africa 4 (6), 244–247.

Novitsky, T.J., Kushner, D.J., 1975. Influences of temperature and salt concentration on the growth of a facultative halophilic "*Micrococcus*" sp. Canadian Journal of Microbiology 21 (1), 107–110.

Ohms, D., McLeod, J.D., Graff, C.J., Frampton, H., Morgan, J.C., Cheung, S.K., Chang, K.T., 2010. Incremental-oil success from waterflood sweep improvement in Alaska. SPE Production & Operations 25 (3), 247–254.

Okeke, T., Lane, R.H., 2012. Simulation and economic screening of improved-conformance oil recovery by polymer flooding and a thermally activated deep diverting gel. In: Proceedings of the SPE Western Regional Meeting, Bakersfield, California, USA, 21–23 March, 2012.

Palaniraj, A., Jayaraman, V., 2011. Production, recovery and application of xanthan gum by *Xanthomonas campestris*. Journal of Food Engineering 106 (1), 1–12.

Patel, J., Borgohain, S., Kumar, M., Rangarajan, V., Somasundaran, P., Sen, R., 2015. Recent developments in microbial enhanced oil recovery. Renewable and Sustainable Energy Reviews 52, 1539–1558.

Paterek, J.R., Smith, P.H., 1983. Isolation of a halophilic methanogenic bacterium from the sediments of great salt Lake and a san Francisco Bay saltern. In: Proceedings of the 83rd Annual Meeting of the American Society for Micobiology, New Orleans, Louisiana, USA, 6–11 March, 1983.

Pinilla, L., Torres, R., Ortiz, C., 2011. Bioethanol production in batch mode by a native strain of *Zymomonas mobilis*. World Journal of Microbiology and Biotechnology 27 (11), 2521–2528.

Poli, A., Donato, P.D., Abbamondi, G.R., Nicolaus, B., 2011. Synthesis, production, and biotechnological applications of exopolysaccharides and polyhydroxyalkanoates by archaea. Archaea 1–13.

Pornsunthorntawee, O., Arttaweeporn, N., Paisanjit, S., Somboonthanate, P., Abe, M., Rujiravanit, R., Chavadej, S., 2008(a). Isolation and comparison of biosurfactants produced by *Bacillus subtilis* PT2 and *Pseudomonas aeruginosa* SP4 for microbial surfactant-enhanced oil recovery. Biochemical Engineering Journal 42 (2), 172–179.

Pornsunthorntawee, O., Wongpanit, P., Chavadej, S., Abe, M., Rujiravanit, R., 2008(b). Structural and physicochemical characterization of crude biosurfactant produced by *Pseudomonas aeruginosa* SP4 isolated from petroleum-contaminated soil. Bioresource Technology 99 (6), 1589–1595.

Portwood, J.T., 1995. A commercial microbial enhanced oil recovery technology: evaluation of 322 projects. In: Proceedings of the SPE Production Operations Symposium, Oklahoma City, Oklahoma, USA, 2–4 April, 1995.

Raleigh, J.T., 1962. The Effect of Rock Properties on Bacteria Plugging in Reservoir Rocks (M.S. thesis). University of Alberta, Edmonton, Alberta, Canada.

Raleigh, J.T., Flock, D.L., 1965. A study of formation plugging with bacteria. Journal of Petroleum Technology 17 (2), 201–206.

Rodriguez-Valera, F., Ruiz-Berraquero, F., Ramos-Cormenzana, A., 1980. Behaviour of mixed populations of halophilic bacteria in continuous culture. Canadian Journal of Microbiology 26 (11), 1259–1263.

Rottava, I., Batesini, G., Silva, M.F., Lerin, L., Oliveira, D., Padilha, F.F., Toniazzo, G., Mossi, A., Cansian, R.L., Luccio, M.D., Treichel, H., 2009. Xanthan gum production and rheological behavior using different strains of *Xanthomonas* sp. Carbohydrate Polymers 77 (1), 65–71.

Rozanova, E.P., Khudyakova, A.I., 1974. A new nonspore-forming thermophilic sulfate-reducing organism, *Desulfovibrio thermophilus* Nov. sp. Mikrobiologiya 43, 1069–1075.

Safdel, M., Anbaz, M.A., Daryasafar, A., Jamialahmadi, M., 2017. Microbial enhanced oil recovery, a critical review on worldwide implemented field trials in different countries. Renewable and Sustainable Energy Reviews 74, 159–172.

Saikia, U., Bharanidharan, B., Vendhan, E., Kumar Yadav, S., Siva Shankar, S., 2013. A brief review on the science, mechanism and environmental constraints of microbial enhanced oil recovery (MEOR). International Journal of Chemical Technology Research 5 (3), 1205–1212.

Salehi, M., Johnson, S.J., Liang, J.T., 2008. Mechanistic study of wettability alteration using surfactants with applications in naturally fractured reservoirs. Langmuir 24 (24), 14099−14107.

Sandvik, E.I., Maerker, J.M., Sandford, P.A., Laskin, A., 1977. Application of xanthan gum for enhanced oil recovery. In: Gould, R.F. (Ed.), Extracellular Microbial Polysaccharides, ACS Symposium Series, vol. 45. American Chemical Society, Washington D.C., pp. 242−264

Santos, M., Teiseira, J., Rodrigues, A., 2000. Production of dextransucrase, dextran and fructose from sucrose using *Leuconostoc mesenteroides* NRRL B512(f). Biochemical Engineering Journal 4 (3), 177−188.

Satyanarayana, T., Johri, B.N., Prakash, A., 2012. Microorganisms in Sustainable Agriculture and Biotechnology. Springer, Berlin, Germany.

Sen, R., 2008. Biotechnology in petroleum recovery: the microbial EOR. Progress in Energy and Combustion Science 34 (6), 714−724.

Speier, J.L., Malek, J.R., 1982. Destruction of microorganisms by contact with solid surfaces. Journal of Colloid and Interface Science 89 (1), 68−76.

Stanier, R.Y., Doudoroff, M., Adelberg, E.A., 1970. The Microbial World, third ed. Prentice-Hall, Englewood Cliffs, New Jersey, USA.

Stanley, S.O., Morita, R.Y., 1968. Salinity effect on the maximal growth temperature of some bacteria isolated from marine environments. Journal of Bacteriology 95 (1), 169−173.

Thimon, L., Peypoux, F., Michel, G., 1992. Interactions of surfactin, a biosurfactant from *Bacillus subtilis*, with inorganic cations. Biotechnology Letters 14 (8), 713−718.

Updegraff, D.M., 1983. Plugging and penetration of petroleum reservoir rock by microorganisms. In: Proceedings of the International Conference on Microbial Enhancement of Oil Recovery, Afton, Oklahoma, USA, 16−21 May, 1982.

Urum, K., Pekdemir, T., 2004. Evaluation of biosurfactants for crude oil contaminated soil washing. Chemosphere 57 (9), 1139−1150.

Walter, V., Syldatk, C., Hausmann, R., 2010. Screening concepts for the isolation of biosurfactant producing microorganisms. In: Sen, R. (Ed.), Biosurfactnats, Advances in Experimental Medicine and Biology, vol. 672. Springer, Berlin, pp. 1−13.

Wang, J., Xu, H., Guo, S., 2007. Isolation and characteristics of a microbial consortium for effectively degrading phenanthrene. Petroleum Science 4 (3), 68−75.

Xavier, J.B., Foster, K.R., 2007. Cooperation and conflict in microbial biofilms. Proceedings of the National Academy of Sciences 104 (3), 876−881.

Yakimov, M.M., Amro, M.M., Bock, M., Boseker, K., Fredrickson, H.L., Kessel, D.G., Timmis, K.N., 1997. The potential of *Bacillus licheniformis* strains for in situ enhanced oil recovery. Journal of Petroleum Science and Engineering 18 (1−2), 147−160.

Yan, P., Lu, M., Yang, Q., Zhang, H.L., Zhang, Z.Z., Chen, R., 2012. Oil recovery from refinery oily sludge using a rhamnolipid biosurfactant-producing *Pseudomonas*. Bioresource Technology 116, 24−28.

Yayanos, A.A., Dietz, A.S., 1982. Thermal inactivation of a deep-sea barophilic bacterium, isolate CNPT-3. Applied and Environmental Microbiology 43 (6), 1481−1489.

Yayanos, A.A., Dietz, A.S., van Boxtel, R., 1981. Obligately barophilic bacterium from the Mariana trench. Proceedings of the National Academy of Sciences United States of America 78 (8), 5212−5215.

Youssef, N., Elshahed, M.S., McInerney, M.J., 2009. Microbial processes in oil fields: culprits, problems, and opportunities. In: Laskin, A.I., Sariaslani, S., Gadd, G.M. (Eds.), Advances in Applied Microbiology, vol. 66. Elsevier, Amsterdam, pp. 141−251.

Youssef, N., Simpson, D.R., McInerney, M.J., Duncan, K.E., 2013. *In-situ* lipopeptide biosurfactant production by *Bacillus* strains correlates with improved oil recovery in two oil wells approaching their economic limit of production. International Biodeterioration & Biodegradation 81, 127−132.

Yu, I.K., Hungate, R.E., 1983. Isolation and characterization of an obligately halophilic methanogenic bacterium. In: Proceedings of the 83rd Annual Meeting of the American Society for Micobiology, New Orleans, Louisiana, USA, 6−11 March, 1983.

Zeikus, J.G., Wolfe, R.S., 1972. *Methanobacterium thermoautotrophicus* sp. n., an anaerobic, autotrophic, extreme thermophile. Journal of Bacteriology 109 (2), 707−713.

Zilling, W., Stetter, K.O., Schäfer, W., Janekovic, D., Wunderl, S., Holz, I., Palm, P., 1981. Thermoproteales: a novel type of extremely thermoacidophilic anaerobic archaebacterial isolated from Icelandic solfataras. Zentralblatt für Bakteriologie, Mikrobiologie und Hygiene 2 (3), 205−227.

Zheng, C., Yu, L., Huang, L., Xiu, J., Huang, Z., 2012. Investigation of a hydrocarbon-degrading strain, *Rhodococcus ruber* Z25, for the potential of microbial enhanced oil recovery. Journal of Petroleum Science and Engineering 81, 49−56.

Zobell, C.E., 1970. Pressure effects on morphology and life processes of bacteria. In: Zimmerman, A.M. (Ed.), High Pressure Effects on Cellular Processes. Academic Press, New York, pp. 85−130.

Zobell, C.E., 1947. Microbial transformation of molecular hydrogen in marine sediments, with particular reference to petroleum. American Association of Petroleum Geologists Bulletin 31 (10), 1709−1751.

Zobell, C.E., Grant, C.W., 1943. Bacterial utilization of low concentrations of organic matter. Journal of Bacteriology 45 (6), 555−564.

Zobell, C.E., Morita, R.Y., 1957. Barophilic bacteria in some deep sea sediments. Journal of Bacteriology 73 (4), 563−568.

Zvyagintsev, D.G., 1970. Growth of microorganisms in thin capillaries and films. Mikrobiologiya 39, 161−165.

Zvyagintsev, D.G., Pitryuk, A.P., 1973. Growth of microorganisms in capillaries of various sizes under continuous flow and static conditions. Mikrobiologiya 42, 60−64.

CHAPTER 2

Microbiology and Microbial Products for Enhanced Oil Recovery

2.1 MICROBIAL ECOLOGY AND ACTIVITIES IN DEEP SUBSURFACE

2.1.1 Microorganisms in Ecosystem

Abundance and diversity of microbial communities in nature are determined by presence of nutrients (foods for microorganisms) and various environmental conditions, such as temperature, pressure, pH, salinity, and oxygen availability (Madigan et al., 1997). Microbial habitats, the environment in which microbial communities thrive, can either be favorable for living for some species or be threatful to other species. In nature, microorganisms metabolite using resources from the ecosystem to keep generating new cells and to adapt to the environment; thereby, they also excrete byproducts back to the environments. Therefore, microorganisms can even change the ecosystem depending on the resource availability and the environmental conditions for metabolism (Madigan et al., 1997). Temperature, pH, salinity, and oxygen availability are the most influential factors for bacterial growth. Oil reservoirs are extreme environments for microbial life associated with high toxicity, high temperature, high salinity, and high pressure (Silva et al., 2013; Cai et al., 2015). This section describes the environment factors affecting growth and survival of microorganisms and explores diverse types of microorganisms that live in hydrocarbon reservoir environments.

A. Temperature

Temperature is one of the most significant factors affecting microorganisms. Generally, the reaction rate of enzymatic activities is accelerated as the temperature increases, such that the growth of microorganisms is faster at the higher temperature. However, microorganisms have their cardinal temperature ranges. There is a minimum temperature that they are not able to grow below as the transport processes are too slow. The maximum temperature is that the protein denatures and the microorganisms are not able to grow above that temperature. The optimal temperature is the temperature at which growth of microorganisms are the fastest

simply because the enzymatic reaction is at the maximal rate. The cardinal temperature range is the characteristics of each microorganism and differs to the species. The range of typical organisms is less than 40°C (Madigan et al., 1997). According to the optimum growth temperature, microorganisms are categorized into several groups: psychrophiles, with low temperature optima (generally under 10°C); mesophiles, with midrange temperature optima (generally under 50°C); thermophiles, with high temperature optima (generally under 70°C); and hyperthermophiles, with extremely high temperature optima (generally over 70°C). Oil reservoirs exhibit a wide range of temperature. Among the other environment factors, the reservoir temperature determines microbial species to survive and exist in the reservoirs.

B. pH

As temperature affects the growth of microorganisms, pH also plays an important role in microbial growth. Similar to the temperature, every microorganism has its own pH range in which it can grow. This pH approximately ranges over 2–3 pH. The microorganisms that grow best in the range of pH 5.5–7.9 are neutrophils. The microorganisms that grow best below pH 5.5 are called acidophiles. Alkaliphiles refer to the microorganisms that show optimal growth above pH 8, and they are frequently found in high pH environments.

During microbial growth in batch cultures, buffers are often required for sustainable growth of microorganisms along with the nutrients. The buffer keeps fluid pH consistent, such that the enzyme activities are minimally affected by any hydroxyl ions generated during metabolic reactions.

C. Salinity

Natural surface or subsurface water is rarely pure. Instead, various solutes, such as salts, sugars, or any other substances, are dissolved in water. In solvent (water)-solutes systems, the solutes readily associate with water molecules; therefore, water becomes less available for microbial growth. This phenomenon is referred to as the reduced

Theory and Practice in Microbial Enhanced Oil Recovery. https://doi.org/10.1016/B978-0-12-819983-1.00002-8

water activity in water-solute systems. The activity of pure water is 1, and the activity of seawater is approximately 0.98. In seawater, halotolerant or halophiles are only found from marine environments. Halotolerant organisms are the microorganisms which are survivable under some level of salinity (less than 10% NaCl concentration). Halophiles are the microorganisms that require some levels of NaCl concentrations for growth. Halophiles are known to grow best 3%−12% of NaCl concentration, and the solutes cannot be replaced by other salts, such as $MgCl_2$, KCl, or $CaCl_2$. An environment with more than 12% NaCl concentration is likely to have a microbial community associated with extreme halophiles. These extreme halophiles can grow under very high concentration of NaCl, typically 12%−30%.

D. Oxygen

Oxygen (O_2) is another essential nutrient for many microorganisms as many of them are unable to grow under anoxic environments. Aerobes refer to the microorganisms that grow in the presence of O_2 by using O_2 for the respiration reaction. Among aerobes, microaerophiles use O_2 only when the oxygen level is lower than the atmospheric concentration (microoxic condition), because they contain oxygen-sensitive components, such as O_2-labile enzymes. Meanwhile, facultative aerobes do not require oxygen; therefore, they can grow in the absence of O_2. Obligate aerobes and microaerophilic aerobes metabolite only via aerobic respiration, whereas facultative aerobes can metabolite through aerobic respiration, anaerobic respiration, and fermentation processes. Because most of oil fields are lack of oxygen, many of facultative aerobes, such as *Bacillus*, *Pseudomonas*, and *Escherichia* species, are most frequently found.

Microorganisms that cannot respire O_2 are referred to as anaerobes. Among anaerobes, aerotolerant microorganisms are able to tolerate O_2 and grow in the presence of O_2 even if they are not respiring under oxic environment. By contrast, obligate anaerobes are unable to survive under oxic environment. *Methanogens* and *Clostridia* are representative obligate anaerobes, which are also frequently found from oil reservoirs.

2.1.2 Reservoir Environments and Their Effects on Microbial Activities

Oil reservoirs have complex environments containing both living and nonliving substances, in which those interact with each other in a complicated dynamic network of nutrients and energy fluxes. As oil reservoirs are geologically heterogeneous and their physicochemical characteristics vary significantly from site to site, the microbial communities in oil reservoirs show a variety of diversity. Living microorganism metabolites through uptaking the substrates (nutrients) and excreting the by-products, such as cells, acids, gases, enzymes, and extracellular polymeric substances (EPS). Those substances alter the physicochemical properties of oils and rocks and hence contribute to the complexity in oil reservoirs. During microbial enhanced oil recovery (MEOR) practices, the physicochemical properties of oil reservoirs, particularly pH, temperature, pressure, salinity, and nutrient availability, are essential because these have direct impacts on the microbial activities. Whether or not the microorganisms to survive in oil fields and produce by-products suitable for MEOR significantly depends on the reservoir physicochemical characteristics.

A. Temperature

Geothermal temperature in subsurface increases with depth according to the geothermal gradient. The temperature generally increases by approximately 2−3°C by every 100 m depth. The temperature of deep oil reservoirs readily exceeds the temperature that usual bacteria can sustain, above 130°C. The reservoir temperature is lowered after the secondary injection because water injection after primary injection cools down the reservoir (Li et al., 2017; Vigneron et al., 2017). Previous studies reveal that the highest concentration of fatty acids are found in the oil reservoirs with the temperature higher than 80°C. This indicates that the biodegradation is unlikely to take place above ~80°C (Fisher, 1987; Barth, 1991; Larter et al., 2003). This observation corroborates the previous finding that in situ oil biodegradation does not take place even when the oil reservoir temperature exceeds 82°C (Philippi, 1977).

However, anaerobic hydrocarbon biodegradation in subsurface oil reservoirs has recently been detected from some samples acquired from oil reservoirs that are as high as 85°C (Aitken et al., 2004). This temperature is similar to the previously reported temperature range, at around ~80°C, although it is slightly higher. Several hyperthermophiles have been found from reservoirs with the temperature higher than 80°C. These include archaeal and bacterial communities from produced water samples from Ekofisk oil field, Norwegian, where the temperature is up to 131°C (Kaster et al., 2009). However, this may need further examination owing to the fact that accurate measurement of the deep reservoirs' temperature is challenging (Kaster et al., 2009; Pannekens et al., 2019). Note that there is an argument that it is unlikely that the organisms survived at such high temperatures.

Table 2.1 shows the bacteria and archaea found from wide ranges of temperature. Diverse microbial species have been found in reservoirs with the temperature of

TABLE 2.1

Bacteria and Archaea Found in Reservoirs With Various Temperatures.

Temperature Optimum	Phylum/Class	Order/Genus
Ubiquitous	Proteobacteria/Epsilonproteobacteria	Campylobacterales/Arcobacter
	Proteobacteria/Epsilonproteobacteria	Campylobacterales/Sulfurospirillum
	Proteobacteria/ Gammaproteobacteria	Pseudomonadales/Pseudomonas
	Proteobacteria/Alphaproteobacteria	Rhizobiales/Rhizobium
	Rhizobiales/	
	Proteobacteria/Alphaproteobacteria	Sphingomonadales/Sphingomonas
	Acinetobacter	
Only > 50°C	Crenarchaeota/Thermoprotei	Fervidicoccales
	Euryarchaeota/Halobacteria	Halobacteriales
	Euryarchaeota/Halobacteria	Haloferacales
	Thaumarchaeota/Nitrososphaeria	Nitrososphaerales
	Nitrospirae/Nitrospira	Nitrospirales/Nitrospira
	Crenarchaeota/Thermoprotei	Sulfolobales
	Proteobacteria/Deltaproteobacteria	Syntrophobacterales/ Thermosulforhabdus
	Euryarchaeota/Thermoplasmata	Thermoplasmatales
	Crenarchaeota/Thermoprotei	Thermoproteales
	Acidobacteria	
	Atribacteria	
Mostly > 50°C	Euryarchaeota/Archaeoglobi	Archaeoglobales
	Firmicutes/Bacilli	Bacillales/Anaerobacillus
	Firmicutes/Bacilli	Bacillales/Bacillus
	Firmicutes/Clostridia Clostridiales	Clostridiales/Thermosyntropha
	Euryarchaeota/Halobacteria	Halobacteriales/Halogeometricum
	Proteobacteria/Hydrogenophilalia	Hydrogenophilales/Tepidiphilus
	Thermotogae/Thermotogae	Kosmotoga
	Euryarchaeota/Methanobacteria	Methanobacteriales/ Methanothermobacter
	Euryarchaeota/Methanomicrobia	Methanocellales/Methanocella
	Euryarchaeota/Methanomicrobia	Methanomicrobiales/ Methanocalculus
	Euryarchaeota/Methanomicrobia	Methanosarcinales/Methanosaeta
	Euryarchaeota/Methanomicrobia	Methanosarcinales/ Methanomethylovorans
	Nitrospirae/Nitrospira Nitrospirales/	Thermodesulfovibrio
	Proteobacteria/Alphaproteobacteria	Rhodospirillales/Tistrella
	Deinococcus−Thermus/Deinococci	Thermales/Thermus
	Firmicutes/Clostridia	Thermoanaerobacterales/ Thermoanaerobacter
	Euryarchaeota/Thermococci	Thermococcales
	Euryarchaeota/Thermoplasmata	Thermoplasmatales/ Thermogymnomonas
	Actinobacteria/Thermoleophilia	
	Bacteroidia/Bacteroidia	
	Deferribacteres/Deferribacteres	
	Firmicutes	
	Proteobacteria/Betaproteobacteria	
	Proteobacteria/Deltaproteobacteria	
	Tenericutes/Mollicutes	
	Thermodesulfobacteria	
	Thermotogae	

Continued

TABLE 2.1
Bacteria and Archaea Found in Reservoirs With Various Temperatures.—cont'd

Temperature Optimum	Phylum/Class	Order/Genus
Mostly < 50°C	Actinobacteria/Actinobacteria	Actinomycetales/Microbacterium
	Actinobacteria/Actinobacteria	Actinomycetales/Dietzia
	Actinobacteria/Actinobacteria	Actinomycetales/Rhodococcus
	Proteobacteria/	Alteromonadales/Marinobacterium
	Gammaproteobacteria	
	Crenarchaeota/Thermoprotei	Desulfurococcales
	Euryarchaeota/Methanobacteria	Methanobacteriales/
		Methanobacterium
	Euryarchaeota/Methanomicrobia	Methanocellales
	Euryarchaeota/Methanococci	Methanococcales/Methanococcus
	Euryarchaeota/Methanomicrobia	Methanomicrobiales
	Euryarchaeota/Methanomicrobia	Methanomicrobiales/
		Methanocorpusculum
	Euryarchaeota/Methanomicrobia	Methanomicrobiales/Methanoculleus
	Euryarchaeota/Methanomicrobia	Methanomicrobiales/Methanolinea
	Euryarchaeota/Methanomicrobia	Methanosarcinales
	Euryarchaeota/Methanomicrobia	Methanosarcinales/Methanolobus
	Proteobacteria/Alphaproteobacteria	Rhodobacterales/Donghicola
	Proteobacteria/Alphaproteobacteria	Rhodobacterales/Hyphomonas
	Proteobacteria/Alphaproteobacteria	Rhodobacterales/Paracoccus
	Bacteroidetes	
	Chloroflexi	
	Planctomycetes	
	Proteobacteria	
	Spirochaetes	
	Synergistetes	

Adopted from Pannekens, M., Kroll, L., Müller, H., Mbow, F.T., Meckenstock, R.U., 2019. Oil reservoirs, an exceptional habitat for microorganisms. New biotechnology 49, 1–9.

~55°C (Pannekens et al., 2019). Diverse indigenous microbial communities have been detected in 10 different locations of Shengli oil field, China, where the temperature varies from 55 to 91°C (Weidong et al., 2014). It is reported that the higher microbial diversity is found as the lower the oil reservoir temperature is (Weidong et al., 2014). Abundant archaebacteria and thermophilic bacteria, such as *Thermus*, *Thermincola*, and *Thermanaeromonas* spp., have been found at the reservoir temperature higher than 90°C, but not as diverse as from the lower reservoir temperature (Lin et al., 2014).

It is expected that the microbial communities in high-temperature oil wells are not diverse, but similar that only thermophilic, thermotolerant, and/or spore-forming families are belonged (Kim et al., 2018). Recent study by Kim et al. (2018) has investigated the indigenous microbial communities of five different high-temperature oil reservoirs near Segno, Texas, United States (~80–85°C), and Crossfield, Alberta, Canada (~75°C). Despite the distance between two reservoirs and the dissimilarities of physicochemical

properties of oil reservoirs, the MiSeq amplicon sequencing and partial 16S rRNA analyses indicated that those different oil reservoirs exhibited unexpectedly high similarity in terms of microbial communities. Methanogens such as *Methanosaetaceae*, *Methanobacterium* and *Methanoculleus*, *Clostridiaceae*, and *Thermotogaceae* were found both from the reservoirs, and it is suggested that the microbial community is predictable to make decision as to which mechanisms should be the major strategy for MEOR practices at such high-temperature reservoirs (Kim et al., 2018).

B. pH

The reservoir pH has a high impact on microbial communities because pH is also one of the most significant factors affecting microbial growth and its metabolic activities (Madigan et al., 1997; Pannekens et al., 2019). While most of bacteria grow best at neutral pH of ~6–8, the pH in oil reservoirs varies from 5 to 8. It is worth noting that the pH measured on surface under ambient pressure and temperature can be different from the in situ pH at deep depth because the gas solubility significantly changes with fluid pressure. In deep reservoirs with high pressure, the in situ pH is expected to be in the range of 3−7 due to the high gas solubility (Magot et al., 2000). The pH of formation water in oil reservoirs is mainly controlled by the CO_2/bicarbonate systems. Production of CO_2 during fermentation processes by microbial activities controls the pH in formation water. In addition, the organic acids, such as acetic, propionic, formic, and butyric acids, can also alter pH in the CO_2/bicarbonate systems. Rock minerals in reservoir formations dissolve, especially more rapidly in lower pH, which in turns alters pH significantly. This buffering reaction of bicarbonate in a brine-mineral system affects the extent of pH change. In summary, pH in oil reservoir is determined by numerous factors, such as microbial fermentation processes and multiphase (solid-gas-liquid) interactions. Table 2.2 lists the bacteria and archaea found from 22 geographically separated oil reservoirs that have diverse range of pH (Gao et al., 2016). In neutral to alkaline pH of ∼7.0−8.2, *Alphaproteobacteria*, *Deltaproteobacteria*, and *Actinobacteria* are found to be most abundant, whereas *Pseudomonas* species are discovered at a low pH regime (∼5.5−7.6). In more acidic environments with the pH of ∼5.5−6.5, *Gammaproteobacteria*, *Betaproteobacteria*, and *Epsilonproteobacteria* have been found (Gao et al., 2016; Pannekens et al., 2019). The fluid pH of the oil reservoir has to be

taken into consideration when operating MEOR practices. It is particularly critical when the target potential bacteria are chosen and when the culture growth media or biostimulation media are designed. Either in case of utilizing indigenous or exogenous bacteria, pH should be taken seriously.

C. Salinity
Salinity in oil reservoirs varies from almost fresh to salt-saturated water (as high as ∼20%). Salinity of oil reservoirs has not been thoroughly investigated compared with the temperature; however, it also has large influences on the microbial growth and metabolic activities as well as the microbial communities of oil reservoir. From 32 producing wells in Halfdan oil field, where most of them were treated with seawater during the secondary flooding operations, *Clostridiales* was found from the wells with the salinity of ∼3.8%, and *Petrotoga* and *Desulfotomaculum* species were commonly found from fluids samples having the salinity of 7.2% (Vigneron et al., 2017). *Desulfotomaculum* species are hydrocarbon-degrading species, and these hydrocarbon degradations appear to occurring even under high-salinity environments (Vigneron et al., 2017). As the secondary oil recovery commonly includes the injection of saline seawater, the MEOR practice, which is the tertiary recovery process, needs to examine the capability of either indigenous or exogenous bacteria whether

TABLE 2.2
Bacteria and Archaea Associated With pH in Petroleum Reservoirs (Pannekens et al., 2019).

pH	Phylum/Class	Order/Genus
Acidic pH	*Euryarchaeota/Methanobacteria*	*Methanobacteriales/Methanothermobacter*
	Proteobacteria/Gammaproteobacteria	*Pseudomonadales/Pseudomonas*
	Proteobacteria/Betaproteobacteria	
	Proteobacteria/Epsilonproteobacteria	
	Proteobacteria/Gammaproteobacteria	
Alkaline pH	*Euryarchaeota/Archaeoglobi*	*Archaeoglobales/Archaeoglobus*
	Proteobacteria/Deltaproteobacteria	*Desulfuromonadales/Desulfuromonas*
	Euryarchaeota/Methanococci	*Methanococcales/Methanococcus*
	Euryarchaeota/Methanomicrobia	*Methanomicrobiales/Methanocorpusculum*
	Euryarchaeota/Methanomicrobia	*Methanomicrobiales/Methanocalculus*
	Euryarchaeota/Methanomicrobia	*Methanomicrobiales/Methanoculleus*
	Euryarchaeota/Methanomicrobia	*Methanomicrobiales/Methanolinea*
	Euryarchaeota/Methanomicrobia	*Methanosarcinales/Methanosaeta*
	Euryarchaeota/Methanomicrobia	*Methanosarcinales/Methanolobus*
	Proteobacteria/Alphaproteobacteria	*Rhodobacterales/Paracoccus*
	Actinobacteria	
	Proteobacteria/Alphaproteobacteria	

or not they are adaptive to the saline environment of the specific potential field. It has been reported that a significantly reduced number of microorganisms present at the reservoirs having the salinity higher than 10% (Röling et al., 2003). Pannekens and coworkers have listed the microorganisms found from low- and high-salinity reservoirs, as listed in Table 2.3 (Pannekens et al., 2019).

2.2 BIOSURFACTANTS
2.2.1 Definitions and Characteristics
Surfactant refers to the surface-active agent that favorably adsorbs on a surface or an interface of the system (Rosen and Kunjappu, 2012). The surfactant alters the interfacial free energy through attachment at the boundary between the immiscible phases. The unique structural characteristics of the surfactant enable its role as a surface-active agent through adsorption; a surfactant molecule consists of both lyophilic and lyophobic parts. A lyophilic part has an affinity to the liquid to be easily dissolved in solvent. On the contrary, a lyophobic part is a liquid-hating part, which is hardly dissolved in solvents. Let us explore water as an example. When a water-soluble surfactant is dissolved in water, the hydrophobic (lyophobic) part in water associates with water molecules in an aqueous (liquid) phase and increases the free energy of the liquid phase. It indicates that the surfactant molecule is thermodynamically more favorable to be located at the interface rather than to be surrounded all around by the water

molecules. Owing to such structural characteristics having both the hydrophilic and hydrophobic parts, surfactants are mostly concentrated at the surface, lowering the energy state and thus modifying the interfacial properties (Fig. 2.1). The presence of surfactants at a surface or an interface reduces the surface free energy per unit area (i.e., surface tension or interfacial tension [IFT]) (Schramm, 2000; Rosen and Kunjappu, 2012). The hydrophobic part of a surfactant is mostly composed of long-chain hydrocarbon residues, whereas the hydrophilic part can have diverse compositions depending on the nature of its electrical charge. Accordingly, depending on the electric nature of the hydrophilic group, surfactants are classified as follows: anionic, cationic, zwitterionic, and nonionic surfactants (Som et al., 2012).

A majority of biosurfactants are amphiphilic surfactant compounds that are naturally produced by microbial activities. Biosurfactants are capable of modifying surface energy at surfaces and interfaces basically via the same mechanisms that chemical surfactants function.

Based on the functional groups, biosurfactants are classified into four types: glycolipids, phospholipids, polymeric, and lipopeptide surfactants. Various types of surfactants are implemented and applied to diverse fields, such as agricultural, food, pharmaceutical, oil recovery, and CO_2 storage. Biosurfactants have attracted much attention as alternatives to the chemical surfactants. Biosurfactants have several advantages over chemical surfactants, as follows:

(1) Biodegradability (ecological suitability): Biosurfactants are readily degradable, which makes biosurfactants a great candidate for a bioremediation agent. By contrast, chemical surfactants are hardly degradable and can remain as a potential threat to environments (Finnerty, 1994).

(2) Low toxicity: Biosurfactant are considered as low toxic materials compared with chemically synthesized surfactants. For an example, rhamnolipids produced by *Pseudomonas aeruginosa* is reported to be four times more effective in hexadecane removal, and glycolipids from *Rhodococcus* species 413A are 50% less toxic than a synthetic surfactant, Tween 80, in naphthalene solubilization tests (Brusseau et al., 1995; Kanga et al., 1997)

(3) Producibility from raw materials: The biosurfactants can be produced by using renewable sources, such as carbon sources from hydrocarbons and carbohydrates (Deziel et al., 1996; Al-Bahry et al., 2013), whereas the chemical surfactants are synthesized with chemicals.

TABLE 2.3
Bacteria and Archaea Associated With Salinity in Petroleum Reservoirs (Pannekens et al., 2019).

Salinity	Phylum/Class	Order/Genus
Higher salinity	Firmicutes/ Clostridia	Clostridiales/ Desulfotomaculum
	Proteobacteria/ Deltaproteobacteria	Desulfovibrionales/ Desulfovermiculus halophilus
	Firmicutes/ Clostridia	Halanaerobiales/ Haloanaerobium
	Euryarchaeota/ Methanococci	Methanococcales/ Methanothermococcus
	Thermotogae/ Thermotogae	Petrotogales/ Petrotoga
Lower salinity	Euryarchaeota/ Methanobacteria	Methanobacteriales/ Methanobacterium
	Euryarchaeota/ Methanomicrobia	Methanomicrobiales/ Methanoplanus

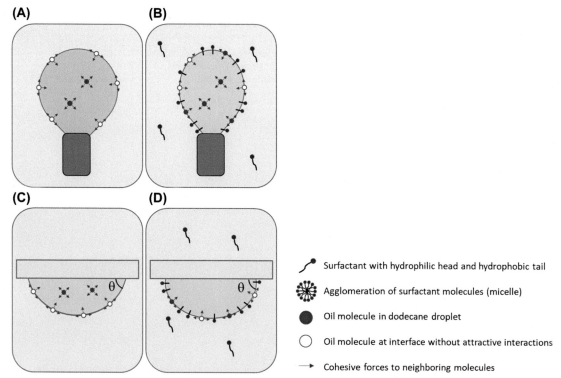

FIG. 2.1 Schematic description of **(A)**, **(B)** interfacial tension alteration and **(C)**, **(D)** contact angle modification by biosurfactant adsorption by time. **(A)** and **(C)** represent the oil droplet in brine without any surfactant, **(B)** and **(D)** represent the surfactant adsorption in the middle of alteration of interfacial tension and contact angle. (Credit: Courtesy of Taehyung Park and Tae-Hyuk Kwon.)

These advantages of biosurfactants over chemical surfactants motivate extensive research to evaluate and compare the efficiency of biosurfactants with chemical surfactants (Finnerty, 1994; Al-Sulaimani et al., 2012; Park et al., 2017). Table 2.4 shows the comparison between biosurfactants and synthetic surfactants for modification of surface tension and IFT, which can effectively change oil-wet minerals to the water-wet status (Fig. 2.2). As shown in Table 2.4, there are several bacteria that can produce diverse biosurfactants, and these biosurfactants can lower the surface tension to the similar range with chemical surfactants. The critical micelle concentration (CMC) of biosurfactants is relatively lower than that of chemical surfactants. This implies the greater efficiency of biosurfactants over chemical surfactants.

Another advantage of using biosurfactants for enhanced oil recovery is their emulsification capability (Fig. 2.1). Emulsification is a process that forms an emulsion in a liquid. An example is forming very small oil droplets suspended in an aqueous phase. High-molecular-weight biosurfactants are effective emulsifiers; these produce a water-in-oil emulsion or an oil-in-water emulsion in the water-oil mixtures.

2.2.2 Biosurfactant Production (Mechanisms, Optimal Conditions, Yields)

Microbial biosurfactants have a wide variety of chemical structures, including glycolipids, lipopeptides, fatty acids, neutral lipids, and phospholipids. Biosurfactants produced by microorganisms are categorized into five major classes, and the list is shown in Table 2.5. Generally, most of biosurfactants include hydrophilic parts composed of amino acids, peptides, or polysaccharides, and at the same time, they have the hydrophobic parts which often consist of saturated, unsaturated, or hydroxylated fatty acids and amphophilic or hydrophobic peptides (Georgiou et al., 1992).

Nutrient availability in the growth media has a pronounced effect on the bacterial growth and the biosurfactant production. In particular, the types and

TABLE 2.4
Comparison Biosurfactants and Chemical Surfactants to Reduce Surface Tension/Interfacial Tension and Their Critical Micelle Concentrations.

	Surface Tension with Air (mN/m)	Interfacial Tension with Crude Oil (mN/m)	Critical Micelle Concentration (mg/L)
BIOSURFACTANTS			
Sophorolipid (produced by *Starmerella* spp.)	40	8	82
Surfactin (produced by *Bacillus* spp.)	30	4.5	20
Rhamnolipid (produced by *Pseudomonas* spp.)	35	1	200
Glycolipid (produced by *Rhodococcus* spp.)	27	6.3	60
CHEMICAL SURFACTANTS			
Linear alkylbenzene sulfonates	31.4	1	1018
Sodium dodecyl sulfate	37	0.02	2120
Cetyltrimethylammonium bromide	30	5	1300
Tween 20	30	4.8	600
Findet 1214N/23	37.9	–	21
Glucopone 215	31.8	–	241
Glucopone 650	32	–	73

Modified from Finnerty, W.R., 1994. Biosurfactants in environmental biotechnology. Current Opinion in Biotechnology 5, 291–295.

concentration of carbon sources (Georgiou et al., 1992), nitrogen substrates (Davis et al., 1999), and ions such as phosphorus (P), magnesium (Mg), iron (Fe), and manganese (Mn) in the medium have huge impacts on the bacterial surfactant production. For this reason, there have been numerous lab tests to optimize biosurfactant production using diverse cultural media.

For an instance of biosurfactant production by *Bacillus* spp., glucose and sucrose are the best carbon sources among diverse carbon sources rather than sodium acetate, fructose, lactose, and tryptone (Makkar and Cameotra, 1997; Abdel-Mawgoud et al., 2008). Use of sodium nitrate and ammonium nitrate as nitrogen sources yields the greatest biosurfactant production,

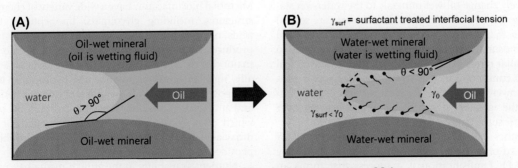

Oil-wet condition ($\theta > 90°$) → Water-wet ($\theta < 90°$)

FIG. 2.2 Enhanced mobilization of biosurfactant treated oil in porous media by transition of (**A**) oil-wet mineral to (**B**) water-wet mineral by biosurfactant activity. (Credit: Courtesy of Taehyung Park and Tae-Hyuk Kwon.)

TABLE 2.5
Biosurfactants Produced by Microorganisms (Safdel et al., 2017).

Biosurfactant Class	Subgroup Class	Producing Microorganism
Glycolipid	Rhamnolipids	*Pseudomonas aeruginosa*
		Pseudomonas sp.
		Pseudomonas chororaphis
		P. aeruginosa UW−1
		P. aeruginosa GL−1
	Trehalolipid/Trehalose lipid	*Rhodococcus erythropolis*
		Mycobacterium sp.
		Norcardia SFC-D
		Rhodococcus sp. *H13 A*
		Rhodococcus sp. *ST−5*
		Norcadiaerithropolis
		Arthobacter sp.
		Pseudomonas sp.
	Sophorolipids	*Torulopsisbombicola*
		Torulopsisbombicola apicola
		Arthrobacter sp.
		Candida bombicola
		Candida apicola
		Candida antarctica
		Torulopispetrophilum
		C. apicola IMET
	Mannosylerythritol lipids (MEL)	*C. antarctica*
		Pseudozymaaphidis
		Pseudozymarugulosa
Lipopeptides and lipoproteins	Peptide-lipid	*B. licheniformis*
	Serrawettin	*Serratiamarcenscens*
	Viscosin	*Pseudomonas fluorescens*
	Surfactin	*Bacillus subtilis*
		B. pumilus A1
		B. subtilis C 9
		Lactobacillus sp.
	Subtilisin	*B. subtilis*
	Gramicidin	*B. subtilis*
	Polymyxin	*Bacillus polymyxia*
	Iturin/fengycin	*B. subtilis*
	Lichenysin	*B. licheniformis*
		B. licheniformis
		B. licheniformis JF−2
Fatty acids, neutral lipids, and phospholipids	Fatty acids	*Corynebacteriumlepus*
	Neutral lipids	*Nocardiaerythropolis*
	Phospholipids	*Thiobacillusthiooxidans*
		Acinetobacter sp.
		Corynebacteriumlepus

but the production is more efficient when those nitrogen sources are not plenty but limited in a cultural medium (Davis et al., 1999; Abdel-Mawgoud et al., 2008). Similarly, both sophorose lipids produced by *Candida* spp. and rhamnolipids produced by *P. aeruginosa* are produced at the high concentrations in a nitrogen-exhausted medium (Göbbert et al., 1984; Ramana and Karanth, 1989). *Rhodococcus* spp. exhibit the maximum growth and biosurfactant production yields with the growth medium containing nitrate as nitrogen sources, which is similar to the other species. Taking into the consideration of the costs of ingredients with the production yields, sucrose and ammonium nitrate show a good promise as nutrients to stimulate bacterial biosurfactant production, which has been revealed through comprehensive tests (Pereira et al., 2013). This supports the MEOR application to be economically feasible. However, the fundamental mechanisms and optimization of biosurfactant production for MEOR remain not fully identified, which warrants further investigation on biosurfactant-producing bacteria for successful implementation of MEOR.

2.2.3 Representative Biosurfactant-Producing Bacteria

A. Bacillus *species*

Among diverse types of biosurfactant-producing bacteria, *Bacillus* species have been regarded most practical microorganisms for its survivability under extreme conditions, such as extreme pH, salinity, high pressure, and high temperature conditions (ZoBell and Johnson, 1949; Yakimov et al., 1995; Willenbacher et al., 2015). Moreover, *Bacillus* spp. are known to be facultative anaerobic bacteria where they are one of the most commonly found bacteria in soil and aquatic environments (Davis et al., 1999; Willenbacher et al., 2015). *Bacillus* spp. sporulate to produce endospores under extreme conditions, which makes these species a great candidate for diverse applications. An endospore is a nonreproductive structure produced by bacteria, which enables bacteria to extend their lives, even for centuries. The morphologies of *Bacillus subtilis* (Fig. 2.3) and the endospores produced by *Bacillus* spp. are shown in Fig. 2.4. When *Bacillus* spp. are exposed to external stimuli, such as extreme temperature, pressure, salinity, pH, ultraviolet radiation, and desiccation and even to the starved environment with the lack of nutrients, they produce endospores and wait for the favorable environment to be available, and the endospore reactivates itself to the vegetative state for growth (Nicholson, 2002). Given the unpredictable complexity of natural environments, this production capability of endospores by *Bacillus* spp. is a unique

and beneficial feature as a candidate for MEOR. Moreover, existence of *Bacillus* spp. endospores have been discovered from unlikely environments such as rocks, dusts, and aquatic environments, which implies the possibility to biostimulate the spore-forming indigenous bacteria (Nicholson et al., 2000).

Possibly due to the aforementioned spore characteristics and survival ability of *Bacillus* spp. under extreme conditions, *Bacillus* spp. have been widely found in oil reservoirs (Kato et al., 2001; Da Cunha et al., 2006; Youssef et al., 2007; Simpson et al., 2011). The temperature of oil reservoirs, where the presence of *Bacillus* spp. is confirmed, varies from 19 to 106°C, the pressure from 4 to 27 MPa, and the salinity from 2% to 24%. It can be seen that *Bacillus* spp. are widely found in many of oil reservoirs and in wide ranges of temperature, pressure, and salinity.

Bacillus spp. produce lipopeptide biosurfactants, which behave as a surface-active agent to modify interfacial properties. *B. subtilis*, one of the most common *Bacillus* spp., produces surfactin. Surfactin is regarded as one of the most powerful lipopeptide biosurfactants to alter the interfacial properties, including surface tension. Surfactin is an amphiphilic bio-chemical compound of which structure contains both hydrophilic and hydrophobic parts (Fig. 2.5). Surfactin consists of a peptide loop of seven hydrophilic amino acids (L-aspartic acid, L-leucine, glutamic acid, L-leucine, L-valine, and two D-leucines), and a hydrophobic fatty acid chain (Ohno et al., 1995; Jacques, 2011; Henkel and Hausmann, 2019). As long as the surfactin concentration is below the CMC, the hydrophobic fatty acid part can extend freely into a solution. This makes surfactin a powerful surfactin. Dissolved surfactin in water can lower the surface tension of water from 72 to 27 mN/m at the concentration as low as 20 μM (Ohno et al., 1995; Abdel-Mawgoud et al., 2008; Shaligram and Singhal, 2010; Jacques, 2011).

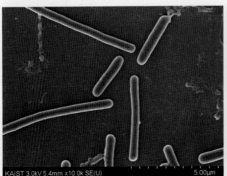
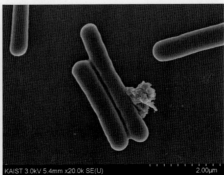

FIG. 2.3 The morphology of the cultured *Bacillus subtilis* cells imaged by using scanning electron microscopy (SEM). (Credit: Courtesy of Taehyung Park and Tae-Hyuk Kwon.)

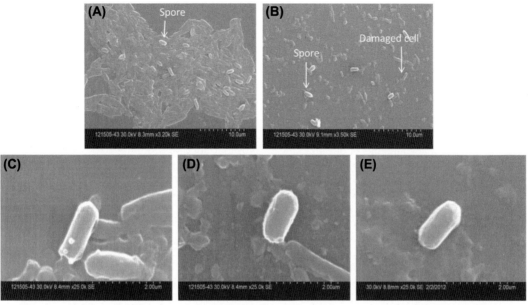

FIG. 2.4 Scanning electron microscopy (SEM) images of *Bacillus coagulans* spores. (Credit: Khanal, S.N., Anand, S., Muthukumarappan, K., 2014. Evaluation of high-intensity ultrasonication for the inactivation of endospores of 3 Bacillus species in nonfat milk. Journal of Dairy Science 97, 5952–5963.)

As the production of biosurfactant is heavily governed by the metabolisms of microbes, such as the respiration or fermentation processes for microbial cells to grow (Davis et al., 1999), the survivability or the growth of microbes under extreme environmental conditions are prerequisites to the application of biosurfactant producing bacteria for MEOR practices.

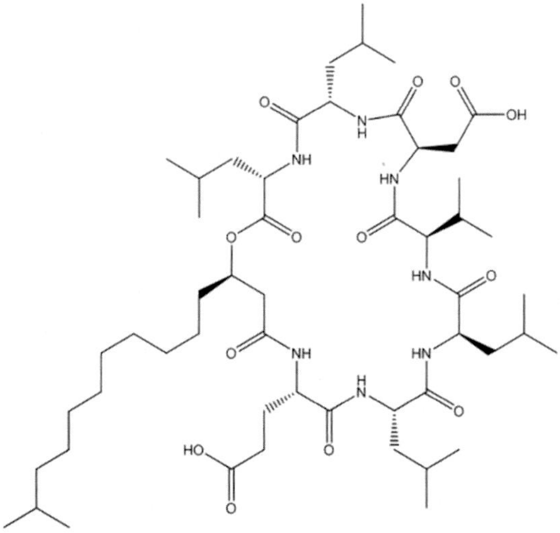

B. Pseudomonas *species*

Rhamnolipids are microbial surfactants that belong to the group of glycolipids, produced by *P. aeruginosa*. They are also known to be one of the most effective biosurfactant regarding to its function to modify the fluids interfacial properties, as well as rock wettability (Amani et al., 2010; Hörmann et al., 2010). Rhamnolipids not only exhibit high surface activity but also have high potential for its high emulsifying activity and solubilization activity. Rhamnolipids have four homologs, where one or two rhamnose units are linked to one or two fatty acids with 8–14 carbon atoms (Hörmann et al., 2010). Rhamnolipids reduce the surface tension of water from 72 to 31 mN/m (Syldatk et al., 1985), which represents relatively higher effectiveness as a biosurfactant. The CMC of rhamnolipid is in the range of 20–225 mg/L in water at 25°C (Syldatk et al., 1985; Dubeau et al., 2009).

Two major factors such as (1) environmental factors such as pH, temperature, agitation, and oxygen availability and (2) nutritional factors such as carbon, nitrogen, and metal ions are known to significantly affect rhamnolipid production (Desai and Banat, 1997; Varjani and Upasani, 2016; Moya-Ramírez et al., 2017). Rhamnolipid production is the maximum in the range of pH from 6.0 to 6.4, and the production drastically drops

FIG. 2.5 Structure of surfactin produced from *Bacillus subtilis*. (Credit: Henkel, M., Hausmann, R., 2019. Diversity and Classification of Microbial Surfactants. Biobased Surfactants. Elsevier, pp. 41–63.)

when pH is higher than 7.2 (Guerra-Santos et al., 1984; Varjani and Upasani, 2016). While *P. aeruginosa* is a facultative anaerobic bacterium, the produced rhamnolipid differs under anaerobic and aerobic environments. The aerobically produced rhamnolipids have shown to have a better surface activity. By contrast, the anaerobically produced rhamnolipids exhibit a better emulsifying activity (Zhao et al., 2016, 2018). Under an anaerobic condition, less rhamnolipids (680 mg/L) are produced compared with that under an aerobic condition (11,650 mg/L) although these amounts of biosurfactant produced by *P. aeruginosa* are sufficient to exceed the CMC (Zhao et al., 2018). There is an additional report that the increased aeration rate led to the increment in rhamnolipid production (Benincasa et al., 2004). The optimum condition for growing and producing rhamnolipids is 37°C and neutral pH with agitation (Moussa et al., 2014). Similar to *Bacillus* spp., the carbon source in a culture medium plays an important role in rhamnolipid production. High production yields of rhamnolipids are observed with glucose with mineral salt medium (Moussa et al., 2014; Varjani and Upasani, 2016). Among diverse nitrogen sources, *P. aeruginosa* produce the highest amount of rhamnolipid with nitrate (Guerra-Santos et al., 1984; Moussa et al., 2014). Furthermore, nitrogen-limited conditions cause the increased production by the factor of $\sim 3-4$. Iron also serves as a promoter of rhamnolipid production (Glick et al., 2010).

Pseudomonas spp. are frequently found in oil reservoirs due to the characteristics that they are facultative anaerobe and can live under microaerobic (microoxic) or anaerobic environments. Numbers of *P. aeruginosa* strains have been isolated from several oil fields (Rocha et al., 1992; Youssef et al., 2007; Amani et al., 2010; Hörmann et al., 2010; Zhao et al., 2016; Khademolhosseini et al., 2019), and this indicates that these strains can adapt to the oil reservoir conditions. The stability of the rhamnolipids has also been confirmed under the variations of pH, temperature, and salinity (Rocha et al., 1992; Varjani and Upasani, 2016; Khademolhosseini et al., 2019). This supports the use of rhamnolipids in MEOR practices.

Pseudomonas spp. also have a hydrocarbon-degrading ability as they utilize hydrocarbons as a carbon source for metabolisms. Therefore, *P. aeruginosa* has a potential to be widely applied for bioremediation or MEOR practices (De Almeida et al., 2016; Chong and Li, 2017).

C. Other species
There are several other microorganisms that can be utilized for MEOR practices. *Rhodococcus* spp. are one of the representative hydrocarbon-degrading strains and grow under a facultative anaerobic condition. They

also produce a biosurfactant when hydrocarbon is used as the carbon source of nutrition. *Rhodococcus ruber* Z25 has been isolates from the formation brine in Daqing oil field, China, and proposed as a potential model microorganism for MEOR application (Zheng et al., 2012). This strain is confirmed to have a capability to produce a biosurfactant, and the produced biosurfactant reduces the IFT of water and oil and has the CMC value of 57 mg/L. Both in aerobic and anaerobic conditions, *R. ruber* degrades crude oil and produces biosurfactants. Most of biosurfactant produced by *Rhodococcus* spp. are glycolipid-type molecules. In addition, *Rhodococcus* sp. strain TA6 shows an emulsifying activity with various hydrocarbons ranging from pentane to light motor oil (Shavandi et al., 2011). Glycolipid, one class of biosurfactants, isolated from Iranian oil–contaminated soils is found to reduce the surface tension of water from 68 to 30 mN/m and shows chemical stability upon the exposures to high salinity (10% NaCl), elevated temperature (120°C for 15 min), and a wide range of pH (4–10). The core-flooding tests on oil-saturated sand packs have demonstrated that the culture broth aid in 70% of residual oil recovery (Shavandi et al., 2011).

Acinetobacter spp. are also frequently found from oil-contaminated soils and oil reservoirs (Chen et al., 2012; Zou et al., 2014; Ohadi et al., 2017a, 2017b). They are not as common as *Pseudomonas* spp.; however, they can also produce rhamnolipid as *Pseudomonas* spp. do. Rhamnolipid, one type of biosurfactants, can reduce surface tension and emulsify crude oil in oil-water system. Rhamnolipid-producing *Acinetobacter junii* BD has been isolated from Xinjiang oil field, northwest China (Dong et al., 2016). This reservoir was 1088 m deep with a temperature of 32°C, which indicates the survivability of *Acinetobacter* spp. under high fluid pressure (Dong et al., 2016).

Arthrobacter spp. are another lipopeptide or trehalolipid biosurfactant producer. They are stable in the temperature of 30–100°C and in the pH range from 2 to 12 (Singh and Cameotra, 2004; Wang et al., 2016). Owing to their survivability under extreme environments, they also have potentials for MEOR practices.

2.3 BIOPOLYMERS
Biopolymers formed by microorganisms are one of widely considered biological products to enhance oil recovery. In particular, a selective plugging strategy can fully exploit the biopolymers through the modification of permeability of reservoir formations. Plugging of high permeability layers or regions in a reservoir by using microbial biopolymers, which is also referred to as "thief

zones," can redirect the water flooding to oil-rich channels, such that it helps the oil recovery (Sen, 2008). This selective plugging strategy can be achieved by two different ways: (A) biostimulation, which stimulates in situ indigenous bacterial community that originally exists within the reservoir and (B) bioaugmentation, which introduces bacterial cells (inoculation) with the nutrients for bacterial growth and the following biopolymer production in targeted highly permeable zones. Once the growth of biopolymer-producing bacteria is confirmed, nutrients can be further fed to facilitate biopolymer production for selective plugging. This section introduces the characteristics of representative biopolymers, which are widely used in various science and engineering practices and have potentials for MEOR.

2.3.1 General Definition
Biopolymer is an assembly of monomers derived from any living organisms including algae, bacteria, and fungi by consuming carbon source during their growth (Kalia, 2016). The biopolymers play various roles, such as energy reserve materials as a carbon source, shelters for microorganisms from the changing environmental conditions, and functioning of cell. Among diverse biopolymers, microbial EPSs specifically refer to the biopolymers that are synthesized by microorganisms and are commonly divided into three main categories: (A) nucleic acids and polynucleotides, which are the genetic materials in all cells, for example DNA and RNA; (B) polypeptides, which consist of amino acids and catalyze biochemical reactions in a specific way; and (C) carbohydrates, which are polysaccharides (Smidsrød et al., 2008). Among the three typical types of biopolymers, polysaccharides are the most commonly used biopolymers in industries (Khatami and O'Kelly, 2013; Vijayendra and Shamala, 2014). The polysaccharide-type biopolymers have chains of polymeric carbohydrate with monosaccharide units, and they form viscous hydrogels in the presence of water due to the abundant surface hydroxyl groups (OH^-) (Chang et al., 2016).

2.3.2 Formation of Biopolymers
From the perspectives of microorganisms, the biggest purpose of EPS production as microbial metabolites is related to the protective agent. Microorganisms attempt to protect themselves from desiccation or against predation by protozoans through surrounding them with EPS (Suresh Kumar et al., 2007). There are two different ways of EPS production. Most EPS are exported to the outside of bacterial cells wall after synthesized intracellularly. However, there are some exceptions such as

xanthan and dextran, which are polymerized and synthesized outside of the bacterial cells by the activity of secreted enzymes that convert substrate to the polymer (Rehm, 2009). Although the mechanism of biosynthesis varies from EPS to EPS, there are common steps of a bacterial biosynthesis pathway. After the uptake of resources such as carbon substrates, it is assimilated by the central metabolite pathway; thereafter, the decomposed substrates are synthesized as polysaccharides by specific enzymes (Vandamme et al., 2002). Biopolymers are formed by condensation reactions which dehydrate to link monomers (Fig. 2.6), or they are synthesized by phosphorylated intermediates that are catalyzed by divalent cation-dependent motors (Fig. 2.7).

2.3.3 Representative Biopolymers and Their Properties
A variety of microorganisms can produce EPS that have potentials to be used for enhanced oil recovery. Among those biopolymers, some representative biopolymers are introduced.

A. Curdlan produced by Alcaligenes spp. and Agrobacterium spp.
a. Microbial production. Curdlan is a neutral, water-insoluble, and alkali-soluble EPS produced by the fermentation of pathogenic bacteria, e.g., *Agrobacterium biobar*, *Alcaligenes faecalis*, under nitrogen-limited conditions (Harada et al., 1968).

The synthesis of curdlan takes place inside of bacterial cells with several enzymes. Curdlan can be effectively formed from glucose and many carbon compounds in a simple, defined medium in the post-stationary growth phase during the conditions of N-starvation. The yield of curdlan shows the maximum at the temperature of $30-32°C$ (Saito et al., 1968; PHILLIPS and Lawford, 1983; Nakanishi et al., 1992). The pH needs to be maintained in the optimum range of $6.5-7.5$ for polysaccharide production during incubation. In batch fermentation, the cell growth rate is the maximum at pH 7.0; on the other hand, curdlan synthesis is optimal at pH 5.5 (Lee et al., 1999). Bacterial cells can be removed by centrifugation after dissolving the curdlan in an alkaline condition medium. Thereafter, by neutralizing the medium with acid, the purified form of curdlan is precipitated and can be separated (Sandford, 1979).

b. Chemical and physicochemical characteristics. Curdlan is composed of linear polysaccharides consisting mainly of β-D-$(1 \rightarrow 3)$-linked

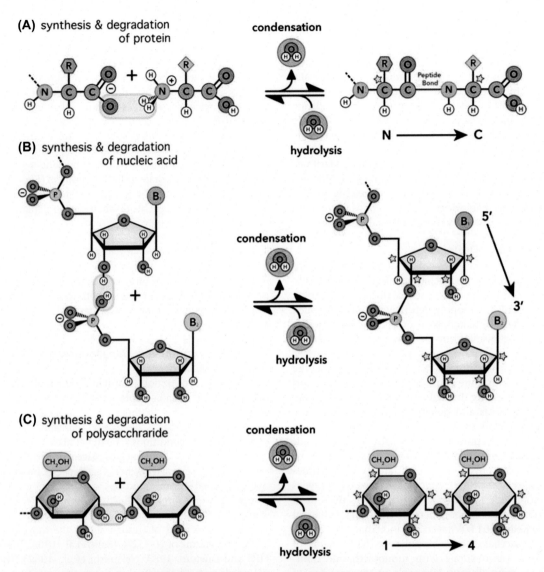

FIG. 2.6 Net reactions for biopolymer formation by condensation reactions. (Credit: Runnels, C.M., Lanier, K.A., Williams, J.K., Bowman, J.C., Petrov, A.S., Hud, N.V., Williams, L.D., 2018. Folding, assembly, and persistence: the essential nature and origins of biopolymers. Journal of Molecular Evolution 86, 598–610.)

D-glucose residues. The curdlan molecules are made up of approximately 12,000 glucose units, which have the molecular weight of $5 \times 10^4 - 2 \times 10^6$ g/mol, as shown in Fig. 2.8 (Futatsuyama et al., 1999; Rehm, 2009; Zhang et al., 2020). Curdlan has a very low solubility in alcohols, most organic solvents, and water in ambient temperature. By contrast, it can be fully dissolved in alkaline solutions such as dilute bases (0.25 M NaOH), dimethylsulfoxide (DMSO), formic acid, and aprotic reagents without heating (Cho, 2001).

The advantage of curdlan is that two different types of curdlan gels can be formed only by the heating process. After heating to the temperature higher than $\sim 80°C$, it produces a high-set, strong thermoirreversible gel, whereas a low-set, reversible gel, which would melt if it is reheated, is produced on heating to the temperature at $\sim 60°C$ (Mcintosh et al., 2005).

c. Application. Due to its unusual rheological properties, curdlan has been mainly used in foods industry as a

(A) Phosphorylated intermediate in protein synthesis

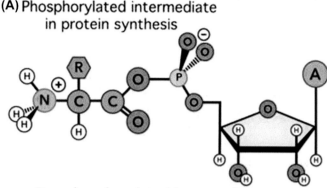

(B) Pyrophosphorylated intermediate in polysaccharide synthesis

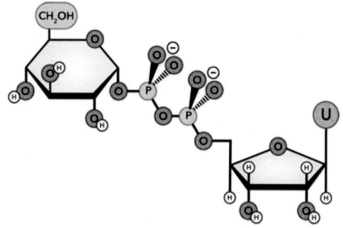

FIG. 2.7 Intermediates in the (**A**) protein and (**B**) polysaccharide synthesis. (Credit: Runnels, C.M., Lanier, K.A., Williams, J.K., Bowman, J.C., Petrov, A.S., Hud, N.V., Williams, L.D., 2018. Folding, assembly, and persistence: the essential nature and origins of biopolymers. Journal of Molecular Evolution 86, 598–610.)

gelling material for jelly products and a food additive in baking (Harada, 1977). However, curdlan has been also applied in nonfood areas. In civil engineering, curdlan is used for the formulation of enhanced fluidity concrete to prevent small aggregates from segregating. It also has been proposed for organic binding agent for ceramics (Harada, 1992).

→3)-β-D-Glc*p*-(1→3)-β-D-Glc*p*-(1→

FIG. 2.8 Structure of curdlan. (Credit: Zhang, H., Zhang, F., Yuan, R., 2020. Applications of natural polymer-based hydrogels in the food industry. Hydrogels Based on Natural Polymers. Elsevier, 357–410.)

B. Pullulan produced by Aureobasidium spp.

a. Microbial production. The yeast-like fungus *Aureobasidium pullulans*, which is one of the commonest and most widespread of fungi, usually produces an extracellular, water-soluble, and neutral EPS known as pullulan. Although pullulan is EPS, it is secreted out of the cell surface after the intracellular synthesis at the cell wall membrane to form a loose, slimy layer (Simon et al., 1993). However, the mechanism of pullulan biosynthesis remains not fully understood yet. According to the conditions and duration of fermentation, and the strain used, the molecular weight and the rheological properties of pullulan can be differed (Catley, 1972; Leduy et al., 1974).

Pullulan can be synthesized from a variety of carbon sources such as sucrose, glucose, fructose, maltose, mannose, and even agricultural waste by saccharification of plant fibers into glucose with its multiple enzyme systems (Bender et al., 1959; Sandford, 1979; Duan et al., 2008). However, at high concentrations of carbon source (i.e., above 5%), the production of pullulan can be inhibited due to the suppression effect of sugars on enzymes involved in pullulan production (Shin et al., 1989; Duan et al., 2008). At optimized concentrations of sucrose and nitrogen source, a pullulan yield reaches up to 60.7% (Cheng et al., 2010). However, it is also reported that pullulan cannot be produced with acetate, D-galactose, glycerol, lactose, or D-mannitol as the carbon source.

Nitrogen sources, especially ammonium ion (NH_4^+), also promote the production of pullulan by influencing on protein synthesis. The depletion of nitrogen is considered as the signal for the start of pullulan formation (Bulmer et al., 1987). It is reported that a 10:1 carbon/nitrogen ratio is the most favorable condition for the maximal pullulan production (Suresh Kumar et al., 2007). The concentration of mineral salts in the growth medium also has to be considered for the production of pullulan (Gao et al., 2010). It has been suggested that the optimal medium for pullulan production contains 75 g/L of sucrose, 3 g/L of yeast extract, and 5 g/L of ammonium sulfate (Chang et al., 2016). The pH and temperature also have huge effects on pullulan production. The optimal pH for pullulan synthesis is in the range of 5.5–7.5 (Lee and Yoo, 1993).

b. Chemical and physiochemical characteristics. Pullulan has the chemical formulation $(C_6H_{10}O_5)_n$, and it has a linear polysaccharide made up of maltotriose-repeating units linked through α-D-(1,6) bonds and small number of α-D-(1,4) bonds, as shown in Fig. 2.9 (Shingel, 2004). The molecular weight of pullulan ranges from 4.5×10^4 to 6×10^5 g/mol, and it varies with various cultivation parameters such as strain, pH, and phosphate content of the medium during fermentation (Kato and Shiosaka, 1975).

c. Application. Pullulan has a wide range of commercial and industrial applications in various fields such as adhesive, cosmetic, especially food and pharmaceutical industry. Due to the excellent physical properties of its films, which are water-soluble, biodegradable and

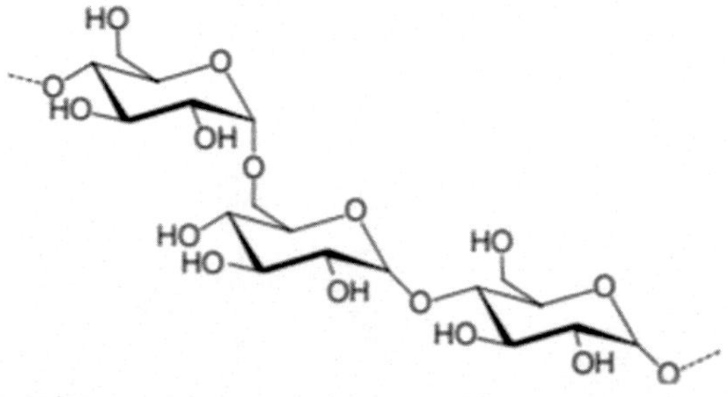

FIG. 2.9 Chemical structure of pullulan. (Credit: Shingel, K.I., 2004. Current knowledge on biosynthesis, biological activity, and chemical modification of the exopolysaccharide, pullulan. Carbohydrate Research 339, 447–460.)

impervious to oxygen, formed from pullulan, it is suitable for coating or packaging foods pharmaceuticals (Cheng et al., 2011). Pullulan has potential to be used for environmental remediation and oil industry. It is found that pullulan-producing strains can remove heavy metal ions such as Cu, Fe, Zn, Mn, Pb, Cd, Ni, and Cr in aqueous environment (Radulović et al., 2008). The pullulan is also candidate for MEOR from the polluted field due to its high viscosity (Cheng et al., 2011).

C. Dextran produced by Leuconostoc spp.
a. Microbial production. Dextran is the neutral EPS which is produced by *Leuconostoc* spp. and *Streptococcus* spp. when fed with sucrose (Jeanes et al., 1954). In specific, dextran is produced by growing *L. mesenteroides* in media containing sucrose, nitrogen source, certain trace mineral, and phosphate. Among many carbon sources, *Leuconostoc* spp. produce a particular class of the enzyme called "dextransucrase" under only the growth medium containing sucrose during their metabolism. This enzyme dextransucrase is glucansucrase belonging to the glycoside hydrolase superfamily, and it synthesizes dextran from glucose. The optimal conditions typically suggested for dextran production include the pH of 6.7–7.2, the temperature of ~25°C, the initial sucrose concentration of ~2%, and the fermentation time of ~24–48 h (Vandamme et al., 2002).

b. Chemical and physicochemical characteristics. Dextran, which has the chemical formulation of $(C_6H_{10}O_6)_n$, has α-(1,2)/α- (1,3)/α-(1,4)-branched and α-(1,6)-linked homopolymer, and the main component is glucose, as shown in Fig. 2.10 (Rehm, 2010). Among different kinds of linkages, dextran has α-(1,6)-linkages more than 50% of the total linkages. The high portion of (1,6)-linkage enables its unusual flexibility.

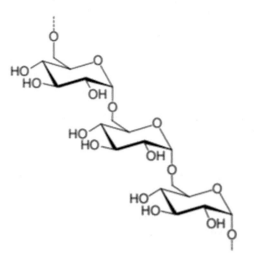

FIG. 2.10 Chemical structure of dextran. (Credit: Shingel, K.I., 2004. Current knowledge on biosynthesis, biological activity, and chemical modification of the exopolysaccharide, pullulan. Carbohydrate Research 339, 447−460.)

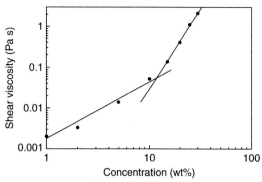

FIG. 2.11 Shear viscosity changes as a function of dextran concentration. (Credit: Tirtaatmadja, V., Dunstan, D.E., Boger, D.V., 2001. Rheology of dextran solutions. Journal of Non-newtonian Fluid Mechanics 97, 295−301.)

The dextran produced by *L. mesenteroides* spp. shows porous and heterogeneous nature due to the high percentage of consecutive α-1,3 linkages (Wilham et al., 1955; Naessens et al., 2005; Jeon et al., 2017). It is worth noting that the exact structure of each type of dextran relies on the producing microbial strain. The length of chain determines the molecular weight, which ranged from 10^6 to 10^9 g/mol (Kim and Fogler, 1999).

Dextran is a porous material, and the environmental scanning electron microscopy (ESEM) shows that the radii of pores in dextran range from 1 to 100 μm and ∼95% of the radius is less than 20 μm (Jeon et al., 2017). Dextran shows different physical properties with its solubility and aqueous concentration. The storage modulus and viscosity of insoluble dextran are estimated to be ∼0.1 Pa and ∼1−10 Pa s, respectively (Jeon et al., 2017). In addition, the shear viscosity of dextran solutions, in which dextran has the nominal molecular weight of 2 million g/mol, has been measured at the different concentrations of dextran (Tirtaatmadja et al., 2001). The shear viscosity of dextran solution increases with the dextran concentration, and it shows two distinct regions with a critical overlap concentration at ∼12 wt.%, as shown in Fig. 2.11 (Tirtaatmadja et al., 2001).

c. Application. Different application strategies can be taken according to the molecular weight. Dextran with relatively low molecular weight (e.g., 40,000−100,000 g/mol) has been used as a blood volume expander (Naessens et al., 2005). In addition, the use of dextran in MEOR has been proposed as a selective clogging material. A highly permeable zone can be clogged by bacterial dextran accumulation in pores, and its permeability can be lowered down close to the value of lower permeable zones (Lappan and Fogler, 1996). Meanwhile, the use of dextran for drilling muds and viscous water-flooding has been successfully tested (Sandford, 1979).

D. Scleroglucan produced by Sclerotium spp.
a. Microbial production. Scleroglucan is a neutral EPS secreted in the forms of powder and liquid by the fermentation of plant pathogen fungi genus *Sclerotium* (Liang et al., 2019). The production of scleroglucan is markedly affected by the type and concentration of carbon sources. It shows the highest yield at the media with 150 g/L sucrose for *Sclerotium rolfsii*. However, those high concentrations decrease the yield of biopolymer with *Sclerotium glucanicum*. The type of nitrogen source also affects the scleroglucan production, which is higher in the media containing nitrate (NO_3^-) than the media containing ammonium (NH_4^+)- or ammonia (NH_3)-based N-sources. By contrast, different factors, such as phosphate and initial pH, have insignificant effects on production of scleroglucan (Wang and McNeil, 1994; Farina et al., 1998; Lee, 1999).

b. Chemical and physicochemical characteristics. Scleroglucan is a water-soluble, neutral biopolymer composed of a linear chain of β-D-(1→3)-linked glucose residues with single β-D-(1→6)-linked glucose attached to every third residue of the main chain, which gives rigid, rodlike, and triple helical structure, as shown in Fig. 2.12 (Li et al., 2020). Due to the existence of 1,6-β-D-glucopyranosyl groups that

Scleroglucan

FIG. 2.12 structure of scleroglucan (Courtesy of Dr. C. Lawson, Carbosynth, Compton UK). (Credit: Li, X., Lu, Y., Adams, G.G., Zobel, H., Ballance, S., Wolf, B., Harding, S.E., 2020. Characterisation of the molecular properties of scleroglucan as an alternative rigid rod molecule to xanthan gum for oropharyngeal dysphagia. Food Hydrocolloids 101, 105446.)

increase the solubility of the polysaccharide, scleroglucan is easily dissolved in water at room temperature (Sandford, 1979). The molecular weight of commercial scleroglucan shows a range of $\sim 5–6 \times 10^6$ g/mol although it varies with the cultivation media and strains used (Lecacheux et al., 1986).

The stable structure of scleroglucan causes a highly viscoelastic behavior like xanthan gum (Palleschi et al., 2005). The viscosity of scleroglucan is slightly affected only by the ambient temperature change. With the addition of chromium salts and borax at pH of 10–11, scleroglucan even forms stable gels (Coviello et al., 2003). Because of the totally nonionic character of the molecule, the viscosity of scleroglucan solutions has tolerance to electrolytes concentration. Due to the rigid structures, it has tolerance to high temperature up to $\sim 130°C$, a broad range of pH, and shear stress. When shear stress is applied, the scleroglucan becomes less viscous, and when the stress is relaxed, it becomes more viscous (Saxman and Crull, 1984; Rivenq et al., 1992).

c. Application. Scleroglucan has been used as viscosifying agents in food, cosmetic, and personal care industries for decades. Meanwhile, the largest potential application of scleroglucan is in the area of oil industry for thickening, discharge of drilling muds, and enhanced oil recovery. The use of polymer solution injection is conducted in North Sea oil reservoir at the field scale. Among more than 140 polymers, scleroglucan has shown the best EOR performance in terms of its viscosifying ability and mobility reduction in porous media under high temperature (90°C), high salinity, and high pressure. The

scleroglucan-containing solution maintains 90% of its original viscosity for 500 days at 90°C of seawater. However, use of polymer to enhance seawater injection and waterflooding processes may be challenging because of cost input, ease of biodegradation and oxidation, and poor filterability (Davison and Mentzer, 1982; Pu et al., 2018). Nowadays, Cargill started program to develop an EOR grade scleroglucan, which can maintain its viscosity or injectivity at high temperature and high salinity, with high purity for use of scleroglucan in harsh reservoir conditions (Jensen et al., 2018).

E. Xanthan gum produced by Xanthomonas spp.

a. Microbial production. Xanthan gum is one of the most versatile polysaccharides, which is produced through the fermentation of glucose by *Xanthomonas campestris* bacterium. The rigid polysaccharide chain of xanthan molecules enhances tolerance over not only the mechanical shear but also against wide ranges of temperature (up to 90°C), pH (2–11), salinity (up to 150 g/L NaCl), and divalent ion concentration (Pollock et al., 1994; Wei et al., 2014).

Xanthan gum is usually produced by the fermentation of *X. campestris* with sucrose, glucose, or corn syrup as the major carbon source (Becker et al., 1998). When the fermentation takes place in a sucrose-rich condition, the use of nitrate ion as a nitrogen source shows the greater yield in xanthan production than ammonium ion (Letisse et al., 2001). In addition, the xanthan gum produced at the temperature of 25°C is found to have the heaviest molecular weight (Casas et al., 2000).

b. Chemical and physicochemical characteristics. Xanthan contains diverse carbohydrates, such as glucose, mannose, glucuronic acid, acetal-linked pyruvic acid, and acetate, as shown in Fig. 2.13 (Ghoumrassi-Barr and Aliouche, 2015; Pu et al., 2018). The molecule has an anionic charge and the molecular weight of $2–50 \times 10^6$ g/mol (Becker et al., 1998).

Because of hydrocolloidal characteristics of xanthan, its solution shows a pronounced shear-thinning behavior when xanthan is mixed with water. This is possibly attributable to the molecule aggregation through hydrogen bonding and polymer entanglements. Shear thinning behavior exhibits a high viscosity at a low shear rate. However, with an increasing shear rate, viscosity decreases rapidly due to the breakage of formed association (Tanner and Yoshida, 1993). Therefore, the xanthan solution exhibits the shear rate–dependent viscosity, which depends on the polymer

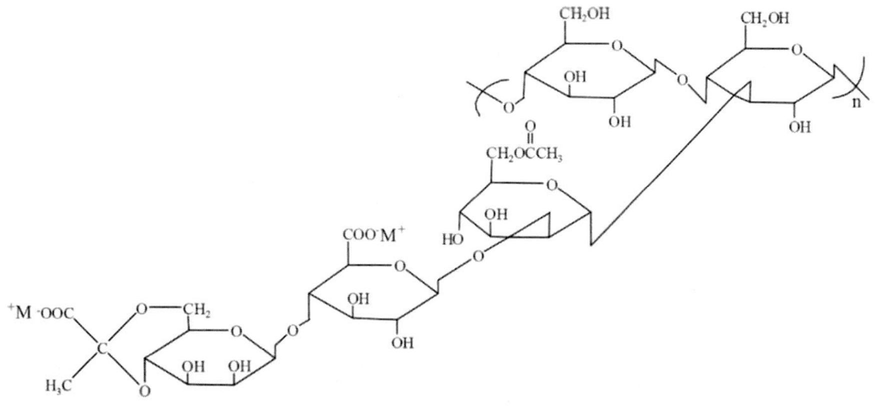

FIG. 2.13 Structure of xanthan. (Credit: Pu, W., Shen, C., Wei, B., Yang, Y., Li, Y., 2018. A comprehensive review of polysaccharide biopolymers for enhanced oil recovery (EOR) from flask to field. Journal of Industrial and Engineering Chemistry 61, 1–11.)

concentration. Milas et al. show that the extent of shear thinning or the relative reduction in viscosity with a shear rate increment decreases with an increases in xanthan concentration, in which xanthan has the molecular weight of 7×10^6 g/mol and the concentration of 0.0625–2 g/L (Milas et al., 1985).

c. Application. Due to its special rheological properties, xanthan is widely used in various industries, such as food, pharmaceutical, cosmetics, toiletries, chemicals, paint, textile, agricultural products, and petroleum industry, especially for the EOR application (Becker et al., 1998). In the petroleum industry, xanthan gum is used in flooding, drilling, fracturing, pipeline cleaning, and work-over and completion as water-flood thickening agents (Katzbauer, 1998). Increasing the viscosity of the injected water, xanthan gum helps to achieve the higher oil recovery than traditional waterflooding. Generally, xanthan gum is added directly to injected water due to the excellent compatibility with salt and resistance to thermal degradation. The thermal stability of xanthan is excellent, and therefore, the rate of viscosity loss is very slow (Katzbauer, 1998). It also can be produced by stimulating the indigenous microbes in the reservoir. As an example, the result of a pilot-scale test conducted from 1994 in China shows a favorable response of water cut and oil production rate upon EOR application with xanthan.

However, xanthan is presumed to have limited applicability for a harsh environment, such as a reservoir warmer than $\sim 80°C$. Also, the impurity of xanthan product can be another limiting factor in practices (Ash et al., 1983; Wellington, 1983; Ryles, 1988; Kalpakci et al., 1990; Seright and Henrici, 1990).

F. Levan produced by Bacillus spp.

a. Microbial production. Levan is one of two typical types of fructan biopolymers synthesized from the fermentation of sucrose by the action of enzyme levansucrase (Rehm, 2009). Levan is produced by a broad range of microbial species such as *Bacillus, Halomonas, Leuconostoc, Rahnella, Lactobacillus, Pseudomonas,* and *Zymomonas* (Han, 1990; Bahl et al., 2010). Levan has high solubility in water and oil due to the presence of β-2,6 linkage. However, it is insoluble in most organic solvents except for dimethyl sulfoxide (DMSO) like other biopolymers (Ullrich, 2009).

The conditions such as carbon source, pH, temperature for microorganism cultivation, and levan formation differ from species to species. The submerged fermentation technique is usually performed for the microbial production of EPS (Han, 1990).

Zymomonas mobilis ATCC 31812 bacterium is proposed as a potential candidate for large-scale levan production (Ananthalakshmy and Gunasekaran, 1999). In batch culture experiments, the production of levan reaches the maximum yield with sucrose, more than that with molasses and sugar cane syrup. The concentrations of yeast extract and monobasic potassium phosphate (KH_2PO_4) show significant impacts on levan production in sucrose media (de Oliveira et al., 2007).

b. Chemical and physicochemical characteristics. The backbone of levan is a β-2,6 polyfructan with β-2,1 bonding at branch points, which occupies a total of 12% branching in the levan polysaccharide chains, as shown in Fig. 2.14 (Venugopal, 2011). Like other polysaccharidic biopolymers, levan has a wide range of molecular weight depending on the microorganisms used. Levan from *B. subtilis* has two different ranges of

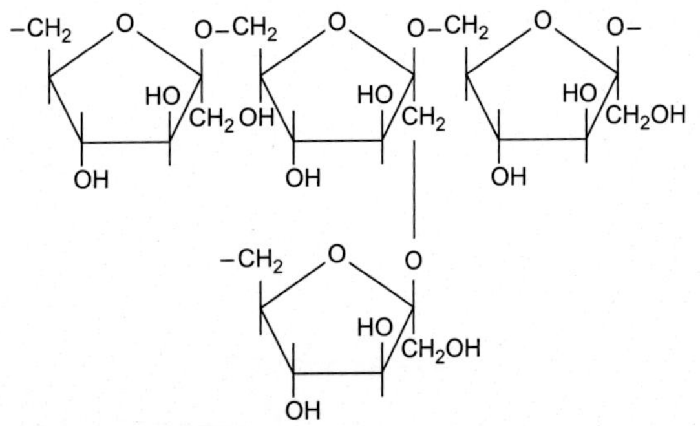

FIG. 2.14 Chemical structure of levan. (Credit: Manandhar, S., Vidhate, S., D'Souza, N., 2009. Water soluble levan polysaccharide biopolymer electrospun fibers. Carbohydrate Polymers 78, 794–798.)

molecular weight; the low-molecular-weight regime is measure as $\sim 1.1 \times 10^4$ g/mol; on the other hand, the high-molecular-weight regime is around 1.8×10^6 g/mol. Levan has a very low viscosity, its inherent viscosity is ~ 0.14 dL/g, and this is attributed to its spherical shape (Arvidson et al., 2006).

Meanwhile, the branches of levan have an influence on its cohesive strength. The large number of hydroxyl groups makes adhesive bonds with a variety of substrates. The tensile strength of levan on aluminum is estimated up to 10.3 MPa (Rehm, 2009).

c. **Application.** The low oxygen permeability of levan provides merits as an edible surface-coating agent, and

thus, levan is widely used in the food and pharmaceuticals industry (Han, 1990). Meanwhile, biopolymer levan is also considered to have a potential to be used for EOR application in specific oil reservoirs condition (Ramsay et al., 1989).

G. Alginate
a. **Microbial production.** Alginate is an EPS produced by various microorganisms such as brown algae and bacteria (Gorin and Spencer, 1966; Painter, 1983). Several bacteria, including *Azotobacter* spp. and *Pseudomonas* spp., can produce alginate, and it is the main constituent of biofilms to protect the bacteria from the hostile environments (Campos et al., 1996). Alginate is mostly produced by extracting from brown algae, mainly from *Laminaria hyperborea*, *Macrocystis pyrifera*, *Laminaria digitata*, and *Ascophyllum nodosum*. Therefore, the composition of alginate in different sources varies according to seasonal and growth conditions (Andresen et al., 1977). Alginate is widely used in food and medical industries for biomedical products and drug delivery.

b. **Chemical and physicochemical characteristics.** Main components of alginate are mannuronic acid and gluronic acid. These materials link each other with β-(1,4) by glycosyltransferase which is polymerizing enzyme, as shown in Fig. 2.15 (Liang et al., 2015). This polymer is a nonrepeating heteropolymer. The molecular weight of alginates ranges $\sim 2 \times 10^5 - 5 \times 10^5$ g/mol (Rehm, 2010).

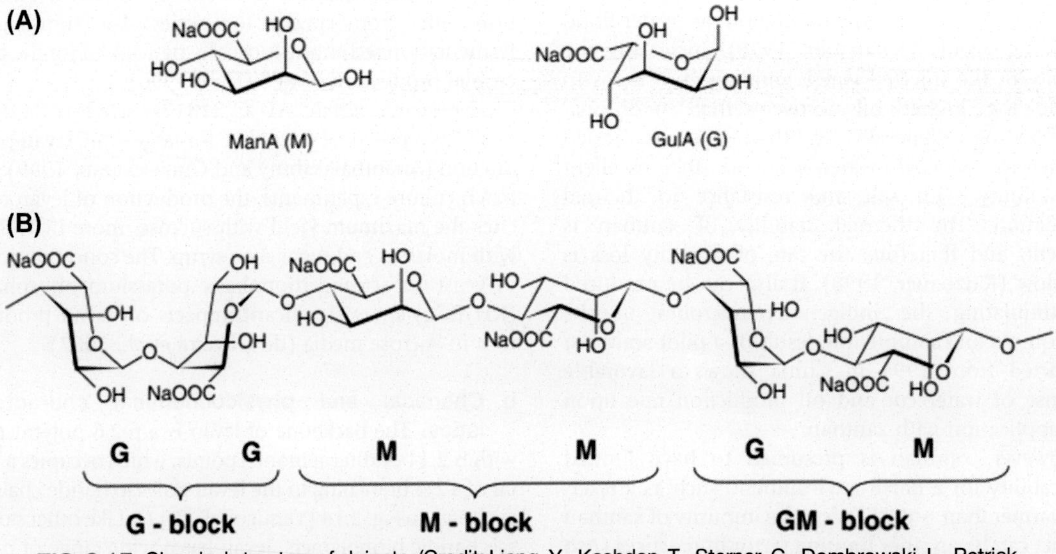

FIG. 2.15 Chemical structure of alginate. (Credit: Liang, Y., Kashdan, T., Sterner, C., Dombrowski, L., Petrick, I., Kröger, M., Höfer, R., 2015. Algal Biorefineries. Industrial Biorefineries & White Biotechnology. Elsevier, pp. 35–90.)

The elastic modulus of alginate shows a dependency on the molecular weight (Draget et al., 1993). Young's modulus is computed to range 25–100 kPa for a certain class of alginate with the G-block length less than 20. Meanwhile, the enzymatic elongation in the polysaccharide chain leads to young's modulus increment (Draget et al., 2000). Alginate shows a linear stress-strain response which at maximum extends up to a deformation strain level of 10%–12%. For saturated alginate gels, the stress–strain linearity holds up to a maximum deformation of 6%–8%. After a linear behavior, a nonlinear stress-strain curve appears, which is common observed in biopolymers. Meanwhile, a plastic behavior is found in calcium-saturated alginates (Mørch et al., 2008).

2.3.4 Applications of EPS for Selective Plugging Strategy of Microbial Enhanced Oil Recovery

The EPSs can alter the flow behavior by clogging of highly permeable zones. However, the success requires several conditions, as follows (Jenneman et al., 1984; Lappin-Scott and Costerton, 1989):

(a) The EPS must be resistant to degradation by the indigenous microbes during the period of bio-augmentation and the following oil production. Thereby, the EPS accumulated at the reservoir behaves a long-lasting flow barrier.

(b) The organisms augmented (or inoculated or introduced from exterior) must be able to survive at the reservoir conditions, such as high temperature, high salinity, and high fluid pressure.

(c) The added organisms should be able to compete against the preexisting microorganisms, such that the bioaugmentation is unaffected by the indigenous microorganisms.

(d) The reservoir is highly heterogeneous, and the low permeable zones contain economically reasonable amounts of crude oils for water flooding.

The selective plugging strategy has been implemented at the field scale in Fuyu oil field, China. According to the report, the microbes used in the tests were isolated for their adaptations at the reservoir (Nagase et al., 2002). The isolated microbes were confirmed to produce cellulous by consuming molasses as the main carbon resources, and the cellulous can effectively clog pores and reduce permeability of the reservoir formation. The mixture of the isolated microorganism and molasses was injected at the production wells, and the huff-n-puff tests were carried out at those production wells to investigate how microorganism would behave under the reservoir condition and to optimize the scheme of injection strategy. Upon introduction of the microbes with

nutrients, the production well shut in 10 days for cultivation and production of biopolymers. The result of production tests reveals that the water cut decreased from 99% to 75%, and the oil production rate increased from 0.25 to 2.0 ton/d. This field-scale demonstration reports a successful implementation of selective plugging using bacterial biopolymers and EPS.

2.4 BIOFILMS AND EXTRACELLULAR POLYMERIC SUBSTANCES

2.4.1 Definitions

As bacterial cells grow on the mineral surface, they usually form biofilms, which are the mixtures of bacterial cells and surrounded adhesive matrix secreted by the cells. This matrix is composed of polysaccharides, proteins, and nucleic acids produced by bacteria. The biofilms help to prevent cell detachment from a solid surface under a high-speed water flow like river, as well as aid in keeping nutrients required in cell growth such as carbon sources. Therefore, bacteria produce biofilms as a microbial self-defense to protect cells from physical forces and toxic molecules and to increase the survival rate of the cells. Biofilms also provide aid in the attachment of cells to nutrient-rich surfaces of interest and increase cell viability through cell-to-cell interactions.

The biofilm growth is much more general growth pattern in a water-saturated condition, such as below the water level in a water reservoir or below the ground water table in subsurface than the planktonic growth of each cells suspended in an aqueous phase. It is because the biofilm growth has several advantages in growth of cells by sharing nutrients, sharing electrons, protecting from physical and chemical attacks, and so on. Naturally occurring biofilms are generally composed of several bacteria species, two to several hundreds of species. Biofilm is not merely a colony of cells trapped in sticky materials, but a society of various bacterial cells and materials helps each other to survive in nature.

One of the main advantages of biofilm growth is high tolerance to antibiotic materials. Cells in biofilms have much higher tolerance to antibiotics, around 1000 times higher than the planktonic cells by blocking antibiotic penetration; biofilms induce genetic variance by cell-to-cell genetic exchange and slower cell growth (Stewart, 2002; Madigan et al., 1997).

There are four main reasons why bacteria generally form biofilm, rather than planktonic growth.

A. Self-defense system

Biofilms help to protect cells from external physical and chemical stimulations. Biofilms provide physical barrier and thus prevent cell removal. Also, these enhance the immune system of the cell-biofilm mixture by protecting from toxic materials such as antibiotics. This cell community system extremely increases survival of cells comparing with the planktonic cell growth.

B. Microbial nutrient niche
By attaching the biofilm to the favorite nutrient condition, the enclosed cell community becomes a nutrient-rich condition. The cells can remain in the nutrient niche and consume the favorite nutrients. Besides, biofilms further accommodate the sharing of nutrients among cells.

C. Cell-to-cell communication
Biofilm structure can serve as a pathway of electrons. Using this pathway, bacterial cells can transfer electrons and ions. These communications help to enhance survival chances of the community.

D. Genetic exchange
As the biofilm causes aggregation of cells, physically being close each other, the cells in the biofilm have more chance of genetic exchange. It can be one of the most powerful reasons that increase survival chances of the cell community.

For these reasons, biofilm formation from prokaryote cells is a general case and "default mode" of their growth. Except under an extremely low nutrient (or starved) condition, not enough energy to form biofilms, in most of cases, bacteria choose biofilm formation rather than planktonic cell growth (Madigan et al., 1997).

2.4.2 Biofilm in Microbial Enhanced Oil Recovery
MEOR is a process to enhance oil production rate from oil reservoirs using one or multiple microbial reactions. There are three general processes to enhance oil recovery rate using microbial reactions: gas production, wettability alteration, and permeability control by selective plugging. Biofilm can have a positive influence on oil recovery in both wettability alteration and permeability control. Since a lot of biofilm-producing bacteria generate biogas during metabolism for bacterial growth, biofilm can be associated with all of three processes in either direct or indirect ways.

A. Wettability modification
Wettability modification is a method to control capillarity of oil at the saturated pore space. A reduction in capillary pressure of oil at pore throats indicates a decrease in the resisting force which prevents oil from being displaced. Typically, chemical surfactants and biosurfactants are considered to alter IFT and wettability in

oil reservoir (Al-Sulaimani et al., 2012; Al-Bahry et al., 2013; Al-Wahaibi et al., 2014; Geetha et al., 2018; Park et al., 2019). However, biofilms also show similar reactions by physical and chemical surface modification (Polson et al., 2010; Karimi et al., 2012). Therefore, the reduction in capillarity of oil by microbial biofilms can lead to an increase in oil recovery efficiency.

B. Permeability control
Another way that biofilm helps MEOR processes is the permeability control in the reservoir. It is widely known that biomasses, such as bacterial cells, biofilms, biopolymers, and biogas accumulating in pores, cause permeability reduction of porous media, referred to as bioclogging (Taylor and Jaffé, 1990; Noh et al., 2016; Kim et al., 2019). Similar to the biopolymers that are mentioned in previous sections, the biofilm-induced bioclogging may have possibility to be used for selective plugging in MEOR. Highly permeable formations in a heterogeneous reservoir can be selectively plugged with biomasses, particularly with biofilms, as shown in Fig. 2.16. Such permeability control by microbial biofilm formation can be used to facilitate water flooding into unswept, oil-rich zones, enhancing oil sweeping efficiency (Raiders et al., 1989; Suthar et al., 2009; Klueglein et al., 2016).

2.4.3 Formation of Bacterial Biofilms
The production of biofilms consists of five steps, as shown in Fig. 2.17: (1) initial cell attachment to a surface, (2) irreversible attachment of cells, (3) colonization, (4) development of bacterial cell and polysaccharide, and (5) active dispersal (Monroe, 2007; Aslam, 2008). Initial cell attachment is initiated from the random movement of a cell at a surface. During the collision, the attraction between biomass and interface causes the adhesion and attachment to the surface. Biomass includes bacterial cells, proteins, and polysaccharides. Biofilm-generating genes, which manage intercellular communications and synthesize encoding proteins, then receive a signal that the cell is attached on the surface. After the biofilm formation process begins, the planktonic cells lose their flagella and hence their mobility.

The detailed mechanism remains to be more revealed; bacteria can find the appropriate surface and coordinate the events, leading to the biofilm growth mode. Thereafter, a switch from planktonic growth to biofilm formation is commenced by producing cyclic dimeric guanosine monophosphate (c-di-GMP; Fig. 2.18). The c-di-GMP is a kind of second messenger, which transmits signals from the environment to the intercellular structure, and these signals serve as a trigger to begin forming

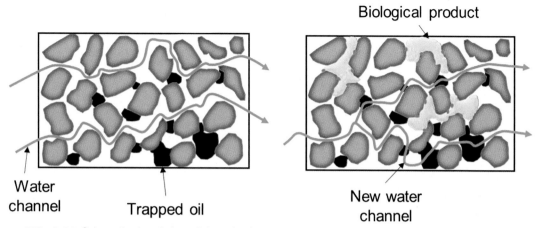

FIG. 2.16 Schematic description of the selective plugging mechanism during MEOR practices. MEOR, microbial enhanced oil recovery. (Credit: Courtesy of Tae-Hyuk Kwon and Taehyung Park.)

biofilms. During the initial phase with the planktonic and sessile growth, the c-di-GMP binds to proteins and enzymes that control the activity of flagella and produce biofilm matrix. In addition, it is found that the c-di-GMP process is controlled by the direct interaction with riboswitches regulatory messenger RNAs (Boyd and O'Toole, 2012; Madigan et al., 1997).

2.4.4 Biofilm Architecture

All biofilms produced by microbial process are unique communities even if they have some general structure. Paradoxically, bio-"film" is highly heterogeneous and contains several porous layers with several bacterial microcolonies of cells, EPS, and flow channels for nutrient, oxygen, nitrogen, and water, as shown in Fig. 2.19 (Iñiguez-Moreno et al., 2018). This heterogeneity is seen in biofilms produced by pure bacteria culture as well as in those by mixed bacteria culture from nature. Even though the structure of biofilm is highly heterogeneous, there are certain criteria that can represent general components, a base film, and sparsely distributed cell layers with water channels (Stoodley et al., 1997; Tolker-Nielsen and Molin, 2000; Donlan, 2002).

Table 2.6 summarizes the factors influencing the formation and structure of biofilms. One of the

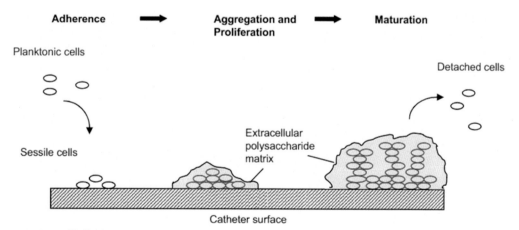

FIG. 2.17 Biofilm formation steps: adherence, aggregation and proliferation, and maturation. (Credit: Aslam, S., 2008. Effect of antibacterials on biofilms. American Journal of Infection Control 36, S175. e179–S175. e111.)

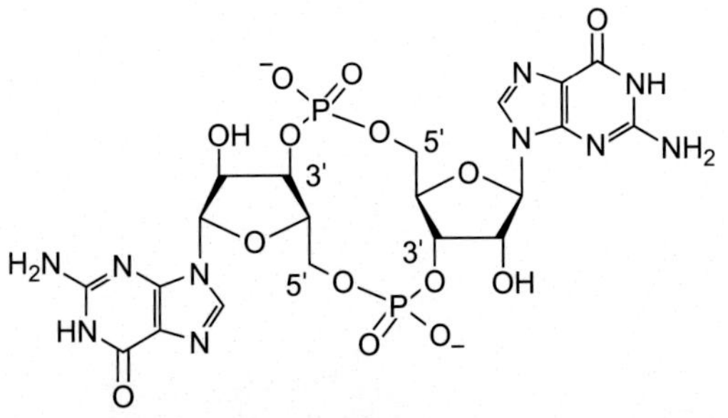

c-di-GMP

FIG. 2.18 Molecular structure of cyclic dimeric guanosine monophosphate. (Credit: Wang, C., Hao, M., Qi, Q., Chen, Y., Hartig, J.S., 2019. Chemical synthesis, purification, and characterization of 3′-5′-linked canonical cyclic dinucleotides (CDNs). Methods in Enzymology 625, 41–59.)

influencing factors, which can affect the biofilm structure, is the cell-to-cell interaction. Generally, mixed culture with multispecies produces thicker biofilms than pure culture with monospecies. Each bacterial species provides support for the other biofilm structure to become more stable, e.g., *Klebsiella pneumoniae-P. aeruginosa*, *Pseudomonas fluorescens-Aeromonas hydrophila-Xanthomonas maltophilia* (Cowan et al., 1991; Siebel and Characklis, 1991; James et al., 1995). However, it is worth noting that mixed culture does not "always" produce enhanced biofilm. For an example, when *P. aeruginosa* and *Pseudomonas putida* are cultured together, it is found that they do not help each other. At first, when these two species are cultured in a flow system together, they start to organize each

microcolony. Once they form each colony, the colonies start to mix by cell migration with time. These phenomena cause the biofilms to become looser than before because of mobility of cells inside of the colonies. Finally, the biofilms are dispersed. This research represents that the mobility of cells is also a major influencing factor to determine biofilm architecture (Tolker-Nielsen and Molin, 2000).

2.4.5 Biofilm Structure (Extracellular Polymeric Substance)

Extracellular polymeric substance (EPS) is the main component of biofilm. The majority of organic carbon in biofilm, approximately 50%–90%, is composed of EPS. Therefore, EPS is considered as the primary matrix material of biofilm. EPS has various physical and chemical properties, and it is mainly composed of polysaccharides. Polysaccharides contribute to ionic characteristics (or electric characteristics) of the biofilm, usually neutral to anionic property due to the presence of several acidic materials, such as uranic acids and ketal-linked pyruvates. This indicates that most of biofilms are electrically neutral to negatively charged. This characteristic causes strong ionic bond with divalent cations such as calcium and magnesium when biofilms are developed (Flemming et al., 2000; Sutherland, 2001). On the other hand, some of gram-positive bacteria (e.g., staphylococci) show cationic charge property.

Biofilms generally show nonuniform structures. The composition of polysaccharides determines the initial biofilm formation. Various structures of polysaccharides determine the rigidity and solubility of EPS. For example, 1,3- or 1,4-β-linked hexose causes more rigid and less

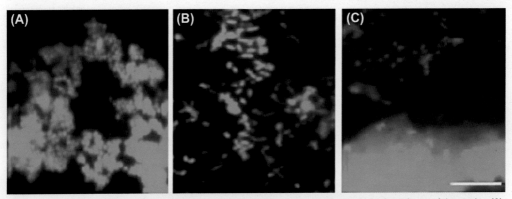

FIG. 2.19 Epifluorescence microscopy image of biofilm from single and dual surface of bacteria; **(A)** *Staphylococcus aureus* 4E, **(B)** *Salmonella* spp., and **(C)** dual species biofilm. (Credit: Iñiguez-Moreno, M., Gutiérrez-Lomelí, M., Guerrero-Medina, P.J., Avila-Novoa, M.G., 2018. Biofilm formation by Staphylococcus aureus and *Salmonella* spp. under mono and dual-species conditions and their sensitivity to cetrimonium bromide, peracetic acid and sodium hypochlorite. Brazilian Journal of Microbiology 49, 310–319.)

TABLE 2.6
Influencing Factors Affecting in Biofilm Formation (Donlan, 2002).

Properties	Bulk Fluid	Cell Properties
Roughness	Flow velocity	Cell surface hydrophobicity
Hydrophobicity	pH	Fimbriae
Biofilm condition	Temperature Cations Antibacterial agents presence	Flagella EPS

soluble EPS. In other cases, EPS becomes highly soluble in water. Different microorganisms cause different amounts of EPS production. Furthermore, the age of biofilm also affects the EPS production amount (Leriche et al., 2000). As the biofilm becomes older, the more amount of EPS is produced. Growth rate of bacteria has an impact to the EPS production; the slower the bacterial growth, the more the amount of EPS production by preventing the biofilm desiccation (Sutherland, 2001).

Hydrophobicity is the unique nature of EPS. Basically, EPS is highly hydratable because it has a significant amount of water molecules with hydrogen bonding. EPS is generally hydrophobic; however, it can be both hydrophobic and hydrophilic in specific conditions.

2.4.6 Representative Biofilm-Bacterial Species

Heterogeneity of permeability in oil reservoir causes a fingering phenomenon of the displacing fluids in high permeability zones and eventually results in high water production and low oil recovery during EOR practices. Aforementioned, the selective plugging technique as one way of MEOR practices involves the use of microbial by-products to selectively lower the permeability of high permeable zones of an oil reservoir. Thereby, this allows displacing fluids to flow through the low permeability zones and leads to oil recovery enhancements.

Although various microorganisms are known to form biofilm, there are several conditions to satisfy for implementation of selective plugging (Jenneman et al., 1984). The cells should be able to transport through porous media. The nutrients should be properly supplied to cultivate the microbes. The microbes must produce adequate by-products that can plug the high permeable zone. And, the growth rate should be limited not to block the wellbore. Oil reservoirs are generally in

oxygen-poor conditions with relatively high temperature and salinities; therefore, aerobic microorganisms that form biofilms are inadequate in situ MEOR.

In this section, the characteristics of several representative species, such as morphology of bacterium, the chemical and mechanical properties of produced EPS, and the optimum growth conditions, are addressed.

A. Pseudomonas aeruginosa

P. aeruginosa is a gram-negative bacterium, and it has a rod shape and monoflagella. The optimum temperature for growth ranges from 25 to 37°C though *P. aeruginosa* can survive at higher temperature (~42°C), which differs from other *Pseudomonas* species (Wu et al., 2015). *P. aeruginosa* also forms biofilms, and the characteristics of biofilm formation differ with strains. Some alginate-producing *P. aeruginosa* strains build highly structured biofilms, which have high resistance to water flows, whereas other strains form normal biofilms that are mainly composed of polysaccharides. Bacterial alginate is the major component of EPS in mucoid strains, which acts as a potential cross-linking site (Körstgens et al., 2001). *P. aeruginosa* also produces biosurfactant called rhamnolipids, which can enhance the oil mobility by reducing the IFT. The high viability in various environmental conditions, resilient biofilm formation, and biosurfactant production make *P. aeruginosa* adequate and attractive for in situ MEOR application. The laboratory observation using HATH strain, which produces both biomass and rhamnolipids, supports the feasibility of use of *P. aeruginosa* as a potential microorganism to increase oil recovery, as shown in Fig. 2.20 (Amani, 2015).

One of the strains, PAO1, which has been extensively studied, changes the cell morphology depending on the growth condition. In a denitrifying condition, the cells elongate (Yawata et al., 2008; Yoon et al., 2011). Under an oxic condition, mushroom-like microcolony is formed in biofilms. By contrast, flat matlike structure is developed under a denitrifying condition. Flow turbulence affects the morphology of biofilm produced by strain PANO67 (Stoodley et al., 2002). Mushroom-like microcolonies appear under laminar flows (Reynolds number $Re < 1200$), whereas filamentous microcolonies grow under turbulent flows ($Re > 1200$). These morphology changes may help the survivals in certain conditions (Justice et al., 2008).

The mechanical properties of EPS can be used as an indicator to the mechanical stability of biomass attached to the surfaces, and this is an important factor to determine the success of MEOR. The stress-stain curves of several strains, PAO1, PAO-JP1, PAN067, and FRD1, reveal the elastic responses under elevated

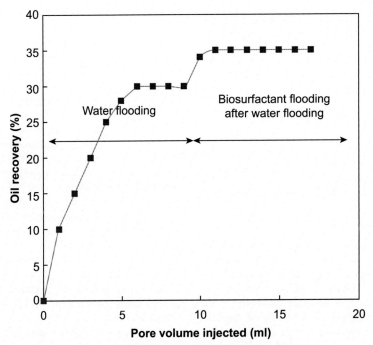

FIG. 2.20 The effect of rhamnolipid produced by *Pseudomonas aeruginosa* HATH on core flooding test. (Credit: Amani, H., 2015. Study of enhanced oil recovery by rhamnolipids in a homogeneous 2D micromodel. Journal of Petroleum Science and Engineering 128, 212–219.)

shear stress in a few seconds (Stoodley et al., 2002). The apparent elastic modulus increases with increased shear strain (Stoodley et al., 2002). This result represents that biofilms behave like elastic solids in a short time scale. The apparent elastic modulus is linearly proportional to the shear stress exerted at which the biofilm was grown. In a longer time scale, the biofilm behaves like linear viscous fluids from creep test (Stoodley et al., 2002). This viscoelastic behavior of biofilms is same as associated polymeric system, suggesting that EPS determines the cohesive strength (Stoodley et al., 2002).

The elastic modulus and yield stress of biofilms are heavily affected by the presence of calcium ions. The stiffness of biofilm from strain SG81 increases as the calcium ion concentration increases up to the critical concentration and thereafter is kept constant in the higher concentration (Körstgens et al., 2001; Orgad et al., 2011). It is because the mechanical properties increase proportional to the number of cross-linking densities caused by calcium ions.

B. *Bacillus licheniformis*
Bacillus licheniformis has been isolated from several high-temperature oil fields, including the oil fields in China, German, and Oman (Al-Hattali et al., 2013; Yunita

Halim et al., 2015). *B. licheniformis* is an adequate microorganism for selective plugging because it produces biofilms and it is a facultative anaerobe unlike most other bacilli, halotolerant, and thermotolerant and has an endospore-forming capability (Yakimov et al., 1997; Rey et al., 2004; Suthar et al., 2009; Yunita Halim et al., 2015). Furthermore, *B. licheniformis* can produce various by-products, including solvents, acids, gases, and surfactant, which can enhance the oil recovery.

The purified EPSs from *B. licheniformis* TT33 are reported to be composed of 26% carbohydrate and 3% proteins according to Fourier transform infrared (FT-IR) spectroscopy and thin-layer chromatography (TLC) analysis. The viscometer test on the purified EPS solution indicates that the EPS exhibits a pseudoplastic behavior associated with shear thinning, where the viscosity decreases with increasing shear rate (Suthar et al., 2009). The EPS production of *B. licheniformis* TT33 is affected by salinity, temperature, pH, and growth condition (Suthar et al., 2009). *B. licheniformis* TT33 produces the maximum EPS in the presence of 10% NaCl, at 50°C, pH 7, and aerobic conditions.

The laboratory cultivation tests using four other strains, BNP29, BNP36, BAS50, and Mep132, which were isolated from Northern German oil reservoirs, show the similar results. These strains can survive at the temperature up to 55°C and at the salinity up to 12% NaCl. However, the maximal metabolic activities of these strains are confirmed at 7% NaCl and 50°C (Yakimov et al., 1997). *B. licheniformis* BNP 29 is presumed to be the most suitable strain among the four strains isolated from Northern German oil reservoirs, due to its highest production rate at the same growth condition in both EPS and biosurfactant (Fig. 2.21). Furthermore, this strain can be transported in low-permeability cores (Yakimov et al., 1997). This strain has a better selective plugging effect than sulfur-reducing bacteria, which release harmful products to MEOR operations. Therefore, it is important to selectively stimulate microbes for selective plugging rather than to stimulate all kinds of microorganisms in the reservoir (Patel et al., 2015).

Meanwhile, there have been several core testing studies with various strains of *B. licheniformis*. The core flooding study of *B. licheniformis* 421 on chalk rocks has demonstrated the increase of oil recovery efficiency from 1.0% to 8.8% OOIP, and it has also shown that the selective plugging is more efficient in heterogeneous porous media (Yunita Halim et al., 2015). The core flooding test with *B. licheniformis* BNP29 has been performed at 30 and 50°C; the similar EOR efficiency was observed in the both temperatures. However, the selective plugging effect was more dominant at 30°C (Yakimov et al., 1997). It has been also reported that the injection of *B. licheniformis* TT33 into a sand pack column increases the additional oil recovery by 23% OOIP

(Suthar et al., 2009). On the other hand, the analyses of effluents collected from the cores in those studies reveal that the effluents had only minimal amount of the inoculated cells, and this implies that most of the bacteria were retained the cores (Yakimov et al., 1997; Suthar et al., 2009; Yunita Halim et al., 2015). The field test has been implemented by Lee et al. (1998), and confirmed the potential of selective plugging using *B. licheniformis*. The sequential injection of spore solution and nutrients into the low permeability carbonate reservoir in Lea County, Mexico, successfully plugged the thief zones, and the selective plugging effect was still maintained after 1 year (Lee et al., 1998).

C. Bacillus cereus

Bacillus cereus is ubiquitous in the environment. *B. cereus* is a gram-positive, rod-shaped, facultative anaerobe, motile, β-hemolytic, spore-forming bacterium commonly found in soil and food (Majed et al., 2016). The spores generated by *B. cereus* are highly resistant and adhesive, which enable to grow in extreme conditions. *B. cereus* produces the various types of biofilms depending on the growth circumstances with a variety of metabolites including surfactants, bacteriocins, enzymes, and toxins (Majed et al., 2016).

Similar to other biofilms, the components of EPS are mostly proteins and carbohydrates. Proteins, which are most important component controlling the adhesion to surface, consist of proteases, lipases, toxins, and cell wall proteins (Gohar et al., 2002; Karunakaran and Biggs, 2011). Amino acids are predominant in the extracellular proteins produced by both strains ATCC 10987

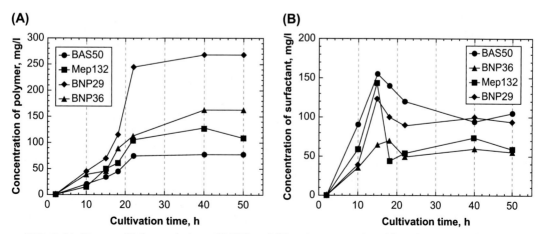

FIG. 2.21 The quantitative analysis on **(A)** EPS and **(B)** surfactant, produced from four strains of *Bacillus licheniformis*. (Credit: Yakimov, M.M., Amro, M.M., Bock, M., Boseker, K., Fredrickson, H.L., Kessel, D.G., Timmis, K.N., 1997. The potential of Bacillus licheniformis strains for in situ enhanced oil recovery. Journal of Petroleum Science and Engineering 18, 147–160.)

and ATCC 14579. This implies that the isoelectric point (IEP) is basic. In other words, the EPS proteins are positively charged at physiological pH of 6.5–7.4 (Karunakaran and Biggs, 2011). Bacterial cell surface assay reveals that biofilms are more hydrophilic than their planktonic cells, where ATCC 10987 is hydrophobic and ATCC 14579 is hydrophilic (Karunakaran and Biggs, 2011). The electrostatic forces between negatively charged bacterial cell surfaces and the positively charged proteins have a significant effect on the cell adhesion to surfaces. Therefore, the biofilm formation is heavily affected by pH, due to the electrostatic interactions among the cells, EPS proteins, and surface. The adhesion of biofilm on a surface is also related to the extracellular DNA (eDNA), which is known as the requirement component for biofilm formation in other species (Vilain et al., 2009; Majed et al., 2016).

Again, biofilms produced by *B. cereus* differ with strains; for an instance, the biofilm-forming habits by *B. cereus* ATCC 10987 and ATCC 14579 show different features. The strain ATCC 10987 forms biofilms on polystyrene plates, whereas the strain ATCC 14579 does not form biofilm on those. In hydrophilic glass wool, both strains form biofilms, but the biofilm by ATCC 14579 contains there fewer cells (Auger et al., 2006; Karunakaran and Biggs, 2011). The EPS production rate by *B. cereus* JBS10 GU812900 in basalt salt solutions increases as the sucrose concentration increases. The EPS production is seen the greatest when the nitrogen source is from ammonium sulfate and the phosphate concentration is 1 mg/mL (Bragadeeswaran et al., 2011).

B. cereus has a capability to produce biosurfactant. *B. cereus*, isolated from Microbial Type Culture collection, University of Alzahra (Tehran, Iran), shows the maximum biosurfactant production when the carbon source is sucrose rather than oil and whey (Amani et al., 2011). The optimum condition for biosurfactant production has been reported when the phosphorus and nitrogen concentrations are 13.53 g/L (Na_2HPO_4, NaH_2PO_4), 1 g/L ($(NH_4)_2SO_4$) (Amani et al., 2011), respectively, and with 25 g/L NaCl. The salt concentration increases from 25 to 75 g/L, the less surface tension reduction is observed, which indicates that the salt has a negative impact on the biosurfactant production and efficiency.

Furthermore, *B. cereus* can cause a sort of the biooxidation reaction, which can degrade crude oil from one long-chain hydrocarbons to two short-chain fatty monoacids or alcohols. This has been confirmed in the field test where both *B. cereus* and *Brevibacillus brevis* were used as microbes (Hou et al., 2005). The injection of *B. cereus* and *B. brevis* into Daqing oil field enhances the oil production (Hou et al., 2005).

D. Shewanella oneidensis *MR-1*

Shewanella oneidensis MR-1 is a facultative anaerobe and gram-negative bacterium, which is widely found in saturated soil, such as marine sediments (Venkateswaran et al., 1999). It has a rod shape with the length of 2–3 μm. The bacterium can swim in water with its single polar flagellum (Venkateswaran et al., 1999). They produce pink and thin biofilms and transport electrons via electrically conductive nanowires and flavins (Gorby et al., 2006; Roy et al., 2013). *S. oneidensis* has a unique capability of reducing metals, such as iron, manganese, mercury, and uranium, and accordingly, this bacterium is popularly used in bioremediation areas (Myers and Nealson, 1990; Xiong et al., 2006; Icopini et al., 2009; Lee et al., 2018).

Biofilm growth process of *S. oneidensis* MR-1 can be summarized as (1) cell growth, (2) attachment colonization, (3) colonies fusion, and (4) towering, which is a general process biofilm formation process, mentioned earlier. Such a biofilm growth habit by *S. oneidensis* MR-1 has been confirmed during reduction of iron and manganese in a flow channel (Thormann et al., 2004). In particular, iron and manganese particles interact with cells in biofilm by electron transport and finally become reduced. *S. oneidensis* MR-1 can produce biofilms thicker than 20 μm and often form a mushroom-like structure of biofilms with a thickness more than 200 μm when sufficient time is allowed, e.g., longer than 5 days (Thormann et al., 2004).

Two main genes, exeM and exeS, affect the biofilm formation by *S. oneidensis* MR-1. The exeM mutants produce less biofilms than wild type, approximately 28% of the amount produced by wild type. On the other hand, the exeS mutants cause more biofilm production, by 1.6 times more than wild type do. The exeM and exeS double mutant secretes less biofilms than wild type, approximately a half of the amount produced by wild type (Thormann et al., 2004).

2.5 BIOGENIC GASES

During EOR, gas injection, either miscible or immiscible, improves displacement efficiency of oil by reducing the capillarity of oils in porous media. Even during water flooding, use of gas aids in reestablishing a pathway for the remaining residual oil to be swept, which can eventually increase the oil recovery rate. In this manner, biogenically produced gases, such as CO_2, N_2, H_2, and CH_4, can facilitate oil recovery process. The idea of using microbial or biogenic gases for MEOR practices is proposed at first by Zobell and Johnson (ZoBell and Johnson, 1949). They proposed to inject specific bacteria or stimulate indigenous bacteria to produce CO_2 or CH_4 and thus restore the pressure in the reservoir. There is a secondary effect to be expected, in which CO_2 gas promotes further dissolution of calcite and siderite in carbonate reservoirs and enhances the oil recovery.

There are several microorganisms that can produce gases, and these produced gases are expected to increase the reservoir pore pressure, reduce the viscosity of oil, swell the oil, and/or improve the displacement efficiency of oils, all of which help the oil recovery. However, it is advised that production of sufficient amounts of gases to achieve such occasions is unlikely in natural anaerobic conditions (Bachmann et al., 2017; Nikolova and Gutierrez, 2020); thus, further investigations to optimize the gas production are needed.

2.5.1 Gas-Bacteria and Microbial Gas Production Mechanisms

A. CO2-producing bacteria via fermentation

The most common microbial mechanism to produce CO_2 is the fermentation process. The fermentation is an energy-yielding, anaerobic process, in which substrates are sequentially transformed through reduction-oxidation processes. It involves enzymatic decomposition of carbohydrates by microorganisms. No external electron acceptor is involved, whereas the redox levels of the substrate and the metabolites remain constant. Therefore, the fermentation requires a relatively low amount of energy.

Typical fermenting bacteria are facultative anaerobes capable of oxidative phosphorylation in the presence of O_2 (e.g., *Escherichia*), aerotolerant anaerobes (e.g., *Lactobacillus*), or they may be strict (O_2-sensitive) anaerobes (e.g., *Clostridium*) (Fenchel et al., 2012). During the fermentation of glucose to lactate or ethanol and CO_2, pyruvate is used to reoxidize the reduced nicotinamide adenine dinucleotide (NADH) that was produced during glycolysis. Thereby, it restores redox balance and produces lactate or ethanol and CO_2 as end products (Fenchel et al., 2012). In particular, the fermentation becomes significant when a high concentration of sugars presents. In such environments, *Lactobacillus*, an acid-tolerant bacterium, for example, has superiority for the fermentation processes over other bacterial species (Fenchel et al., 2012).

B. N2-producing bacteria via denitrification

Denitrifying bacteria are a group of bacteria that can perform denitrification as part of the nitrogen cycle. Denitrifying bacteria metabolize nitrogenous compounds using their enzymes, reducing nitrogen oxides such as nitrite (NO_2^-) and nitrate (NO_3^-) to nitrogen gas (N_2) or nitrous oxide (N_2O). Heterotrophic bacteria including *Paracoccus denitrificans*, *P. aeruginosa*, and *Pseudomonas stutzeri*, and autotrophic bacteria such as *Thiobacillus denitrificans* and *Micrococcus denitrificans* are representative denitrifying bacteria that reduce nitrate and produce nitrogen in an anaerobic condition by the reaction: $NO_3^- \rightarrow NO_2^- \rightarrow NO \rightarrow N_2O \rightarrow N_2$ (van Spanning et al., 2007).

Denitrification typically occurs in anoxic environments, where the concentration of oxygen is limited. Denitrifying bacteria use nitrate or nitrite as a substitute terminal electron acceptor instead of oxygen. The terminal electron acceptor is a compound undergoing reduction reactions by receiving electrons. Oil reservoirs are mostly regarded as anoxic environment; thus, oil reservoirs have a favorable condition to stimulate a denitrification process from oxygen perspectives. Meanwhile, there are several of microorganisms that can denitrify under oxic environments. *P. denitrificans* is one of the representative species engaged in denitrification under both oxic and anoxic environments. Those aerobic denitrifiers are mostly gram-negative bacteria, and under oxic environments, those species utilize nitrous oxide reductase, an enzyme that catalyzes the denitrification process. It is often suggested to inject nitrate to stimulate nitrate-reducing bacteria in the oil reservoir, such that indigenous nitrogen-producing bacteria inside the reservoir produces nitrogen gas in situ (Nuryadi et al., 2011; Fida et al., 2017).

C. Representative H2-producing species

Biohydrogen is hydrogen gas (H_2) that is biologically produced via fermentation. The main reactions involve the fermentation of sugars, which start with glucose that is converted to acetic acid and then produces hydrogen gas (Thauer, 1998). H_2 production is catalyzed by two hydrogenases: the one is called [FeFe]-hydrogenase and another one is called [NiFe]-hydrogenase. Representative examples of hydrogen producing genera are *Clostridium*, *Desulfovibrio*, *Ralstonia*, and *Helicobacter* (Casalot and Rousset, 2001; Vignais and Colbeau, 2004). H_2 is necessary for in situ CH_4 production by hydrogenotrophic methanogens in oil reservoirs during carbon dioxide reduction to methane (Madigan et al., 1997).

D. Representative CH4-producing species

Methanogens are microorganisms that produce methane as a metabolic by-product in oxygen-limited environments. Methanogens are mostly anaerobic organisms that cannot function under aerobic conditions. However, several species, for example, *Methanothrix paradoxum*, can generate methane under aerobic environments. Methanogens are known to produce methane from substrates such as H_2/CO_2, acetate, formate, methanol, and methylamines in a process called methanogenesis (Blaut, 1994). Different methanogenic reactions are catalyzed by unique sets of enzymes and coenzymes. While the reaction mechanism and energetics vary between one reaction and another, all of these reactions contribute to net positive energy production by creating ion concentration gradients that are used to drive ATP synthesis (Dybas and Konisky, 1992). There are over 50 species of methanogens, but well-known microorganisms that produce methane via H_2/CO_2 methanogenesis are *Methanococcus*, *Methanosarcina*, *Methanobacterium*, *Methanobrevibacter*, *Methanothermobacter*, and *Methanoculleus* species (Balch et al., 1979; Boone et al., 1993). Under anaerobic conditions, degradation of hydrocarbons to methane by methanogenic microbes is also a well-known process in the geosphere, and this methanogenic biodegradation mechanism is commonly occurring in subsurface oil reservoirs (Jones et al., 2008).

2.6 SOLVENTS, ACIDS

Microbially produced solvent has been considered as a potential way to facilitate MEOR practices. Solvents that have good miscibility with hydrocarbons dissolve in crude oils and reduce the oil viscosity, which hence increases the mobility of oils. Liquid solvents such as acetone, butanol, and ethanol are produced by *Clostridium* species and *Zymomonas* species (Van Hamme et al., 2003). The acetone-butanol-ethanol process occurs during bacterial fermentation translating carbohydrates such as glucose to butyrul-CoA, as shown in Fig. 2.22 (Kumar et al., 2015). The process is associated with the yeast fermenting sugars to produce ethanol. The acetone-butanol-ethanol fermentation is a strictly anaerobic process, which can be satisfied in most of deep subsurface reservoirs. Among *Clostridium* species, which is one of the most well-known species to produce solvents (Behlulgil et al., 1992; Marsh et al., 1995), *Clostridium acetobutylicum* has been tested for oil recovery

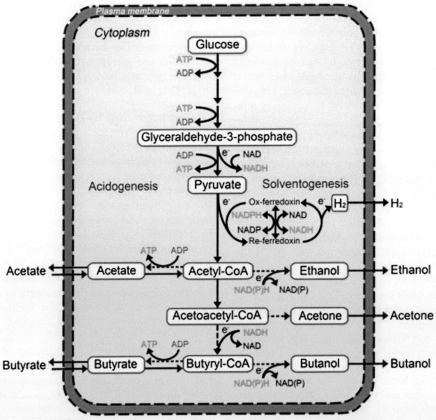

FIG. 2.22 Redox reactions of acetone-butanol-ethanol fermentation process by *Clostridium* species. (Credit: Liu, C.-G., Xue, C., Lin, Y.-H., Bai, F.-W., 2013. Redox potential control and applications in microaerobic and anaerobic fermentations. Biotechnology Advances 31, 257—265.)

improvements, and it is found to be effective in terms of its functionalities and stability under reservoir environments (Behlulgil et al., 1992; Ishizaki et al., 1999).

Acid-producing bacteria, which are capable of producing organic or inorganic acids, also can assist in MEOR. Similar to solvents, acids can reduce the oil viscosity, improve the permeability by dissolving rock minerals and increasing the porosity, and help emulsification of crude oils. As acids dissolve carbonate minerals in from carbonate-rich formations, acid-producing bacteria are considered to be promising for carbonate or carbonaceous reservoirs. Representative candidates of acid-producing bacteria for MEOR practices are *Bacillus*, *Clostridium*, and *Enterobacter* species, and their main products are acetic, lactic, formic, and succinic acids. It is reported that *B. subtilis* isolated from the production water of Rio-Itaúnas formation oil field, Conceição da Barra, Brazil, produces both organic acids and solvents (Fernandes et al., 2016). *B. subtilis* produces 2,3-butanediol during fermentation (Shariat et al., 1995) and organic acids such as lactate and succinate, which are known to be commonly produced during the fermentation metabolisms of *Bacillus* species.

However, one study has pointed out that the use of acids/solvent-producing bacteria would be economically attractive only when the viscosity difference between the oil and the solvent is large enough to reduce oil viscosity by a significant level. If not, it would take vast volumes of solvent to be injected in the reservoir to achieve even the slightest increase in oil recovery (Gray et al., 2008). Therefore, it warrants further investigation on the characteristics of solvents or acids produced by microorganisms in reservoir environment and to figure out if it is economically feasible or not.

REFERENCES

Abdel-Mawgoud, A.M., Aboulwafa, M.M., Hassouna, N.A.-H., 2008. Characterization of surfactin produced by *Bacillus subtilis* isolate BS5. Applied Biochemistry and Biotechnology 150, 289–303.

Aitken, C.M., Jones, D.M., Larter, S., 2004. Anaerobic hydrocarbon biodegradation in deep subsurface oil reservoirs. Nature 431, 291–294.

Al-Bahry, S., Al-Wahaibi, Y., Elshafie, A., Al-Bemani, A., Joshi, S., Al-Makhmari, H., Al-Sulaimani, H., 2013. Biosurfactant production by *Bacillus subtilis* B20 using date molasses and its possible application in enhanced oil recovery. International Biodeterioration & Biodegradation 81, 141–146.

Al-Hattali, R., Al-Sulaimani, H., Al-Wahaibi, Y., Al-Bahry, S., Elshafie, A., Al-Bemani, A., Joshi, S.J., 2013. Fractured carbonate reservoirs sweep efficiency improvement using microbial biomass. Journal of Petroleum Science and Engineering 112, 178–184.

Al-Sulaimani, H., Al-Wahaibi, Y., Al-Bahry, S., Elshafie, A., Al-Bemani, A., Joshi, S., 2012. Residual-oil recovery through injection of biosurfactant, chemical surfactant, and mixtures of both under reservoir temperatures: induced-wettability and interfacial-tension effects. SPE Reservoir Evaluation and Engineering 15, 210–217.

Al-Wahaibi, Y., Joshi, S., Al-Bahry, S., Elshafie, A., Al-Bemani, A., Shibulal, B., 2014. Biosurfactant production by *Bacillus subtilis* B30 and its application in enhancing oil recovery. Colloids and Surfaces B: Biointerfaces 114, 324–333.

Amani, H., 2015. Study of enhanced oil recovery by rhamnolipids in a homogeneous 2D micromodel. Journal of Petroleum Science and Engineering 128, 212–219.

Amani, H., Haghighi, M., Sarrafzadeh, M., Mehrnia, M., Shahmirzaee, F., 2011. Optimization of the production of biosurfactant from iranian indigenous bacteria for the reduction of surface tension and enhanced oil recovery. Petroleum Science and Technology 29, 301–311.

Amani, H., Sarrafzadeh, M.H., Haghighi, M., Mehrnia, M.R., 2010. Comparative study of biosurfactant producing bacteria in MEOR applications. Journal of Petroleum Science and Engineering 75, 209–214.

Ananthalakshmy, V., Gunasekaran, P., 1999. Optimization of levan production by *Zymomonas* mobilis. Brazilian Archives of Biology and Technology 42, 291–298.

Andersen, I.-L., Skipnes, O., Smidsrod, O., Ostgaard, K., Hemmer, P.C., 1977. Some Biological Functions of Matrix Components in Benthic Algae in Relation to Their Chemistry and the Composition of Seawater. ACS Publications.

Arvidson, S.A., Rinehart, B.T., Gadala-Maria, F., 2006. Concentration regimes of solutions of levan polysaccharide from *Bacillus* sp. Carbohydrate Polymers 65, 144–149.

Ash, S., Clarke-Sturman, A., Calvert, R., Nisbet, T., 1983. Chemical stability of biopolymer solutions. In: SPE Annual Technical Conference and Exhibition. Society of Petroleum Engineers.

Aslam, S., 2008. Effect of antibacterials on biofilms. American Journal of Infection Control 36. S175. e179-S175. e111.

Auger, S., Krin, E., Aymerich, S., Gohar, M., 2006. Autoinducer 2 affects biofilm formation by *Bacillus cereus*. Applied and Environmental Microbiology 72, 937–941.

Bachmann, H., Molenaar, D., Branco dos Santos, F., Teusink, B., 2017. Experimental evolution and the adjustment of metabolic strategies in lactic acid bacteria. FEMS Microbiology Reviews 41, S201–S219.

Bahl, M.A., Schultheis, E., Hempel, D.C., Nörtemann, B., Franco-Lara, E., 2010. Recovery and purification of the exopolysaccharide PS-EDIV from *Sphingomonas pituitosa* DSM 13101. Carbohydrate Polymers 80, 1037–1041.

Balch, W., Fox, G., Magrum, L., Woese, C., Wolfe, R., 1979. Methanogens: reevaluation of a unique biological group. Microbiological Reviews 43, 260.

Barth, T., 1991. Organic acids and inorganic ions in waters from petroleum reservoirs, Norwegian continental shelf: a multivariate statistical analysis and comparison with

American reservoir formation waters. Applied Geochemistry 6, 1–15.

Becker, A., Katzen, F., Pühler, A., Ielpi, L., 1998. Xanthan gum biosynthesis and application: a biochemical/genetic perspective. Applied Microbiology and Biotechnology 50, 145–152.

Behlulgil, K., Mehmetoglu, T., Donmez, S., 1992. Application of microbial enhanced oil recovery technique to a Turkish heavy oil. Applied Microbiology and Biotechnology 36, 833–835.

Bender, H., Lehmann, J., Wallenfels, K., 1959. Pullulan, ein extracelluläres Glucan von Pullularia pullulans. Biochimica et Biophysica Acta 36, 309–316.

Benincasa, M., Abalos, A., Oliveira, I., Manresa, A., 2004. Chemical structure, surface properties and biological activities of the biosurfactant produced by Pseudomonas aeruginosa LBI from soapstock. Antonie van Leeuwenhoek 85, 1–8.

Blaut, M., 1994. Metabolism of methanogens. Antonie van Leeuwenhoek 66, 187–208.

Boone, D.R., Whitman, W.B., Rouvière, P., 1993. Diversity and Taxonomy of Methanogens. Methanogenesis. Springer, pp. 35–80.

Boyd, C.D., O'Toole, G.A., 2012. Second messenger regulation of biofilm formation: breakthroughs in understanding c-di-GMP effector systems. Annual Review of Cell and Developmental Biology 28, 439–462.

Bragadeeswaran, S., Jeevapriya, R., Prabhu, K., Rani, S.S., Priyadharsini, S., Balasubramanian, T., 2011. Exopolysaccharide production by Bacillus cereus GU812900, a fouling marine bacterium. African Journal of Microbiology Research 5, 4124–4132.

Brusseau, M.L., Miller, R.M., Zhang, Y., Wang, X., Bai, G.-Y., 1995. Biosurfactant-and Cosolvent-Enhanced Remediation of Contaminated Media. ACS Publications.

Bulmer, M.A., Catley, B.J., Kelly, P.J., 1987. The effect of ammonium ions and pH on the elaboration of the fungal extracellular polysaccharide, pullulan, by Aureobasidium pullulans. Applied Microbiology and Biotechnology 25, 362–365.

Cai, M., Nie, Y., Chi, C.-Q., Tang, Y.-Q., Li, Y., Wang, X.-B., Liu, Z.-S., Yang, Y., Zhou, J., Wu, X.-L., 2015. Crude oil as a microbial seed bank with unexpected functional potentials. Scientific Reports 5, 16057.

Campos, M., Martínez-Salazar, J.M., Lloret, L., Moreno, S., Núñez, C., Espín, G., Soberón-Chávez, G., 1996. Characterization of the gene coding for GDP-mannose dehydrogenase (algD) from Azotobacter vinelandii. Journal of Bacteriology 178, 1793–1799.

Casalot, L., Rousset, M., 2001. Maturation of the [NiFe] hydrogenases. Trends in Microbiology 9, 228–237.

Casas, J., Santos, V., Garcıa-Ochoa, F., 2000. Xanthan gum production under several operational conditions: molecular structure and rheological properties. Enzyme and Microbial Technology 26, 282–291.

Catley, B., 1972. Pullulan elaboration, an inducible system of Pullularia pullulans. FEBS Letters 20, 174–176.

Chang, I., Im, J., Cho, G.-C., 2016. Introduction of microbial biopolymers in soil treatment for future environmentally-friendly and sustainable geotechnical engineering. Sustainability 8, 251.

Chen, J., Huang, P., Zhang, K., Ding, F., 2012. Isolation of biosurfactant producers, optimization and properties of biosurfactant produced by Acinetobacter sp. from petroleum-contaminated soil. Journal of Applied Microbiology 112, 660–671.

Cheng, K.-C., Demirci, A., Catchmark, J.M., 2010. Enhanced pullulan production in a biofilm reactor by using response surface methodology. Journal of Industrial Microbiology and Biotechnology 37, 587–594.

Cheng, K.-C., Demirci, A., Catchmark, J.M., 2011. Pullulan: biosynthesis, production, and applications. Applied Microbiology and Biotechnology 92, 29.

Cho, S.S., 2001. Handbook of Dietary Fiber. CRC Press.

Chong, H., Li, Q., 2017. Microbial production of rhamnolipids: opportunities, challenges and strategies. Microbial Cell Factories 16, 137.

Coviello, T., Coluzzi, G., Palleschi, A., Grassi, M., Santucci, E., Alhaique, F., 2003. Structural and rheological characterization of scleroglucan/borax hydrogel for drug delivery. International Journal of Biological Macromolecules 32, 83–92.

Cowan, M.M., Warren, T.M., Fletcher, M., 1991. Mixed-species colonization of solid surfaces in laboratory biofilms. Biofouling 3, 23–34.

Da Cunha, C.D., Rosado, A.S., Sebastián, G.V., Seldin, L., Von der Weid, I., 2006. Oil biodegradation by Bacillus strains isolated from the rock of an oil reservoir located in a deep-water production basin in Brazil. Applied Microbiology and Biotechnology 73, 949–959.

Davis, D., Lynch, H., Varley, J.J.E., Technology, M., 1999. The production of surfactin in batch culture by Bacillus subtilis ATCC 21332 is strongly influenced by the conditions of nitrogen metabolism. Enzyme and Microbial Technology 25, 322–329.

Davison, P., Mentzer, E., 1982. Polymer flooding in North Sea reservoirs. Society of Petroleum Engineers Journal 22, 353–362.

De Almeida, D.G., Soares Da Silva, R.d.C.F., Luna, J.M., Rufino, R.D., Santos, V.A., Banat, I.M., Sarubbo, L.A., 2016. Biosurfactants: promising molecules for petroleum biotechnology advances. Frontiers in Microbiology 7, 1718.

de Oliveira, M.R., da Silva, R.S.S.F., Buzato, J.B., Celligoi, M.A.P.C., 2007. Study of levan production by Zymomonas mobilis using regional low-cost carbohydrate sources. Biochemical Engineering Journal 37, 177–183.

Desai, J.D., Banat, I.M., 1997. Microbial production of surfactants and their commercial potential. Microbiology and Molecular Biology Reviews 61, 47–64.

Deziel, E., Paquette, G., Villemur, R., Lepine, F., Bisaillon, J., 1996. Biosurfactant production by a soil Pseudomonas strain growing on polycyclic aromatic hydrocarbons. Applied and Environmental Microbiology 62, 1908–1912.

Dong, H., Xia, W., Dong, H., She, Y., Zhu, P., Liang, K., Zhang, Z., Liang, C., Song, Z., Sun, S., 2016. Rhamnolipids produced by indigenous Acinetobacter junii from petroleum reservoir and its potential in enhanced oil recovery. Frontiers in Microbiology 7, 1710.

Donlan, R.M., 2002. Biofilms: microbial life on surfaces. Emerging Infectious Diseases 8, 881.

Draget, K., Simensen, M., Onsøyen, E., Smidsrød, O., 1993. Gel strength of Ca-limited alginate gels made in situ. In: Fourteenth International Seaweed Symposium. Springer, pp. 563–569.

Draget, K.I., Strand, B., Hartmann, M., Valla, S., Smidsrød, O., Skjåk-Bræk, G., 2000. Ionic and acid gel formation of epimerised alginates; the effect of AlgE4. International Journal of Biological Macromolecules 27, 117–122.

Duan, X., Chi, Z., Wang, L., Wang, X., 2008. Influence of different sugars on pullulan production and activities of α-phosphoglucose mutase, UDPG-pyrophosphorylase and glucosyltransferase involved in pullulan synthesis in *Aureobasidium pullulans* Y68. Carbohydrate Polymers 73, 587–593.

Dubeau, D., Déziel, E., Woods, D.E., Lépine, F., 2009. *Burkholderia thailandensis* harbors two identical rhl gene clusters responsible for the biosynthesis of rhamnolipids. BMC Microbiology 9, 263.

Dybas, M., Konisky, J., 1992. Energy transduction in the methanogen *Methanococcus* voltae is based on a sodium current. Journal of Bacteriology 174, 5575–5583.

Farina, J., Sineriz, F., Molina, O., Perotti, N., 1998. High scleroglucan production by *Sclerotium rolfsii*: influence of medium composition. Biotechnology Letters 20, 825–831.

Fenchel, T., Blackburn, H., King, G.M., Blackburn, T.H., 2012. Bacterial Biogeochemistry: The Ecophysiology of Mineral Cycling. Academic press.

Fernandes, P., Rodrigues, E., Paiva, F., Ayupe, B., McInerney, M., Tótola, M., 2016. Biosurfactant, solvents and polymer production by *Bacillus subtilis* RI4914 and their application for enhanced oil recovery. Fuel 180, 551–557.

Fida, T.T., Gassara, F., Voordouw, G., 2017. Biodegradation of isopropanol and acetone under denitrifying conditions by *Thauera* sp. TK001 for nitrate-mediated microbially enhanced oil recovery. Journal of Hazardous Materials 334, 68–75.

Finnerty, W.R., 1994. Biosurfactants in environmental biotechnology. Current Opinion in Biotechnology 5, 291–295.

Fisher, J.B., 1987. Distribution and occurrence of aliphatic acid anions in deep subsurface waters. Geochimica et Cosmochimica Acta 51, 2459–2468.

Flemming, H.-C., Wingender, J., Griegbe, T., Mayer, C., 2000. Physico-chemical properties of biofilms. In: Biofilms: Recent Advances in Their Study and Control. Harwood Academic Publishers, Amsterdam, pp. 19–34.

Futatsuyama, H., Yui, T., Ogawa, K., 1999. Viscometry of curdlan, a linear $(1 \rightarrow 3)$-β-D-glucan, in DMSO or alkaline solutions. Bioscience Biotechnology and Biochemistry 63, 1481–1483.

Gao, P., Tian, H., Wang, Y., Li, Y., Li, Y., Xie, J., Zeng, B., Zhou, J., Li, G., Ma, T., 2016. Spatial isolation and environmental factors drive distinct bacterial and archaeal communities in different types of petroleum reservoirs in China. Scientific Reports 6, 20174.

Gao, W., Kim, Y.-J., Chung, C.-H., Li, J., Lee, J.-W., 2010. Optimization of mineral salts in medium for enhanced production of pullulan by *Aureobasidium pullulans* HP-2001 using an orthogonal array method. Biotechnology and Bioprocess Engineering 15, 837–845.

Geetha, S., Banat, I.M., Joshi, S.J., 2018. Biosurfactants: production and potential applications in microbial enhanced oil recovery (MEOR). Biocatalysis and Agricultural Biotechnology 14, 23–32.

Georgiou, G., Lin, S.-C., Sharma, M.M., 1992. Surface–active compounds from microorganisms. Biotechnology 10, 60–65.

Ghoumrassi-Barr, S., Aliouche, D., 2015. Characterisation and rheological study of xanthan polymer for enhanced oil recovery (EOR) application. In: Offshore Mediterranean Conference and Exhibition. Offshore Mediterranean Conference.

Glick, R., Gilmour, C., Tremblay, J., Satanower, S., Avidan, O., Déziel, E., Greenberg, E.P., Poole, K., Banin, E., 2010. Increase in rhamnolipid synthesis under iron-limiting conditions influences surface motility and biofilm formation in *Pseudomonas aeruginosa*. Journal of Bacteriology 192, 2973–2980.

Göbbert, U., Lang, S., Wagner, F., 1984. Sophorose lipid formation by resting cells of *Torulopsis bombicola*. Biotechnology Letters 6, 225–230.

Gohar, M., Økstad, O.A., Gilois, N., Sanchis, V., Kolst⊘, A.B., Lereclus, D., 2002. Two-dimensional electrophoresis analysis of the extracellular proteome of *Bacillus cereus* reveals the importance of the PlcR regulon. Proteomics 2, 784–791.

Gorby, Y.A., Yanina, S., McLean, J.S., Rosso, K.M., Moyles, D., Dohnalkova, A., Beveridge, T.J., Chang, I.S., Kim, B.H., Kim, K.S., 2006. Electrically conductive bacterial nanowires produced by *Shewanella oneidensis* strain MR-1 and other microorganisms. Proceedings of the National Academy of Sciences 103, 11358–11363.

Gorin, P., Spencer, J., 1966. Exocellular alginic acid from *Azotobacter vinelandii*. Canadian Journal of Chemistry 44, 993–998.

Gray, M., Yeung, A., Foght, J., Yarranton, H.W., 2008. Potential microbial enhanced oil recovery processes: a critical analysis. In: SPE Annual Technical Conference and Exhibition. Society of Petroleum Engineers.

Guerra-Santos, L., Käppeli, O., Fiechter, A., 1984. *Pseudomonas aeruginosa* biosurfactant production in continuous culture with glucose as carbon source. Applied and Environmental Microbiology 48, 301–305.

Han, Y.W., 1990. Microbial levan. In: Advances in Applied Microbiology. Elsevier, pp. 171–194.

Harada, T., 1977. Production, Properties, and Application of Curdlan. ACS Publications.

Harada, T., 1992. The story of research into curdlan and the bacteria producing it. Trends in Glycoscience and Glycotechnology 4, 309–317.

Harada, T., Misaki, A., Saito, H., 1968. Curdlan: a bacterial gelforming β-1, 3-glucan. Archives of Biochemistry and Biophysics 124, 292–298.

Henkel, M., Hausmann, R., 2019. Diversity and Classification of Microbial Surfactants. In: Biobased Surfactants. Elsevier, pp. 41–63.

Hörmann, B., Müller, M.M., Syldatk, C., Hausmann, R., 2010. Rhamnolipid production by Burkholderia plantarii DSM 9509T. European Journal of Lipid Science and Technology 112, 674–680.

Hou, Z., Wu, X., Shi, M., Han, P., Wang, Y., Xu, Y., Jin, R., 2005. The mechanism and application of MEOR by Brevibacillus brevis and Bacillus cereus in daqing oilfield. In: SPE International Improved Oil Recovery Conference in Asia Pacific. Society of Petroleum Engineers, Kuala Lumpur, Malaysia, p. 9.

Icopini, G.A., Lack, J.G., Hersman, L.E., Neu, M.P., Boukhalfa, H., 2009. Plutonium (V/VI) reduction by the metal-reducing bacteria Geobacter metallireducens GS-15 and Shewanella oneidensis MR-1. Applied and Environmental Microbiology 75, 3641–3647.

Iñiguez-Moreno, M., Gutiérrez-Lomelí, M., Guerrero-Medina, P.J., Avila-Novoa, M.G., 2018. Biofilm formation by Staphylococcus aureus and Salmonella spp. under mono and dual-species conditions and their sensitivity to cetrimonium bromide, peracetic acid and sodium hypochlorite. Brazilian Journal of Microbiology 49, 310–319.

Ishizaki, A., Michiwaki, S., Crabbe, E., Kobayashi, G., Sonomoto, K., Yoshino, S., 1999. Extractive acetone-butanol-ethanol fermentation using methylated crude palm oil as extractant in batch culture of Clostridium saccharoperbutylacetonicum N1-4 (ATCC 13564). Journal of Bioscience and Bioengineering 87, 352–356.

Jacques, P., 2011. Surfactin and Other Lipopeptides from Bacillus Spp. Biosurfactants. Springer, pp. 57–91.

James, G.A., Beaudette, L., Costerton, J.W., 1995. Interspecies bacterial interactions in biofilms. Journal of Industrial Microbiology 15, 257–262.

Jeanes, A., Haynes, W.C., Wilham, C., Rankin, J.C., Melvin, E., Austin, M.J., Cluskey, J., Fisher, B., Tsuchiya, H., Rist, C., 1954. Characterization and classification of dextrans from ninety-six strains of bacteria. Journal of the American Chemical Society 76, 5041–5052.

Jenneman, G.E., Knapp, R.M., McInerney, M.J., Menzie, D., Revus, D., 1984. Experimental studies of in-situ microbial enhanced oil recovery. Society of Petroleum Engineers Journal 24, 33–37.

Jensen, T., Kadhum, M., Kozlowicz, B., Sumner, E., Malsam, J., Muhammed, F., Ravikiran, R., 2018. Chemical EOR under harsh conditions: scleroglucan as A viable commercial solution. In: SPE Improved Oil Recovery Conference. Society of Petroleum Engineers.

Jeon, M.-K., Kwon, T.-H., Park, J.-S., Shin, J.H., 2017. In situ viscoelastic properties of insoluble and porous polysaccharide biopolymer dextran produced by Leuconostoc mesenteroides using particle-tracking microrheology. Geomechanics and Engineering 12, 849–862.

Jones, D., Head, I., Gray, N., Adams, J., Rowan, A., Aitken, C., Bennett, B., Huang, H., Brown, A., Bowler, B., 2008. Crude-oil biodegradation via methanogenesis in subsurface petroleum reservoirs. Nature 451, 176–180.

Justice, S.S., Hunstad, D.A., Cegelski, L., Hultgren, S.J., 2008. Morphological plasticity as a bacterial survival strategy. Nature Reviews Microbiology 6, 162–168.

Kalia, V.C., 2016. Microbial Factories: Biodiversity, Biopolymers, Bioactive Molecules. Springer.

Kalpakci, B., Jeans, Y., Magri, N., Padolewski, J., 1990. Thermal stability of scleroglucan at realistic reservoir conditions. In: SPE/DOE Enhanced Oil Recovery Symposium. Society of Petroleum Engineers.

Kanga, S.A., Bonner, J.S., Page, C.A., Mills, M.A., Autenrieth, R.L., 1997. Solubilization of naphthalene and methyl-substituted naphthalenes from crude oil using biosurfactants. Environmental Science & Technology 31, 556–561.

Karimi, M., Mahmoodi, M., Niazi, A., Al-Wahaibi, Y., Ayatollahi, S., 2012. Investigating wettability alteration during MEOR process, a micro/macro scale analysis. Colloids and Surfaces B: Biointerfaces 95, 129–136.

Karunakaran, E., Biggs, C.A., 2011. Mechanisms of Bacillus cereus biofilm formation: an investigation of the physicochemical characteristics of cell surfaces and extracellular proteins. Applied Microbiology and Biotechnology 89, 1161–1175.

Kaster, K.M., Bonaunet, K., Berland, H., Kjeilen-Eilertsen, G., Brakstad, O.G., 2009. Characterisation of culture-independent and -dependent microbial communities in a high-temperature offshore chalk petroleum reservoir. Antonie van Leeuwenhoek 96, 423–439.

Kato, K., Shiosaka, M., 1975. Process for the Production of Pullulan. U.S. Patent 3,912,591, issued October 14, 1975.

Kato, T., Haruki, M., Imanaka, T., Morikawa, M., Kanaya, S., 2001. Isolation and characterization of long-chain-alkane degrading Bacillus thermoleovorans from deep subterranean petroleum reservoirs. Journal of Bioscience and Bioengineering 91, 64–70.

Katzbauer, B., 1998. Properties and applications of xanthan gum. Polymer Degradation and Stability 59, 81–84.

Khademolhosseini, R., Jafari, A., Mousavi, S.M., Hajfarajollah, H., Noghabi, K.A., Manteghian, M., 2019. Physicochemical characterization and optimization of glycolipid biosurfactant production by a native strain of Pseudomonas aeruginosa HAK01 and its performance evaluation for the MEOR process. RSC Advances 9, 7932–7947.

Khanal, S.N., Anand, S., Muthukumarappan, K., 2014. Evaluation of high-intensity ultrasonication for the inactivation of endospores of 3 Bacillus species in nonfat milk. Journal of Dairy Science 97, 5952–5963.

Khatami, H.R., O'Kelly, B.C., 2013. Improving mechanical properties of sand using biopolymers. Journal of Geotechnical and Geoenvironmental Engineering 139, 1402–1406.

Kim, D.-S., Fogler, H.S., 1999. The effects of exopolymers on cell morphology and culturability of Leuconostoc mesenteroides during starvation. Applied Microbiology and Biotechnology 52, 839–844.

Kim, D.D., O'Farrell, C., Toth, C.R., Montoya, O., Gieg, L.M., Kwon, T.H., Yoon, S., 2018. Microbial community analyses of produced waters from high-temperature oil reservoirs reveal unexpected similarity between geographically distant oil reservoirs. Microbial biotechnology 11, 788–796.

Kim, Y.-M., Park, T., Kwon, T.-H., 2019. Engineered bio-clogging in coarse sands by using fermentation-based bacterial biopolymer formation. Geomechanics and engineering 17, 485–496.

Klueglein, N., Kögler, F., Adaktylou, I.J., Wuestner, M.L., Mahler, E., Scholz, J., Herold, A., Alkan, H., 2016. Understanding selective plugging and biofilm formation of a halophilic bacterial community for MEOR application. In: SPE Improved Oil Recovery Conference. Society of Petroleum Engineers.

Körstgens, V., Flemming, H.-C., Wingender, J., Borchard, W., 2001. Influence of calcium ions on the mechanical properties of a model biofilm of mucoid *Pseudomonas aeruginosa*. Water Science and Technology 43, 49–57.

Kumar, A., Singh, J., Baskar, C., Ramakrishna, S., 2015. Bioenergy: biofuels process technology. In: Advances in Bioprocess Technology. Springer, pp. 165–207.

Lappan, R.E., Fogler, H.S., 1996. Reduction of porous media permeability from in situ *Leuconostoc mesenteroides* growth and dextran production. Biotechnology and Bioengineering 50, 6–15.

Lappin-Scott, H.M., Costerton, J.W., 1989. Bacterial biofilms and surface fouling. Biofouling 1, 323–342.

Larter, S., Wilhelms, A., Head, I., Koopmans, M., Aplin, A., Di Primio, R., Zwach, C., Erdmann, M., Telnaes, N., 2003. The controls on the composition of biodegraded oils in the deep subsurface—part 1: biodegradation rates in petroleum reservoirs. Organic Geochemistry 34, 601–613.

Lecacheux, D., Mustiere, Y., Panaras, R., Brigand, G., 1986. Molecular weight of scleroglucan and other extracellular microbial polysaccharides by size-exclusion chromatography and low angle laser light scattering. Carbohydrate Polymers 6, 477–492.

Leduy, A., Marsan, A., Coupal, B., 1974. A study of the rheological properties of a non-Newtonian fermentation broth. Biotechnology and Bioengineering 16, 61–76.

Lee, H.O., Bae, J.H., Hejl, K., Edwards, A., 1998. Laboratory design and field implementation of microbial profile modification process. In: SPE Annual Technical Conference and Exhibition. Society of Petroleum Engineers, New Orleans, Louisiana, p. 12.

Lee, J.H., Lee, I.Y., Kim, M.K., Park, Y.H., 1999. Optimal pH control of batch processes for production of curdlan by *Agrobacterium* species. Journal of Industrial Microbiology and Biotechnology 23, 143–148.

Lee, K., 1999. Characterization of Scleroglucan Fermentation by *Sclerotium Rolfsii* in Terms of Cell, Scleroglucan and By-Product, Oxalic Acid Concentrations, Viscosity and Molecular Weight Distribution.

Lee, K.Y., Yoo, Y.J., 1993. Optimization of pH for high molecular weight pullulan. Biotechnology Letters 15, 1021–1024.

Lee, S., Kim, D.-H., Kim, K.-W., 2018. The enhancement and inhibition of mercury reduction by natural organic matter in the presence of *Shewanella oneidensis* MR-1. Chemosphere 194, 515–522.

Leriche, V., Sibille, P., Carpentier, B., 2000. Use of an enzyme-linked lectinsorbent assay to monitor the shift in polysaccharide composition in bacterial biofilms. Applied and Environmental Microbiology 66, 1851–1856.

Letisse, F., Chevallereau, P., Simon, J.-L., Lindley, N.D., 2001. Kinetic analysis of growth and xanthan gum production with *Xanthomonas campestris* on sucrose, using sequentially consumed nitrogen sources. Applied Microbiology and Biotechnology 55, 417–422.

Li, X.-X., Mbadinga, S.M., Liu, J.-F., Zhou, L., Yang, S.-Z., Gu, J.-D., Mu, B.-Z., 2017. Microbiota and their affiliation with physiochemical characteristics of different subsurface petroleum reservoirs. International Biodeterioration & Biodegradation 120, 170–185.

Li, X., Lu, Y., Adams, G.G., Zobel, H., Ballance, S., Wolf, B., Harding, S.E., 2020. Characterisation of the molecular properties of scleroglucan as an alternative rigid rod molecule to xanthan gum for *oropharyngeal dysphagia*. Food Hydrocolloids 101, 105446.

Liang, K., Han, P., Chen, Q., Su, X., Feng, Y., 2019. Comparative study on enhancing oil recovery under high temperature and high salinity: polysaccharides versus synthetic polymer. ACS Omega 4, 10620–10628.

Liang, Y., Kashdan, T., Sterner, C., Dombrowski, L., Petrick, I., Kröger, M., Höfer, R., 2015. Algal Biorefineries. Industrial Biorefineries & White Biotechnology. Elsevier, pp. 35–90.

Lin, J., Hao, B., Cao, G., Wang, J., Feng, Y., Tan, X., Wang, W., 2014. A study on the microbial community structure in oil reservoirs developed by water flooding. Journal of Petroleum Science and Engineering 122, 354–359.

Liu, C.-G., Xue, C., Lin, Y.-H., Bai, F.-W., 2013. Redox potential control and applications in microaerobic and anaerobic fermentations. Biotechnology Advances 31, 257–265.

Madigan, M.T., Martinko, J.M., Parker, J., 1997. Brock Biology of Microorganisms. Prentice hall, Upper Saddle River, NJ.

Magot, M., Ollivier, B., Patel, B.K., 2000. Microbiology of petroleum reservoirs. Antonie Van Leeuwenhoek 77, 103–116.

Majed, R., Faille, C., Kallassy, M., Gohar, M., 2016. *Bacillus cereus* biofilms—same, only different. Frontiers in Microbiology 7, 1054.

Makkar, R., Cameotra, S.S., 1997. Biosurfactant production by a thermophilic *Bacillus subtilis* strain. Journal of Industrial Microbiology and Biotechnology 18, 37–42.

Manandhar, S., Vidhate, S., D'Souza, N., 2009. Water soluble levan polysaccharide biopolymer electrospun fibers. Carbohydrate Polymers 78, 794–798.

Marsh, T., Zhang, X., Knapp, R., McInerney, M., Sharma, P., Jackson, B., 1995. Mechanisms of Microbial Oil Recovery by *Clostridium Acetobutylicum* and *Bacillus* Strain JF-2. BDM Oklahoma, Inc., Bartlesville, OK (United States).

Mcintosh, M., Stone, B., Stanisich, V., 2005. Curdlan and other bacterial (1→3)-β-D-glucans. Applied Microbiology and Biotechnology 68, 163–173.

Milas, M., Rinaudo, M., Tinland, B., 1985. The viscosity dependence on concentration, molecular weight and shear rate of xanthan solutions. Polymer Bulletin 14, 157–164.

Monroe, D., 2007. Looking for chinks in the armor of bacterial biofilms. PLoS Biology 5.

Mørch, Y.A., Holtan, S., Donati, I., Strand, B., Skjåk-Bræk, G., 2008. Mechanical properties of C-5 epimerized alginates. Biomacromolecules 9, 2360–2368.

Moussa, T., Mohamed, M., Samak, N., 2014. Production and characterization of di-rhamnolipid produced by Pseudomonas aeruginosa TMN. Brazilian Journal of Chemical Engineering 31, 867–880.

Moya-Ramírez, I., Garcia-Roman, M., Fernandez-Arteaga, A., 2017. Rhamnolipids: highly compatible surfactants for the enzymatic hydrolysis of waste frying oils in microemulsion systems. ACS Sustainable Chemistry & Engineering 5, 6768–6775.

Myers, C.R., Nealson, K.H., 1990. Respiration-linked proton translocation coupled to anaerobic reduction of manganese (IV) and iron (III) in Shewanella putrefaciens MR-1. Journal of Bacteriology 172, 6232–6238.

Naessens, M., Cerdobbel, A., Soetaert, W., Vandamme, E.J., 2005. Leuconostoc dextransucrase and dextran: production, properties and applications. Journal of Chemical Technology and Biotechnology: International Research in Process, Environmental & Clean Technology 80, 845–860.

Nagase, K., Zhang, S., Asami, H., Yazawa, N., Fujiwara, K., Enomoto, H., Hong, C., Liang, C., 2002. A successful field test of microbial EOR process in Fuyu Oilfield, China. In: SPE/DOE Improved Oil Recovery Symposium. Society of Petroleum Engineers.

Nakanishi, I., Kimura, K., Kanamaru, T., 1992. Studies on curdlan-type polysaccharide. I. Industrial production of curdlan-type polysaccharide. Journal of Takeda Research Laboratories 51, 99–108.

Nicholson, W., 2002. Roles of Bacillus endospores in the environment. Cellular and Molecular Life Sciences CMLS 59, 410–416.

Nicholson, W.L., Munakata, N., Horneck, G., Melosh, H.J., Setlow, P., 2000. Resistance of Bacillus endospores to extreme terrestrial and extraterrestrial environments. Microbiology and Molecular Biology Reviews 64, 548–572.

Nikolova, C., Gutierrez, T., 2020. Use of microorganisms in the recovery of oil from recalcitrant oil reservoirs: current state of knowledge, technological advances and future perspectives. Frontiers in Microbiology 10, 2996.

Noh, D.H., Ajo-Franklin, J.B., Kwon, T.H., Muhunthan, B., 2016. P and S wave responses of bacterial biopolymer formation in unconsolidated porous media. Journal of Geophysical Research: Biogeosciences 121, 1158–1177.

Nuryadi, A., Kishita, A., Watanabe, N., Vilcaez Perez, J., Kawai, N., 2011. EOR simulation by in situ nitrogen production via denitrifying bacteria and performance improvement by nitrogen alternating surfactant injection. In: SPE Asia Pacific Oil and Gas Conference and Exhibition. Society of Petroleum Engineers.

Ohadi, M., Dehghannoudeh, G., Shakibaie, M., Banat, I.M., Pournamdari, M., Forootanfar, H., 2017a. Isolation, characterization, and optimization of biosurfactant production by an oil-degrading Acinetobacter junii B6 isolated from an Iranian oil excavation site. Biocatalysis and Agricultural Biotechnology 12, 1–9.

Ohadi, M., Forootanfar, H., Rahimi, H.R., Jafari, E., Shakibaie, M., Eslaminejad, T., Dehghannoudeh, G., 2017b. Antioxidant potential and wound healing activity of biosurfactant produced by Acinetobacter junii B6. Current Pharmaceutical Biotechnology 18, 900–908.

Ohno, A., Ano, T., Shoda, M., 1995. Production of a lipopeptide antibiotic, surfactin, by recombinant Bacillus subtilis in solid state fermentation. Biotechnology and Bioengineering 47, 209–214.

Orgad, O., Oren, Y., Walker, S.L., Herzberg, M., 2011. The role of alginate in Pseudomonas aeruginosa EPS adherence, viscoelastic properties and cell attachment. Biofouling 27, 787–798.

Painter, T.J., 1983. Algal Polysaccharides. The Polysaccharides. Elsevier, pp. 195–285.

Palleschi, A., Bocchinfuso, G., Coviello, T., Alhaique, F., 2005. Molecular dynamics investigations of the polysaccharide scleroglucan: first study on the triple helix structure. Carbohydrate Research 340, 2154–2162.

Pannekens, M., Kroll, L., Müller, H., Mbow, F.T., Meckenstock, R.U., 2019. Oil reservoirs, an exceptional habitat for microorganisms. New biotechnology 49, 1–9.

Park, T., Jeon, M.-K., Yoon, S., Lee, K.S., Kwon, T.-H., 2019. Modification of interfacial tension and wettability in oil-brine-quartz system by in situ bacterial biosurfactant production at reservoir conditions: implications to microbial enhanced oil recovery. Energy & Fuels 33, 4909–4920.

Park, T., Joo, H.-W., Kim, G.-Y., Kim, S., Yoon, S., Kwon, T.-H., 2017. Biosurfactant as an enhancer of geologic carbon storage: microbial modification of interfacial tension and contact angle in carbon dioxide/water/quartz systems. Frontiers in Microbiology 8, 1285.

Patel, J., Borgohain, S., Kumar, M., Rangarajan, V., Somasundaran, P., Sen, R., 2015. Recent developments in microbial enhanced oil recovery. Renewable and Sustainable Energy Reviews 52, 1539–1558.

Pereira, J.F., Gudiña, E.J., Costa, R., Vitorino, R., Teixeira, J.A., Coutinho, J.A., Rodrigues, L.R., 2013. Optimization and characterization of biosurfactant production by Bacillus subtilis isolates towards microbial enhanced oil recovery applications. Fuel 111, 259–268.

Philippi, G., 1977. On the depth, time and mechanism of origin of the heavy to medium-gravity naphthenic crude oils. Geochimica et Cosmochimica Acta 41, 33–52.

Phillips, K.R., Lawford, H., 1983. Curdlan: its properties and production in batch and continuous fermentations. Progress in Industrial Microbiology 18, 201–229.

Pollock, T.J., Thorne, L., Yamazaki, M., Mikolajczak, M.J., Armentrout, R.W., 1994. Mechanism of bacitracin resistance in gram-negative bacteria that synthesize exopolysaccharides. Journal of Bacteriology 176, 6229–6237.

Polson, E.J., Buckman, J.O., Bowen, D.G., Todd, A.C., Gow, M.M., Cuthbert, S.J., 2010. An environmental-scanning-electron-microscope investigation into the effect of biofilm on the wettability of quartz. SPE Journal 15, 223–227.

Pu, W., Shen, C., Wei, B., Yang, Y., Li, Y., 2018. A comprehensive review of polysaccharide biopolymers for enhanced oil recovery (EOR) from flask to field. Journal of Industrial and Engineering Chemistry 61, 1—11.

Radulović, M.Đ., Cvetković, O.G., Nikolić, S.D., Đorđević, D.S., Jakovljević, D.M., Vrvić, M.M., 2008. Simultaneous production of pullulan and biosorption of metals by Aureobasidium pullulans strain CH-1 on peat hydrolysate. Bioresource Technology 99, 6673—6677.

Raiders, R.A., Knapp, R.M., McInerney, M.J., 1989. Microbial selective plugging and enhanced oil recovery. Journal of Industrial Microbiology 4, 215—229.

Ramana, K.V., Karanth, N., 1989. Factors affecting biosurfactant production using Pseudomonas aeruginosa CFTR-6 under submerged conditions. Journal of Chemical Technology and Biotechnology 45, 249—257.

Ramsay, J.A., Cooper, D., Neufeld, R., 1989. Effects of oil reservoir conditions on the production of water-insoluble Levan by Bacillus licheniformis. Geomicrobiology Journal 7, 155—165.

Rehm, B., 2009. Microbial Production of Biopolymers and Polymer Precursors. Caister Academic.

Rehm, B.H., 2010. Bacterial polymers: biosynthesis, modifications and applications. Nature Reviews Microbiology 8, 578—592.

Rey, M.W., Ramaiya, P., Nelson, B.A., Brody-Karpin, S.D., Zaretsky, E.J., Tang, M., Lopez de Leon, A., Xiang, H., Gusti, V., Clausen, I.G., Olsen, P.B., Rasmussen, M.D., Andersen, J.T., Jørgensen, P.L., Larsen, T.S., Sorokin, A., Bolotin, A., Lapidus, A., Galleron, N., Ehrlich, S.D., Berka, R.M., 2004. Complete genome sequence of the industrial bacterium Bacillus licheniformis and comparisons with closely related Bacillus species. Genome Biology 5. R77-R77.

Rivenq, R., Donche, A., Nolk, C., 1992. Improved scleroglucan for polymer flooding under harsh reservoir conditions. SPE Reservoir Engineering 7, 15—20.

Rocha, C., San-Blas, F., San-Blas, G., Vierma, L., 1992. Biosurfactant production by two isolates of Pseudomonas aeruginosa. World Journal of Microbiology and Biotechnology 8, 125—128.

Roy, J.N., Garcia, K.E., Luckarift, H.R., Falase, A., Cornejo, J., Babanova, S., Schuler, A.J., Johnson, G.R., Atanassov, P.B., 2013. Applied electrode potential leads to Shewanella oneidensis MR-1 biofilms engaged in direct electron transfer. Journal of The Electrochemical Society 160, H866—H871.

Röling, W.F., Head, I.M., Larter, S.R., 2003. The microbiology of hydrocarbon degradation in subsurface petroleum reservoirs: perspectives and prospects. Research in Microbiology 154, 321—328.

Rosen, M.J., Kunjappu, J.T., 2012. Surfactants and Interfacial Phenomena. John Wiley & Sons.

Runnels, C.M., Lanier, K.A., Williams, J.K., Bowman, J.C., Petrov, A.S., Hud, N.V., Williams, L.D., 2018. Folding, assembly, and persistence: the essential nature and origins of biopolymers. Journal of Molecular Evolution 86, 598—610.

Ryles, R., 1988. Chemical stability limits of water-soluble polymers used in oil recovery processes. SPE Reservoir Engineering 3, 23—34.

Safdel, M., Anbaz, M.A., Daryasafar, A., Jamialahmadi, M., 2017. Microbial enhanced oil recovery, a critical review on worldwide implemented field trials in different countries. Renewable and Sustainable Energy Reviews 74, 159—172.

Saito, H., Misaki, A., Harada, T., 1968. A comparison of the structure of curdlan and pachyman. Agricultural & Biological Chemistry 32, 1261—1269.

Sandford, P.A., 1979. Exocellular, microbial polysaccharides. In: Advances in Carbohydrate Chemistry and Biochemistry. Elsevier, pp. 265—313.

Saxman, D., Crull, A., 1984. Biotechnology and enhanced petroleum production. In: SPE Annual Technical Conference and Exhibition. Society of Petroleum Engineers.

Schramm, L.L., 2000. Surfactants: Fundamentals and Applications in the Petroleum Industry. Cambridge University Press.

Sen, R., 2008. Biotechnology in petroleum recovery: the microbial EOR. Progress in Energy and Combustion Science 34, 714—724.

Seright, R., Henrici, B., 1990. Xanthan stability at elevated temperatures. SPE Reservoir Engineering 5, 52—60.

Shaligram, N.S., Singhal, R.S., 2010. Surfactin—a review on biosynthesis, fermentation, purification and applications. Food Technology and Biotechnology 48, 119—134.

Shariat, P., Mitchell, W.J., Boyd, A., Priest, F.G., 1995. Anaerobic metabolism in Bacillus licheniformis NCIB 6346. Microbiology 141, 1117—1124.

Shavandi, M., Mohebali, G., Haddadi, A., Shakarami, H., Nuhi, A., 2011. Emulsification potential of a newly isolated biosurfactant-producing bacterium, Rhodococcus sp. strain TA6. Colloids and Surfaces B: Biointerfaces 82, 477—482.

Shin, Y.C., Kim, Y.H., Lee, H.S., Cho, S.J., Byun, S.M., 1989. Production of exopolysaccharide pullulan from inulin by a mixed culture of Aureobasidium pullulans and Kluyveromyces fragilis. Biotechnology and Bioengineering 33, 129—133.

Shingel, K.I., 2004. Current knowledge on biosynthesis, biological activity, and chemical modification of the exopolysaccharide, pullulan. Carbohydrate Research 339, 447—460.

Siebel, M.A., Characklis, W.G., 1991. Observations of binary population biofilms. Biotechnology and Bioengineering 37, 778—789.

Silva, T., Verde, L., Neto, E.S., Oliveira, V., 2013. Diversity analyses of microbial communities in petroleum samples from Brazilian oil fields. International Biodeterioration & Biodegradation 81, 57—70.

Simon, L., Caye-Vaugien, C., Bouchonneau, M., 1993. Relation between pullulan production, morphological state and growth conditions in Aureobasidium pullulans: new observations. Microbiology 139, 979—985.

Simpson, D.R., Natraj, N.R., McInerney, M.J., Duncan, K.E., 2011. Biosurfactant-producing Bacillus are present in produced brines from Oklahoma oil reservoirs with a wide range of salinities. Applied Microbiology and Biotechnology 91, 1083.

Singh, P., Cameotra, S.S., 2004. Enhancement of metal bioremediation by use of microbial surfactants. Biochemical and Biophysical Research Communications 319, 291—297.

Smidsrød, O., Moe, S., Moe, S.T., 2008. Biopolymer Chemistry. Tapir Academic Press.

Som, I., Bhatia, K., Yasir, M., 2012. Status of surfactants as penetration enhancers in transdermal drug delivery. Journal of Pharmacy & Bioallied Sciences 4, 2.

Stewart, P.S., 2002. Mechanisms of antibiotic resistance in bacterial biofilms. International journal of medical microbiology 292, 107–113.

Stoodley, P., Boyle, J.D., Dodds, I., Lappin-Scott, H.M., 1997. Consensus model of biofilm structure.

Stoodley, P., Cargo, R., Rupp, C.J., Wilson, S., Klapper, I., 2002. Biofilm material properties as related to shear-induced deformation and detachment phenomena. Journal of Industrial Microbiology and Biotechnology 29, 361–367.

Suresh Kumar, A., Mody, K., Jha, B., 2007. Bacterial exopolysaccharides—a perception. Journal of Basic Microbiology 47, 103–117.

Suthar, H., Hingurao, K., Desai, A., Nerurkar, A., 2009. Selective plugging strategy-based microbial-enhanced oil recovery using *Bacillus licheniformis* TT33. Journal of Microbiology and Biotechnology 19, 1230–1237.

Sutherland, I.W., 2001. Biofilm exopolysaccharides: a strong and sticky framework. Microbiology 147, 3–9.

Syldatk, C., Lang, S., Wagner, F., Wray, V., Witte, L., 1985. Chemical and physical characterization of four interfacial-active rhamnolipids from *Pseudomonas* spec. DSM 2874 grown on n-alkanes. Zeitschrift für Naturforschung C 40, 51–60.

Tanner, R., Yoshida, T., 1993. Bioproducts and Bioprocesses. Springer.

Taylor, S.W., Jaffé, P.R., 1990. Biofilm growth and the related changes in the physical properties of a porous medium: 1. Experimental investigation. Water Resources Research 26, 2153–2159.

Thauer, R.K., 1998. Biochemistry of methanogenesis: a tribute to marjory Stephenson: 1998 marjory Stephenson prize lecture. Microbiology 144, 2377–2406.

Thormann, K.M., Saville, R.M., Shukla, S., Pelletier, D.A., Spormann, A.M., 2004. Initial phases of biofilm formation in *Shewanella oneidensis* MR-1. Journal of Bacteriology 186, 8096–8104.

Tirtaatmadja, V., Dunstan, D.E., Boger, D.V., 2001. Rheology of dextran solutions. Journal of Non-newtonian Fluid Mechanics 97, 295–301.

Tolker-Nielsen, T., Molin, S., 2000. Spatial organization of microbial biofilm communities. Microbial Ecology 40, 75–84.

Ullrich, M., 2009. Bacterial Polysaccharides: Current Innovations and Future Trends. Horizon Scientific Press.

Van Hamme, J.D., Singh, A., Ward, O.P., 2003. Recent advances in petroleum microbiology. Microbiology and Molecular Biology Reviews 67, 503–549.

van Spanning, R.J., Richardson, D.J., Ferguson, S.J., 2007. Introduction to the Biochemistry and Molecular Biology of Denitrification. Biology of the Nitrogen Cycle. Elsevier, pp. 3–20.

Vandamme, E.J., De Baets, S., Steinbüchel, A., 2002. Biopolymers, Polysaccharides II: Polysaccharides from Eukaryotes. Wiley-Blackwell.

Varjani, S.J., Upasani, V.N., 2016. Core flood study for enhanced oil recovery through ex-situ bioaugmentation with thermo-and halo-tolerant rhamnolipid produced by *Pseudomonas aeruginosa* NCIM 5514. Bioresource Technology 220, 175–182.

Venkateswaran, K., Moser, D.P., Dollhopf, M.E., Lies, D.P., Saffarini, D.A., MacGregor, B.J., Ringelberg, D.B., White, D.C., Nishijima, M., Sano, H., 1999. Polyphasic taxonomy of the genus *Shewanella* and description of *Shewanella oneidensis* sp. nov. International Journal of Systematic and Evolutionary Microbiology 49, 705–724.

Venugopal, V., 2011. Extracellular polysaccharides from marine microorganism. In: Marine Polysaccharides: Food Applications. CRC Press, Taylor & Francis Group, LLC., Florida, USA.

Vignais, P., Colbeau, A., 2004. Molecular biology of microbial hydrogenases. Current Issues in Molecular Biology 6, 159–188.

Vigneron, A., Alsop, E.B., Lomans, B.P., Kyrpides, N.C., Head, I.M., Tsesmetzis, N., 2017. Succession in the petroleum reservoir microbiome through an oil field production lifecycle. The ISME Journal 11, 2141–2154.

Vijayendra, S., Shamala, T., 2014. Film forming microbial biopolymers for commercial applications—a review. Critical Reviews in Biotechnology 34, 338–357.

Vilain, S., Pretorius, J.M., Theron, J., Brözel, V.S., 2009. DNA as an adhesin: *Bacillus cereus* requires extracellular DNA to form biofilms. Applied and Environmental Microbiology 75, 2861–2868.

Wang, C., Hao, M., Qi, Q., Chen, Y., Hartig, J.S., 2019. Chemical synthesis, purification, and characterization of 3′-5′-linked canonical cyclic dinucleotides (CDNs). Methods in Enzymology 625, 41–59.

Wang, H., Liu, Y., Li, J., Lin, M., Hu, X., 2016. Biodegradation of atrazine by Arthrobacter sp. C3, isolated from the herbicide-contaminated corn field. International Journal of Environmental Science and Technology 13, 257–262.

Wang, Y., McNeil, B., 1994. Scleroglucan and oxalic acid formation by *Sclerotium glucanicum* in sucrose supplemented fermentations. Biotechnology Letters 16, 605–610.

Wei, B., Romero-Zerón, L., Rodrigue, D., 2014. Mechanical properties and flow behavior of polymers for enhanced oil recovery. Journal of Macromolecular Science 53, 625–644.

Weidong, W., Junzhang, L., Xueli, G., Jing, W., Ximing, L., Yan, J., Fengmin, Z., 2014. MEOR field test at block Luo801 of Shengli oil field in China. Petroleum Science and Technology 32, 673–679.

Wellington, S.L., 1983. Biopolymer solution viscosity stabilization-polymer degradation and antioxidant use. Society of Petroleum Engineers Journal 23, 901–912.

Wilham, C., Alexander, B.H., Jeanes, A., 1955. Heterogeneity in dextran preparations. Archives of Biochemistry and Biophysics 59, 61–75.

Willenbacher, J., Rau, J.-T., Rogalla, J., Syldatk, C., Hausmann, R., 2015. Foam-free production of Surfactin via anaerobic fermentation of *Bacillus subtilis* DSM 10 T. AMB Express 5, 21.

Wu, W., Jin, Y., Bai, F., Jin, S., 2015. Chapter 41 - *Pseudomonas aeruginosa*. In: Tang, Y.-W., Sussman, M., Liu, D., Poxton, I., Schwartzman, J. (Eds.), Molecular Medical Microbiology, second ed. Academic Press, Boston, pp. 753–767.

Xiong, Y., Shi, L., Chen, B., Mayer, M.U., Lower, B.H., Londer, Y., Bose, S., Hochella, M.F., Fredrickson, J.K., Squier, T.C., 2006. High-affinity binding and direct electron transfer to solid metals by the *Shewanella oneidensis* MR-1 outer membrane c-type cytochrome OmcA. Journal of the American Chemical Society 128, 13978–13979.

Yakimov, M.M., Amro, M.M., Bock, M., Boseker, K., Fredrickson, H.L., Kessel, D.G., Timmis, K.N., 1997. The potential of *Bacillus licheniformis* strains for in situ enhanced oil recovery. Journal of Petroleum Science and Engineering 18, 147–160.

Yakimov, M.M., Timmis, K.N., Wray, V., Fredrickson, H.L., 1995. Characterization of a new lipopeptide surfactant produced by thermotolerant and halotolerant subsurface *Bacillus licheniformis* BAS50. Applied and Environmental Microbiology 61, 1706–1713.

Yawata, Y., Nomura, N., Uchiyama, H., 2008. Development of a novel biofilm continuous culture method for simultaneous assessment of architecture and gaseous metabolite production. Applied and Environmental Microbiology 74, 5429–5435.

Yoon, M.Y., Lee, K.-M., Park, Y., Yoon, S.S., 2011. Contribution of cell elongation to the biofilm formation of *Pseudomonas aeruginosa* during anaerobic respiration. PloS One 6, e16105.

Youssef, N., Simpson, D., Duncan, K., McInerney, M., Folmsbee, M., Fincher, T., Knapp, R., 2007. In situ biosurfactant production by *Bacillus* strains injected into a limestone petroleum reservoir. Applied and Environmental Microbiology 73, 1239–1247.

Yunita Halim, A., Marie Nielsen, S., Eliasson Lantz, A., Sander Suicmez, V., Lindeloff, N., Shapiro, A., 2015. Investigation of spore forming bacterial flooding for enhanced oil recovery in a North Sea chalk Reservoir. Journal of Petroleum Science and Engineering 133, 444–454.

Zhang, H., Zhang, F., Yuan, R., 2020. Applications of natural polymer-based hydrogels in the food industry. In: Hydrogels Based on Natural Polymers. Elsevier, pp. 357–410.

Zhao, F., Shi, R., Ma, F., Han, S., Zhang, Y., 2018. Oxygen effects on rhamnolipids production by *Pseudomonas aeruginosa*. Microbial Cell Factories 17, 39.

Zhao, F., Zhou, J.-D., Ma, F., Shi, R.-J., Han, S.-Q., Zhang, J., Zhang, Y., 2016. Simultaneous inhibition of sulfate-reducing bacteria, removal of H2S and production of rhamnolipid by recombinant *Pseudomonas stutzeri* Rhl: applications for microbial enhanced oil recovery. Bioresource Technology 207, 24–30.

Zheng, C., Yu, L., Huang, L., Xiu, J., Huang, Z., 2012. Investigation of a hydrocarbon-degrading strain, *Rhodococcus ruber* Z25, for the potential of microbial enhanced oil recovery. Journal of Petroleum Science and Engineering 81, 49–56.

ZoBell, C.E., Johnson, F.H., 1949. The influence of hydrostatic pressure on the growth and viability of terrestrial and marine bacteria. Journal of Bacteriology 57, 179.

Zou, C., Wang, M., Xing, Y., Lan, G., Ge, T., Yan, X., Gu, T., 2014. Characterization and optimization of biosurfactants produced by *Acinetobacter baylyi* ZJ2 isolated from crude oil-contaminated soil sample toward microbial enhanced oil recovery applications. Biochemical Engineering Journal 90, 49–58.

CHAPTER 3

Theory and Experiments

3.1 PRINCIPLES OF INTERFACIAL TENSION AND WETTABILITY

3.1.1 Interfacial Tension or Surface Tension

Porous media containing immiscible fluid phases are inevitably involved in interactions with the fluids. Surface energy prevailing at fluid interfaces affects fluid saturations, fluid mobility, and overall distributions of fluids within porous media (Ingham and Pop, 1998). Therefore, the interfacial forces among phases have a direct influence on fluid flow behavior. The interfacial force is quantified in terms of interfacial tension (IFT) or interfacial energy. The IFT has the dimension of force per unit length (e.g., N/m), and the interfacial energy has the dimension of energy per area (e.g., J/m^2), but the both dimensions are basically identical, such that the IFT and interfacial energy are interchangeability used. The "*interfacial*" tension refers to the tension existing at an interface between two immiscible fluids or between a liquid and a solid (Berry et al., 2015). Meanwhile, the "*surface*" tension typically refers to the specific case where the interface is placed between a liquid and its vapor or air. However, those two terms, surface tension and IFT, are often interchangeably used.

As graphically described in Fig. 3.1, the interior molecules within a liquid phase have an equal level of intermolecular cohesive forces with other surrounding molecules, which results in zero net force in every direction. However, the boundary molecules at the interface of liquid-liquid (or vapor-liquid and any immiscible liquids) are lack of some of the intermolecular cohesive forces. Therefore, the net molecular adhesive force is not zero and prevalent at an interface, which changes the shape of liquid droplets. Liquids tend to minimize the energy state at the interface by reducing interfacial (or surface) area (Berry et al., 2015). IFT is relevant to the shape of liquid droplets according to the forces acting on the interface (Drelich et al., 2002). In the cases where the adhesive force is stronger than the cohesive force, the liquid has a convex meniscus (Fig. 3.2A, water). By contrast, the stronger cohesive force than the adhesive force causes a convex meniscus (Fig. 3.2B, mercury) (Drelich et al., 2002). Wetting of water is more favorable than that of mercury on a glass surface. This indicates that the glass surface is hydrophilic, in terms of the ability of a liquid to maintain contact with a solid surface; this is referred to as the wettability.

There are various methods to measure IFT or surface tension, and the details are described in Table 3.1 and Fig. 3.3. The simplest way to measure the surface tension of a liquid is to use a capillary tube (Fig. 3.3E, capillary rise method; Harkins and Brown, 1919). With a capillary tube placed in a beaker filled with water, rises of water by capillarity is measured, and hence the surface tension of water with air can be calculated, as follows:

$$\gamma_{wa} = \frac{r_c h (\rho_w - \rho_a) g}{2 \cos \theta}, \qquad (3.1)$$

where γ_{wa} is the surface tension between water and air, ρ is the density of components, g is the gravitational force, r is the radius of a capillary tube, h is the height of raised water, and θ is the contact angle of solid surface with water.

3.1.2 Surface Wettability

The distributions and transport of fluids in porous media are affected not only by the interaction between the fluids (IFT) but also by the interactive forces between the fluids and solid surface, which can be expressed with the wettability (Ingham and Pop, 1998).

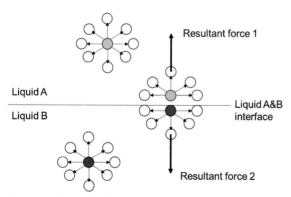

FIG. 3.1 Liquid-liquid interface and balance of forces on molecules of immiscible liquids. (Credit: Courtesy of Taehyung Park and Tae-Hyuk Kwon.)

Theory and Practice in Microbial Enhanced Oil Recovery. https://doi.org/10.1016/B978-0-12-819983-1.00003-X

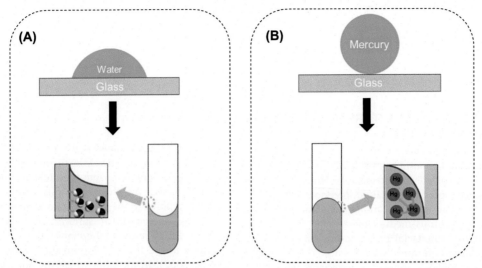

FIG. 3.2 **(A)** Concave-up and **(B)** convex-down menisci of water and mercury in a hydrophilic glass tube. (Credit: Courtesy of Taehyung Park and Tae-Hyuk Kwon.)

Wettability refers to the state of a liquid to maintain contact with a solid surface, due to the intermolecular interactions of liquids with solid surfaces. The degree of tendency of a fluid to spread or adhere to a certain solid surface can be quantitatively evaluated with the wettability, and such degree of wetting of a fluid on a certain solid surface is determined by a force balance between adhesive and cohesive forces (Melrose, 1965). The adhesive forces between a liquid and a solid can cause a liquid droplet to spread over the surface (i.e., strong adhesive forces between water and glass; Fig. 3.2A). By contrast, the cohesive forces of the liquid with the solid hinder a liquid droplet to spread but produce a liquid shape to be round, minimizing the contact area with the surface (i.e., strong cohesive forces within mercury molecules; Fig. 3.2B). The balance between adhesive and cohesive forces determines the contact angle, and the contact angle indicates the wettability of a fluid on a given solid surface in a quantitatively manner (Melrose, 1965). When a contact angle is less than 90 degrees (Fig. 3.4A), wetting of the fluid on that surface is favorable, indicating that the fluid has a high degree of wettability on the very surface. On the contrary, when a contact angle is larger than 90 degrees (Fig. 3.4B), the fluid has a low degree of wettability, and it means the unfavorable wetting of the fluid on the surface (Hui and Blunt, 2000). In a particular case of water, a high wettability surface is referred to as a hydrophilic surface, and a low wettability surface is referred to as a hydrophobic surface.

Reservoir rocks frequently involve multiphase pore fluids, where two or more immiscible fluids are present in pores. In such a condition, one fluid phase can be more favorable and more attracted to a solid (herein, rock mineral surface) than the other fluid phase, and this more favorable phase is referred to the *"wetting phase or wetting fluid."* The other fluid phase that is less attracted to the rock mineral is referred as the *"nonwetting phase or nonwetting fluid"* (Law and Zhao, 2016b). The similar concept can also be applied to describe the wettability of rocks, preference of a solid to a certain fluid. In oil reservoir, if the rock is *oil-wet*, it indicates that the rock mineral has preference to be in contact with oil over brine or water. If the rock is *water-wet*, it is the opposite.

Contact angle hysteresis is defined as the difference between the advancing (θ_a) and receding (θ_r) contact angles. The advancing contact angle is the maximum stable angle when the fluid advances on the surface. The receding contact angle is the minimum stable angle when the fluid recedes on the surface (Law and Zhao, 2016b). No contact angle hysteresis exists, such that the advancing and receding contact angles stay constant and the same, when a surface is physically flat, smooth, and chemically homogeneous. However, the surfaces of natural geologic materials and rocks hardly have perfectly smooth and homogeneous surfaces, and thus, the physicochemical heterogeneity unavoidably causes the contact angle hysteresis (Johnson Jr and Dettre, 1964). As a matter of fact, diverse chemical compositions of rock minerals and pore fluids as well as the

TABLE 3.1
Various Methods to Measure Surface/Interfacial Tensions (Drelich et al., 2002).

Surface Tension Measurement Method	Principles	Governing Equation	Key Parameter	Advantage and Limitation
(a) Wilhelmy plate method (Biswas et al., 2001; Drelich et al., 2002)	• A vertical thin plate used to measure the uplifted liquid meniscus by force applied to the plate. • Uses microbalance for measurements	$\sigma = \frac{F_c}{(2w_p + 2d_p)\cos\theta}$	• Contact angle of liquid with plate	• Accurate determination of surface kinetics on a wide range of timescales • Not suitable for liquids with absorbable
(b) Maximum bubble pressure method (Bendure, 1971; Drelich et al., 2002)	• Measures the maximum pressure to force a gas bubble out of a capillary into a liquid	$\sigma = \frac{\Delta P_{max} r_c}{2}$	• Maximum pressure drop	• Direct and rapid • No requirement of contact angle measurement • Not very accurate
(c) Spinning drop method (Cayias et al., 1975; Drelich et al., 2002)	• Measurement in a rotating horizontal tube with a dense fluid. • Surface tension derived from the shape of the drop at the equilibrium point	$\sigma = \frac{r_c^3 \Delta\rho\omega_v^2}{4}$	• Density difference between the fluids • Rotational velocity	• No need of interface curvature estimation • Measures ultralow surface tensions (up to 10^{-6} mN/m) • Not accurate for the high surface tension measurement
(d) Du Noüy Ring method (du Noüy, 1925; Drelich et al., 2002)	• Involves slowly lifting a ring • Force required to raise the ring to measure surface tension	$\sigma = \frac{F_{max}}{4\pi r_r \cos\theta} f_c$	• Correction factor (f) • Surface wettability • Density difference • Dimensions of ring	• High accuracy • Deformation of ring (delicate) causes measurement error
(e) Capillary rise method (Harkins and Brown, 1919; Drelich et al., 2002)	• Measure height of the meniscus in a round glass tube with known radius	$\sigma = \frac{\Delta\rho g h r_c}{2\cos\theta}$	• Height of meniscus • Inner radius of tube • Density difference • Contact angle of liquid and glass tube	• One of the most accurate techniques • Fabrication of a uniform capillary tube is challenging • Not convenient for measuring the interfacial tension between two immiscible liquids
(f) Pendant drop method (Arashiro and Demarquette, 1999; Drelich et al., 2002)	• Measures the surface tension of liquid drop hangs on a dosing needle	$\sigma = \frac{\Delta\rho g D_E^2}{H_c}$	• Pendant drop profile • Density difference	• Simple and reliable method • Requires extreme cleanliness of needle inside

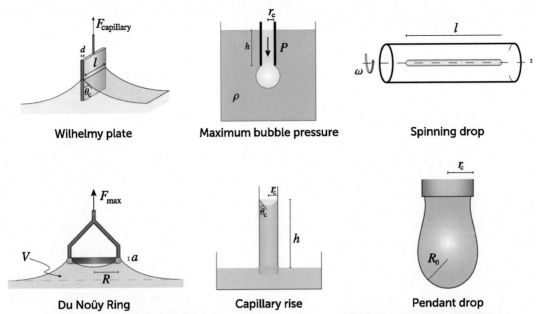

Wilhelmy plate   **Maximum bubble pressure**   **Spinning drop**

Du Noüy Ring   **Capillary rise**   **Pendant drop**

FIG. 3.3 Schematics of various experimental techniques used to determine interfacial tension. (Credit: Berry, J.D., Neeson, M.J., Dagastine, R.R., Chan, D.Y., Tabor, R.F., 2015. Measurement of surface and interfacial tension using pendant drop tensiometry. Journal of Colloid and Interface Science 454, 226–237.)

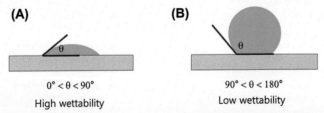

(A)   **(B)**

$0° < \theta < 90°$ $90° < \theta < 180°$

High wettability   Low wettability

FIG. 3.4 Wetting of fluids on solid surface: **(A)** a fluid with smaller contact angle due to the high wettability and **(B)** a fluid with higher contact angle due to the low wettability. (Credit: Courtesy of Taehyung Park and Tae-Hyuk Kwon.)

physicochemical reactions between rocks and fluids lead to complex wettability behaviors in natural environments. For an instance, the rock wettability in an oil reservoir may vary from point to point due to its heterogeneity. The wettability of rock, whether it is water-wet or oil-wet, affects the fluid saturation state within porous medium (Morrow, 1990).

3.1.3 Fluid Flow in Porous Media—Capillary Forces

In a porous medium with multiphase fluid-solid systems, it is essential to understand the thermodynamics of surface energy of fluids in the system because the surface energy governs the intrusion of a fluid phase in a porous medium. Importantly, the capillary pressure,

a pressure existing between two immiscible fluids in a narrow space, attributes to the interactive force between the fluids and solid surfaces. The capillary pressure serves as an opposing and/or driving force for fluids to transport, and it is derived from the forces at the fluid-fluid and fluid-solid interfaces when two immiscible fluids are in contact to each other and to a solid rock (Melrose, 1965). Therefore, the capillary pressure is determined by the IFT between fluids and the rock wettability, and it can be expressed, as follows:

$$p_c = \frac{2\sigma \cos \theta}{r},\tag{3.2}$$

where p_c is the capillary pressure, σ is the IFT between the fluids, θ is the contact angle of fluids with a solid surface, and r is the radius of pore. As shown in Eq. (3.2), the capillary pressure is proportional to the IFT, inversely proportional to the pore radius, and affected by the contact angle, which indicates the contact area between the wetting phase and the rock.

The capillary pressure plays a vital role in prediction of fluid transports and fluid saturations in porous media and, therefore, in determini2ng the relative permeability when two immiscible fluid phases are presents in a porous medium. This has direct relevance to the oil production rate in petroleum engineering and the cap-rock seal integrity in geologic carbon storage applications.

3.2 PORE-SCALE MECHANISMS OF MOBILITY CONTROL BY SURFACTANT ADHESION

3.2.1 Reduction in Interfacial Tension

The molecules at the surface of a liquid have higher potential energies than the molecules interior of the liquid because the attractive forces of molecules at the interface are greater than those in an interior phase. The IFT refers to the surface free energy per unit area, which indicates the minimum force required to bring a sufficient amount of molecules at the interface of immiscible fluids. For an instance, the potential energies of hydrocarbon molecules and water molecules across the interface (meniscus) are different. The similarities of these two facing but immiscible fluids determine the interaction energy across the interface and the IFT. When the two molecules are similar (e.g., water and ethylene alcohol), the interaction energy per unit area across the interface is large and leads to the small IFT. On the contrary to that, for the two fluids very dissimilar each other (e.g., water and hydrocarbon), the interaction energy is small, which leads to the high IFT.

Reduction of IFT between brine and oils trapped in a porous medium is one of the most effective strategies for successful implementation of enhanced oil recovery (EOR). Surfactant molecules replace oil and brine molecules at the oil-brine interface and, thereby, reduce IFT between oil and brine (or oil-brine IFT). A surfactant molecule consists of hydrophilic and hydrophobic parts. When surface-active agents such as surfactants are added to the system of two immiscible fluids, e.g., oil and brine, the surface-active agents are adsorbed at the fluid-fluid interface (Fig. 3.5). Then, the surfactants orient themselves, locating the hydrophilic group toward the brine molecules and the hydrophobic group toward the oil molecules (Fig. 3.5). The hydrophilic parts of surfactants interact with water molecules, and the hydrophobic tails interact with hydrocarbon

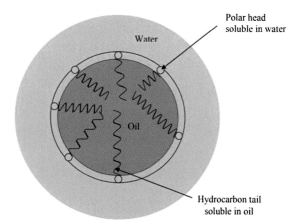

FIG. 3.5 Principles of surfactants that surfactant adsorbs at oil/water interface. (Credit: Olajire, A.A., 2014. Review of ASP EOR (alkaline surfactant polymer enhanced oil recovery) technology in the petroleum industry: prospects and challenges. Energy 77, 963-982.)

on the other side of the interface. These interactions of the hydrophilic and hydrophobic parts with two immiscible fluids significantly increases the interaction energy and thus decreases the IFT. As the structural dissimilarities of two immiscible fluid phases are more significant, the IFT reduction by surfactant becomes greater; one good example is the surfactant effect in oil-water systems.

When surfactant molecules spontaneously are adsorbed at fluid-fluid interfaces, such as air-water or water-hydrocarbon interfaces, an oriented monolayer of surfactants is formed, as shown in Fig. 3.6. Above the critical micelle concentration (CMC), the bulk surfactant molecules are adsorbed at the interfaces and create micelles (Fig. 3.6). The micellization, transition of surfactant molecules to form the micelle structure, happens within a timescale on the order of micro- to nanoseconds. Below the CMC, adsorption of surfactant monomers to interface occurs while it decreases the IFT. Once the surfactant concentration reaches the CMC, additional surfactant above CMC will simply form micelles. Above the CMC, the adsorption of surfactant to interface does not take place any further because the interface is simply saturated, and thus the IFT stays constant thereafter. Simply, the IFT is determined by the surfactant concentration at below CMC but becomes constant above CMC, as shown in Fig. 3.6.

3.2.2 Surfactant Efficiency and Effectiveness

Performance of a particular type of surfactant as a surface-active agent is characterized by using its

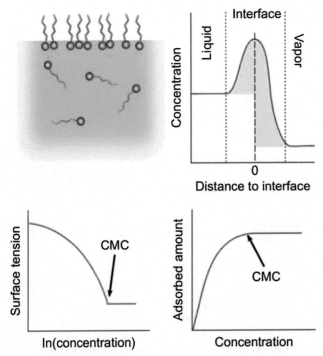

FIG. 3.6 Surfactant adsorption at the air-water interface. (Credit: Eastoe, J., Tabor, R.F., 2014. Surfactants and Nanoscience. Colloidal Foundations of Nanoscience. Elsevier, pp. 135–157.)

adsorption behavior. Most commonly, efficiency of a surfactant is evaluated by measuring the required concentration of surfactant to reduce IFT (or surface tension) to a certain degree (Rosen and Kunjappu, 2012; Eastoe and Tabor, 2014). On the other hand, surfactant effectiveness refers to the maximum reduction in IFT (or surface tension) that the surfactant can achieve, and this does not consider how much surfactant is added or the surfactant concentration. It should be noted that the efficiency and the effectiveness are two different terminologies; the most effective surfactant is not the most efficient, and the most efficient surfactant is not always effective (Eastoe and Tabor, 2014).

The efficiency of a surfactant is determined by its thermodynamic behavior: the preference of a surfactant to be adsorbed at the interface rather than in the bulk phase, in terms of an energy state. The more massive hydrophobic tail a surfactant has, the higher efficiency it tends to show. The surfactant efficiency is evaluated by quantifying the amount of surfactant adsorbed with respect to a given IFT reduction (under CMC condition) at the fluid-fluid interface. For an example, the water-air IFT or surface tension can be used as an indicator to the surfactant efficiency because measuring the surface tension is relatively easier. Therefore, the

surfactant efficiency is widely described in a term called pC_{20}, which is defined as the negative log of the bulk phase surfactant concentration required to reduce the surface tension by 20 mN/m (Rosen and Kunjappu, 2012).

Contrarily, the effectiveness of a surfactant is mainly related to the surfactant packing at the interface and thus determined by the chemical structure and the molecular shape/size of the surfactant (Eastoe and Tabor, 2014). The surfactant effectiveness can be also described in terms of surface tension reduction; a term called Π_{CMC} that is defined as the degree of surface tension reduction at the CMC is often used. The surfactant effectiveness Π_{CMC} equals to $\gamma_0 - \gamma_{CMC}$, where γ is the surface tension.

3.3 INTERFACIAL TENSION AND WETTABILITY MODIFICATION BY BIOSURFACTANT PRODUCERS

3.3.1 Surfactin Produced by *Bacillus* spp.

Due mainly to the ubiquitous occurrence of *Bacillus* spp. in soil and aquatic environments and their survivability under harsh environments such as extreme pH, high salinity (Simpson et al., 2011), high temperature, high pressure (Yakimov et al., 1995), and anoxic conditions (Willenbacher et al., 2015), *Bacillus* spp. have been widely selected and evaluated as a model bacterium for microbial enhanced oil recovery (MEOR) (Al-Bahry et al., 2013; Al-Wahaibi et al., 2014; Peet et al., 2015). Furthermore, they have an ability to sporulate, which guarantees their high resistance against extreme temperature, desiccation, and chemical disinfectants (Nicholson et al., 2000). For their survivability under extreme conditions, *Bacillus* spp. have been often found in oil reservoirs in the United States (Oklahoma, United States; Simpson et al., 2011), Brazil (Atlantic Ocean, offshore Rio de Janeiro, Brazil; Da Cunha et al., 2006), and Japan (Minami-aga Oil field, Japan; Kato et al., 2001). This supports the potential of *Bacillus* strain as the model organism to be used for MEOR. Among Bacillus spp., *Bacillus subtilis* is widely investigated for its highest efficiency as a lipopeptide surfactin producer (Cooper et al., 1981).

In one recent study, both IFT and contact angle changes of brine (mineral salt medium), dodecane and quartz systems, i.e., IFT of brine-dodecane and contact angle of brine-dodecane-quartz, are monitored during the cultivation of *B. subtilis* at 35, 40, and 45°C and 10 MPa (Park et al., 2019). *B. subtilis* are cultivated at such fairly harsh environments in a view cell in which IFT and contract angle are measured with the oil droplet on a substrate (Figs. 3.7 and 3.8).

For the IFT measurement, under a stable temperature-pressure condition, a pendant dodecane droplet is placed on a capillary tube, and the droplet images are acquired over the course of bacterial growth and biosurfactant production (Fig. 3.9). The acquired images are analyzed with the low bond axisymmetric drop shape analysis (LBADSA) method that fits a first-order approximation of the Young–Laplace equation (Berry et al., 2015). This LBADSA method is known to allow accurate estimation on IFT, as long as the bond number, the dimensionless ratio of a buoyance force to a surface force, is well controlled to stay in the range of 0.1–0.9 by managing the volume of droplets (Berry et al., 2015; Park et al., 2019).

The most common method to measure the contact angle is the sessile drop method where a sessile droplet image is analyzed once a droplet is positioned on the substrate (Panel B in Fig. 3.7). Time-lapsed images of the dodecane sessile droplet on the quartz substrate are acquired during bacterial growth and the following biosurfactant production. The contact angles of dodecane-brine-quartz can be determined using the drop-snake plugin for ImageJ software (Stalder et al., 2006; Park et al., 2019). Theoretically, Young's contact angle is irrelevant to the size of the droplet. However, the volume of the droplet should be carefully considered for reliable and consistent estimates because the gravitational force generated by the droplet may not be small enough to neglect. It is generally reported that the volume of a droplet less than 10 μL is considered suitable and appropriate for contact angle measurement (Jung and Wan, 2012; Law and

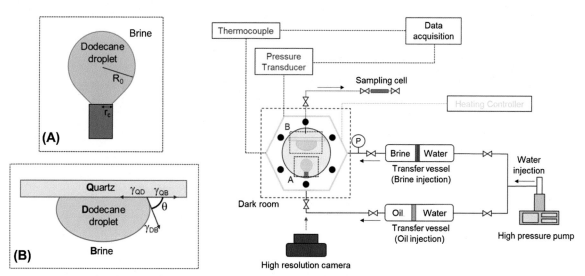

FIG. 3.7 A schematic of the experimental system for interfacial tension (panel A) and contact angle (panel B) measurement. (Credit: Park, T., Jeon, M.-K., Yoon, S., Lee, K.S., Kwon, T.-H., 2019. Modification of interfacial tension and wettability in oil-brine-quartz system by in situ bacterial biosurfactant production at reservoir conditions: implications to microbial enhanced oil recovery. Energy & Fuels 33, 4909–4920.)

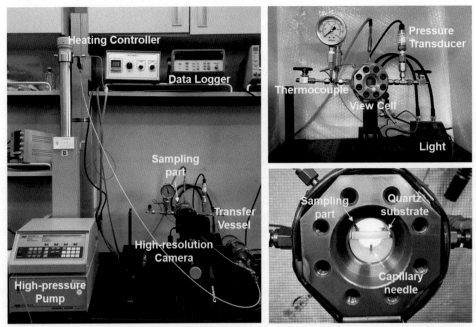

FIG. 3.8 A picture of the experimental systems for interfacial tension and contact angle measurements during microbial cultivation at high-pressure high-temperature conditions. (Credit: Park, T., Jeon, M.-K., Yoon, S., Lee, K.S., Kwon, T.-H., 2019. Modification of interfacial tension and wettability in oil-brine-quartz system by in situ bacterial biosurfactant production at reservoir conditions: implications to microbial enhanced oil recovery. Energy & Fuels 33, 4909–4920.)

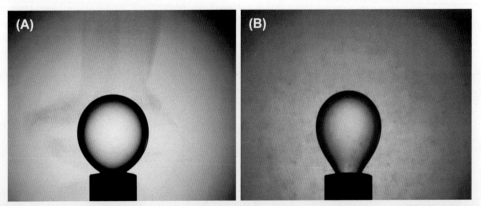

FIG. 3.9 Digital images of oil droplets for interfacial tension measurement **(A)** before and **(B)** after microbial growth and surfactin production. (Credit: Park, T., Jeon, M.-K., Yoon, S., Lee, K.S., Kwon, T.-H., 2019. Modification of interfacial tension and wettability in oil-brine-quartz system by in situ bacterial biosurfactant production at reservoir conditions: implications to microbial enhanced oil recovery. Energy & Fuels 33, 4909–4920.)

Zhao, 2016a). Fig. 3.10 shows the oil droplets in brine before and after surfactin production, which are used for the contact angle measurement.

During bacterial cultivation, the concentrations of carbon, nitrogen sources, and surfactin can be measured with time, as shown in Figs. 3.11 and 3.12. The gradual accumulation in surfactin concentration is one of the direct evidences of the growth of *B. subtilis* and the biosurfactant production. The optimal temperature for *B. subtilis* growth is known to be 37°C (Kim et al.,

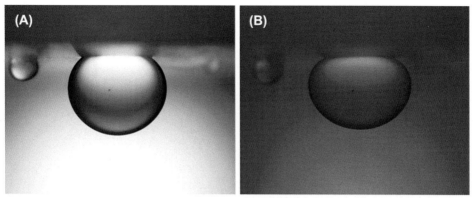

FIG. 3.10 Digital images of oil droplets on a quartz disc for contact angle measurement **(C)** before and **(D)** after microbial growth and surfactin production. (Credit: Park, T., Jeon, M.-K., Yoon, S., Lee, K.S., Kwon, T.-H., 2019. Modification of interfacial tension and wettability in oil-brine-quartz system by in situ bacterial biosurfactant production at reservoir conditions: implications to microbial enhanced oil recovery. Energy & Fuels 33, 4909–4920.)

1997; Davis et al., 1999; Pereira et al., 2013). Accordingly, the onset of surfactin production at 35°C and 40°C is observed when 8 h elapsed, and the onset at 45°C is detected after 12 h. This indicates the slower production of surfactin at the temperature higher than 40°C (Park et al., 2019).

The IFT reductions are clearly observed owing to the surfactin production by *B. subtilis* under the pressure of 10 MPa and the temperatures of 35, 40, and 45°C. *B. subtilis* produce ~110−130 mg/L of surfactin at 35 and 40°C and ~70−80 mg/L of surfactin at 45°C, which exceeds the CMC (Fig. 3.11). At the same time, surfactin reduces the IFT of brine and dodecane by 42−44 mN/m within less than 30 h, from ~50 to 54 mN/m at the beginning to ~8−10 mN/m at the CMC. Fig. 3.12 shows the variations of contact angle in dodecane-brine-quartz systems during the course of *B. subtilis* growth at 10 MPa and 35, 40, and 45°C, respectively. As shown in Figs. 3.11 and 3.12, the final concentrations of surfactin readily exceed the CMC, and accordingly, the alterations of IFT and contact angle are kept consistent after certain level of surfactin concentration in a liquid phase reached. At the CMC, these interfacial properties are altered to its possible maximum extent.

There are several studies that have isolated *B. subtilis* from oil reservoirs that have characterized biosurfactant produced from the isolates (Al-Bahry et al., 2013; Pereira et al., 2013; Veshareh et al., 2019). The studies by Al-Bahry et al. (2013), Pereira et al. (2013), and Veshareh et al. (2019) show the feasibility of using *B. subtilis* and its biosurfactant production for MEOR. *B. subtilis* are isolated from Iranian oil field (Veshareh et al., 2019), from a petroleum-contaminated garage

site (Al-Bahry et al., 2013) and from crude oil samples obtained from Brazilian oil field (Pereira et al., 2013). The biosurfactant produced from the isolated *B. subtilis* is identified as surfactin through examinations of the structural identities (Pereira et al., 2013); this has been confirmed via Fourier transform infrared spectroscopy (FTIR), proton nuclear magnetic resonance (^{1}H NMR), and matrix-assisted laser desorption ionization-time of flight mass spectrometry (MALDI-TOF). The modifications of IFT and wettability are examined by using the biosurfactant called surfactin produced by the isolated *B. subtilis*. The contact angle of brine-oil-calcite decreases from 152 to 119 degrees with 500 ppm of surfactin, and the surface tension is reduced from 73.5 to 30 mN/m by adding 60 mg/L of surfactin (Veshareh et al., 2019). Meanwhile, the production yield of ~229 mg/L of biosurfactant reduces the surface tension from 60 to 25 mN/m and oil-brine IFT from 27 to 5 mN/m (Al-Bahry et al., 2013). The surfactin reduces the IFT of Arabian light crude oil/water from ~18 to ~7 mN/m (Pereira et al., 2013). The details of previous studies on the MEOR potential of *B. subtilis* are summarized in Table 3.2.

The stability of biosurfactant can be tested with the extracted surfactants measuring the surface tension under different environmental conditions, a wide range of temperature, pH, and salinity. Surfactin is stable for the temperature up to 70°C, the pH range of 7−12, and the salinity less than 4%, effectively reducing the surface tension (Al-Bahry et al., 2013). However, the surfactin becomes ineffective at pH 2−4, presumably because of lowered solubility under such acidic environments. On the other hand, the microbial metabolism of *B. subtilis* including surfactin production is

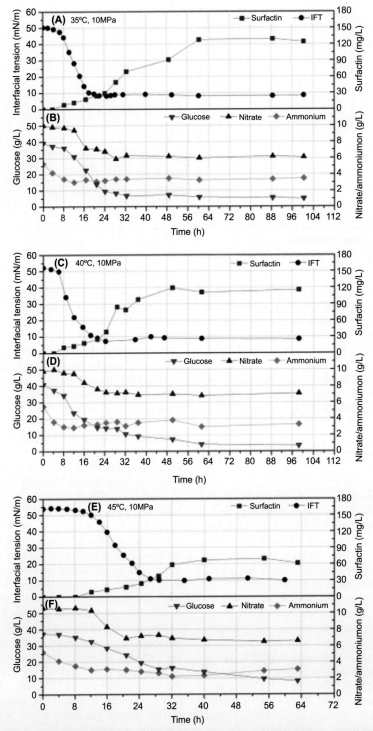

FIG. 3.11 Variations in dodecane/brine IFT and surfactin concentration (35 **(A)**, 40 **(C)**, and 45°C **(E)**) and glucose, nitrate, and ammonium (35 **(B)**, 40 **(D)**, and 45°C **(F)**) during the cultivation of B*acillus subtilis* at various temperatures and 10 MPa. (Credit: Park, T., Jeon, M.-K., Yoon, S., Lee, K.S., Kwon, T.-H., 2019. Modification of interfacial tension and wettability in oil-brine-quartz system by in situ bacterial biosurfactant production at reservoir conditions: implications to microbial enhanced oil recovery. Energy & Fuels 33, 4909—4920.)

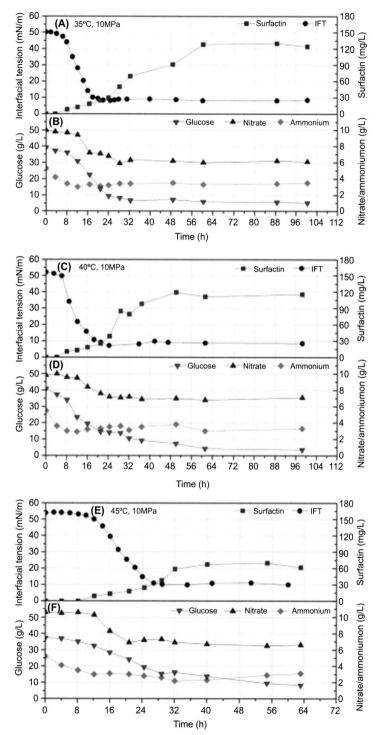

FIG. 3.12 Variations in contact angle of dodecane-brine-quartz and surfactin concentration (35 **(A)**, 40 **(C)**, and 45 °C **(E)**) and glucose, nitrate, and ammonium (35 **(B)**, 40 **(D)**, and 45 °C **(F)**) during the cultivation of B*acillus subtilis* at various temperatures and 10 MPa. (Credit: Park, T., Jeon, M.-K., Yoon, S., Lee, K.S., Kwon, T.-H., 2019. Modification of interfacial tension and wettability in oil-brine-quartz system by in situ bacterial biosurfactant production at reservoir conditions: implications to microbial enhanced oil recovery. Energy & Fuels 33, 4909–4920.)

TABLE 3.2
Investigations of *Bacillus subtilis* for Interfacial Property Modifications.

Reference	Experimental Details	Interfacial Property Modification	Finding
Al-Wahaibi et al. (2014)	ST/IFT (with crude oil) measurement Contact angle (of air on hydrophobic plate) measurement Used different types of carbon sources Oil recovery measurement (core-flooding test)	72 mN/m → 27 mN/m (ST) 35 mN/m → 15 mN/m (IFT) 59 degrees → 27.2 degrees	Either glucose/molasses suitable for *Bacillus subtilis* and surfactin production Surfactin production gave 17%–31% additional crude oil recovery
Pereira et al. (2013)	*B. subtilis* isolated from oil reservoir IFT measurement with crude oil (at room T, P)	18 mN/m → 7 mN/m (IFT with water/Arabian light oil)	*B. subtilis* isolated from Brazilian oil reservoir produced surfactin Better interfacial activity and lower CMC than chemical surfactants
Liu et al. (2015)	Surfactin from *B. subtilis* tested for i) ST reduction (at room T, P) ii) oil-washing efficiency and oil displacement efficiency (at room T, P) Surfactin stability test under extreme conditions (pH, temperature, and salinity)	72 mN/m → 28 mN/m	At 30 mg/L, 88.5% of oil washing efficiency and 13.5% of oil displacement efficiency Strong stability under extreme condition such as pH (3–13), salinity (50 g/L NaCl + 0 −170 g/L CaCl$_2$) and temperature (30–120°C), but lost stability over 50g NaCl + 130g CaCl$_2$ solution
Pornsunthorntawee et al. (2008)	*B. subtilis* isolated from sludge oil ST measurement (at room T, P) Oil recovery measurement (core-flooding test)	72 mN/m → 26 mN/m	*B. subtilis* PT2 isolated from sludge oil–produced surfactin which lowered surface tension Compared with other biosurfactant producing bacteria (*P. aeruginosa*), much lower CMC (25 and 120 mg/L, respectively)
Al-Bahry et al. (2013)	Used low cost materials for *B. subtilis* and measured ST and IFT (with crude oil) (at room T, P) Surfactin stability test under extreme conditions (pH, temperature, and salinity)	60 mN/m → 25 mN/m (ST) 27 mN/m → 5 mN/m (IFT)	Cheap material (molasses-based media) suitable for *B. subtilis* growth and surfactin production Strong surfactin stability under extreme temperature (40–100°C) and salinity under (0–4 w/v%) but lost its stability under extreme salinity (over 6 w/v%) and extreme pH (2, 4, 8, 10, 12)

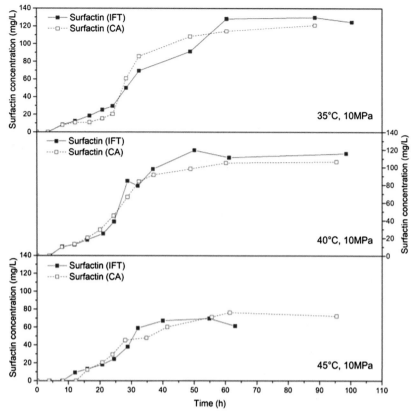

FIG. 3.13 Surfactin production rates under different temperature conditions. Both the surfactin concentrations determined in the IFT and contact angle experiments are plotted. *IFT*, interfacial tension. (Credit: Park, T., Jeon, M.-K., Yoon, S., Lee, K.S., Kwon, T.-H., 2019. Modification of interfacial tension and wettability in oil-brine-quartz system by in situ bacterial biosurfactant production at reservoir conditions: implications to microbial enhanced oil recovery. Energy & Fuels 33, 4909−4920.)

significantly affected by temperature (Fig. 3.13). Therefore, microbial surfactin production can be limited when the temperature is greater than 45°C (Park et al., 2019 (Ohno et al., 1995; Abdel-Mawgoud et al., 2008). The surfactin production is optimized at the temperature in the range of ∼30−35°C, pH in the range of ∼6.5−7, and an aeration percentage of 90% (Abdel-Mawgoud et al., 2008). It is concluded that *B. subtilis* are widely found in oil reservoirs, and they are able to produce biosurfactant even at extreme environments, such as high temperature, high pressure, and under saline environments. However, as the reservoir conditions can significantly affect biosurfactant production as well as its functionality, the environmental conditions of which a reservoir has should be carefully considered.

3.3.2 Rhamnolipid Produced by *Pseudomonas* spp.

Pseudomonas aeruginosa is also one of the best-known bacteria for biosurfactant production. Rhamnolipids, produced by *P. aeruginosa* is known for its high potential for MEOR practices (Wang et al., 2007; Amani et al., 2013; Zhao et al., 2015; Dong et al., 2016; Varjani and Upasani, 2016; Chong and Li, 2017). Owing to the characteristics that *Pseudomonas* spp. are facultative anaerobe and can live under microaerobic or anaerobic environment, *Pseudomonas* spp. are frequently found in oil-related areas (Rocha et al., 1992; Amani et al., 2010; Zhao et al., 2016; Khademolhosseini et al., 2019). This supports the potential of *P. aeruginosa* to adapt at oil reservoir conditions. Stability of rhamnolipid itself is comparatively solid under various pH, temperature,

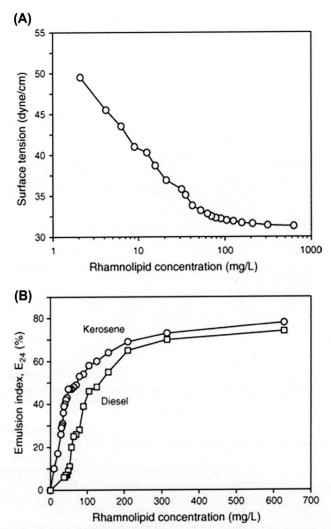

FIG. 3.14 **(A)** Surface tension and **(B)** emulsification index measurements during the bacterial growth of *Pseudomonas aeruginosa* J4 in MS medium containing 2% glucose at 30°C. (Credit: Wei, Y.-H., Chou, C.-L., Chang, J.-S., 2005. Rhamnolipid production by indigenous Pseudomonas aeruginosa J4 originating from petrochemical wastewater. Biochemical Engineering Journal 27, 146-154.)

and salinity conditions (Rocha et al., 1992; Varjani and Upasani, 2016; Khademolhosseini et al., 2019) conditions. As shown in Fig. 3.14, rhamnolipid exhibits a high surface activity that it has high emulsification activity and reduces the surface tension of water from 72 to 31 mN/m (Syldatk et al., 1985; Wei et al., 2005; Gudiña et al., 2015). This represents relatively high effectiveness among many biosurfactants. The CMC of rhamnolipid is reported as ~20–225 mg/L in water at 25°C (Syldatk et al., 1985; Dubeau et al., 2009).

Isolation of *Pseudomonas* spp. from oil reservoirs is not as common as *Bacillus* spp.; however, there have

been several studies that have found *Pseudomonas* spp. from oil reservoirs and have tested their potential as surface-active agents. Rocha et al. (1992) have isolated two strains of *P. aeruginosa* in injection water and crude oil–associated water from a Venezuelan oil field. Both strains of *P. aeruginosa* produce rhamnolipids-resembled biosurfactants which reduce the surface tension of water from 72 to 28 mN/m and the IFT of crude oil-water from 16 to 2.5 mN/m. Emulsions of oil-water mixtures in the presence of rhamnolipids are stable even under high temperature up to 100°C, pH from 1 to 12, and 5000 ppm of salinity (Rocha et al., 1992).

Pseudomonas stutzeri Rhl isolated from produced water, Daqing oil field, is also confirmed to produce rhamnolipid more than its CMC and reduce the surface tension from 72 to 30 mN/m (Zhao et al., 2016). Zhao et al. (2016) also proposed an interesting strategy that *P. stutzeri* Rhl can simultaneously remove S^{2-} and produce rhamnolipids under S^{2-} stress conditions, whereas H_2S production inhibits the growth of *P. stutzeri* and thus reduces the rhamnolipid production yields. Noting that H_2S produced by sulfate-reducing bacteria (SRB) causes corrosion and plugging, thus is considered as an obstacle in the petroleum industry.

P. aeruginosa AP02-1 is also isolated from a hot spring environment in hydrocarbon-containing muddy sediment (Perfumo et al., 2006). This strain shows an optimal growth rate at the temperature of 45°C, where *P. aeruginosa* degrades 99% of oil (crude oil at 1% v/v in a basal mineral medium; diesel oil at 2% v/v in a basal mineral medium). During incubation time of 7 days, *P. aeruginosa* AP02-1gradually lowers the surface tension from 70 to 40 mN/m. Various hydrocarbons that *P. aeruginosa* AP02-1 can use as carbon source are listed in Table 3.3. This indicates that *P. aeruginosa* AP02-1 grows well with most of hydrocarbons including alkanes, aromatic hydrocarbons, PHAs, heterocyclic

TABLE 3.3
Growth of *Pseudomonas aeruginosa* AP02-1 on Various Hydrocarbons (Perfumo et al., 2006).

Hydrocarbons	C Source Utilization
Pentane	+++
Hexane	+++
Cyclohexane	+++
Octane	+++
Dodecane	+++
Hexadecane	+++
Pristane	+++
Benzene	+
Toluene	+
o-Xylene	++
Naphthalene	+
Anthracene	+
Thianaphtene	++
Dibenzothiophene	++
Metyl-tert-butyl-ether	+++

The utilization of hydrocarbons as carbon source was indicated by the color development of a redox indicator dye, and plus signs are based on the absorbance values measured at 590 nm.

sulforate hydrocarbons, and metyl-*tert*-butyl-ether (Perfumo et al., 2006). Owing to its survivability under high temperature, high surface activity, and being able to use diverse hydrocarbon sources for the growth, *P. aeruginosa* AP02-1 shows a great potential to be used for MEOR applications.

3.3.3 Other Biosurfactant-Producing Species

In addition to representative biosurfactant-producing bacteria, *Bacillus* spp. and *Pseudomonas* spp., several of other strains have been also examined for their potential use for MEOR. *Rhodococcus* spp. are one of those candidates, and they can produce biosurfactant in oil reservoirs. *Rhodococcus* spp. show several capabilities advantageous for MEOR, including hydrocarbon degradation, biosurfactant production, and being facultative anaerobe. Zheng et al. (2012) have isolated *Rhodococcus ruber* Z25 from the formation brine in Daqing oil field. *Rhodococcus ruber* is incubated under an anaerobic condition, and it produces biosurfactant while hydrocarbon is used as the sole carbon source (Zheng et al., 2012). Given that the temperature, depth, and salinity of formation brine from Daqing oil field are 35−45°C, 800−1000 m, and 6300−7000 g/L, respectively, *R. ruber* has its survivability under harsh reservoir environments. As shown in Fig. 3.15, the biosurfactant produced by *R. ruber* shows a good emulsification activity, and it lowers the surface tension from 68.6 to 29.5 mN/m and the IFT of water and *n*-hexadecane from 43.6 to 1 mN/m with the CMC value of 133 mg/L (Zheng et al., 2012). In addition, the waterflooding experiment results demonstrate that the biosurfactant produced by *R. ruber* alters wettability, enhances oil mobility, and thus improves the oil recovery rate from 8.88% to 25.78%.

Rhodococcus sp. TA6 isolated from Iranian oil−contaminated soil is found to produce biosurfactants to a concentration greater than CMC and decrease the surface tension from 68 to 30 mN/m (Shavandi et al., 2011). In this study, sucrose and several hydrocarbon substrates are used as sole carbon source. Furthermore, the produced biosurfactant shows stable emulsification formation with various hydrocarbons including pentane to light motor oil. The stability of the biosurfactant is also confirmed at high salinity (10% NaCl), elevated temperatures (120°C), and at a pH range of 4.0−10.0 (Shavandi et al., 2011). The column test result reveals that the cultural broth containing biosurfactant is effective in recovering 65% of the residual oil trapped in sand packs (Shavandi et al., 2011).

There are other additional species that have the potential to use their metabolites for enhanced oil recovery. Sarafzadeh et al. (2014) have investigated

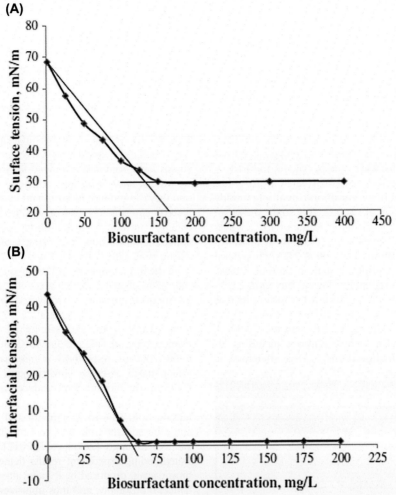

FIG. 3.15 Surface tension and interfacial tension reduction by the biosurfactant produced by *Rhodococcus ruber* Z25. (Credit: Zheng, C., Yu, L., Huang, L., Xiu, J., Huang, Z., 2012. Investigation of a hydrocarbon-degrading strain, *Rhodococcus ruber* Z25, for the potential of microbial enhanced oil recovery. Journal of Petroleum Science and Engineering 81, 49-56.)

Enterobacter cloacae which reduces the IFT of water/crude oil from 30 to 2.7 mN/m within 24 h of cultivation. Owing to the IFT reduction and wettability alterations, approximately 33% more of residual oil in a sand packed column is recovered (Shavandi et al., 2011). Although *E. cloacae* is not isolated from oil reservoir, it does have potential for ex situ application of MEOR because it has produced high surface-active surfactants which can lead to the enhancement in oil recovery.

Fusarium sp. BS-8, an isolate from crude oil—contaminated soil from MissaKeswal oil field, Pakistan, produces a crude biosurfactant which reduces the surface tension from 72 to 32 mN/m, with 71% hydrocarbon emulsifying index (Qazi et al., 2013). Under optimal growth conditions, *Fusarium* sp. produces 5.25 g/L of biosurfactant. With this crude biosurfactant, 46% residual oil is recovered through the core-flooding tests with a sand packed column. The biosurfactant maintains its stability under a wide range of temperature (0—90°C), pH (5—9), and salinity (1%—15%), indicating its suitability for ex situ biosurfactant implications of MEOR (Qazi et al., 2013).

Das, 2018 have tested *Candida tropicalis* MTCC230 to determine its suitability in MEOR. Various analytical methods including near IR, FTIR, high-performance liquid chromatography, and mass spectroscopy characterize the biosurfactant produced by *C. tropicalis* as lipopeptide surfactin. This surfactin reduces the surface tension from 72 to 32 mN/m with the 32.5 mg/L of CMC. The produced surfactin is stable under extreme conditions, which includes the pH of 2–12, the salinity of 2%–10%, and the temperature of 30–90°C. In a sand-packed column, 39.8% of additional oil recovery is achieved with the surfactin produced by *C. tropicalis* which suggests its possible applicability for the use in MEOR (Das, 2018).

3.3.4 Summary
Oil reservoirs exhibit a wide range of environmental conditions which can affect metabolic activities of microorganisms and production of microbial byproducts such as biosurfactants, such as temperature, pressure, pH, salinity, and oxygen levels. These are significant factors that can determine success or failure of biosurfactant-aided MEOR. Table 3.4 shows the strategies to improve the biosurfactant production economics (Geetha et al., 2018). Meanwhile, both *B. subtilis* and *P. aerugionosa*, the most promising microorganisms for MEOR, are expected to be suitable for "cold" reservoirs with the temperature less than 50°C. There are also other candidate species such as *Rhodococcus* spp, *E. cloacae*, *Fusarium* spp, and *C. tropicalis*.

3.4 ADDITIONAL MICROBIAL ENHANCED OIL RECOVERY MECHANISMS
3.4.1 Viscosity Reduction by Biogenic Gas and Solvent Producers
Use of bacterial consortia that produce gases and solvents can be considered to reduce oil viscosity and enhance oil recovery. Bacteria produces gases such as CH_4, CO_2, and H_2 during carbohydrate fermentation processes. These biogenic gases not only contribute to pressure buildup in a pressure-depleted reservoir but also dissolve into crude oils so that the oil viscosity is reduced (Sen, 2008). Viscosity reduction of crude oil leads to the greater oil mobility through porous media so that oil recovery rate is enhanced. Similarly, solvents such as acetone, ethanol, 1-butanol, and butanone are produced during microbially induced fermentation, which can also reduce the oil viscosity.

Clostridium acetobutylicum, an anaerobic bacterium that shows optimal growth at 38°C, is found to produce organic solvents (acetone, butanol, and ethanol) through fermentation of molasses and reduce the oil viscosity of oil (Behlulgil et al., 1992). In a one-dimensional reservoir model containing Turkish heavy oil (Raman oil) in limestone grains, inoculation of *C. acetobutylicum* and a nutrient medium containing 10% of molasses reduces the oil viscosity by 12%

−34%, which in turn leads to additional recovery of 12% of residual oil (Behlulgil et al., 1992). *B. subtilis*, a representative MEOR candidate as a biosurfactant producer, also produces organic acids. In a comprehensive experimental study conducted by Fernandes et al. (2016), *B. subtilis* RI4914 is found to produce 2,3-butanediol during fermentation, as well as other metabolites, including solvents, biosurfactants, and acids such as lactate and succinates, and these enhance the oil recovery (Fernandes et al., 2016).

Castorena-Cortés et al. (2012) have also tested fermentative bacteria for oil recovery enhancements. Anaerobic, thermophilic, halotorent, and fermentative enrichment cultures are obtained from nine oil samples collected from a carbonate oil reservoir in Cordoba Platform, Veracruz, Mexico (Castorena-Cortés et al., 2012). 16S rRNA analysis reveals that the culture contains 99.9% of *Thermoanaerobacter ethanolicus*, which is well known for its thermophilic, anaerobic, ethanol-producing characteristics. Biogenic gases such as CO_2, solvents such as ethanol, acetone, acids (acetic acid), and biosurfactants are all detected after 96 h of cultivation. Those metabolites including biogenic gases and solvents reduce oil viscosity and improve the oil recovery rate by 12% in a granular carbonate porous medium (Castorena-Cortés et al., 2012).

3.4.2 Permeability Increase by Solvent- and Acid-Producing Microorganisms
Bacterially induced fermentation includes the production of acids and solvents, and these by-products can effectively help the oil recovery by improving the permeability. Especially in carbonate rocks, the solvents, such as acetone, butanol, and propan-2-diol, dissolve carbonate rock minerals and hence increase the porosity and permeability of the reservoir rock. Kalish et al. (1964) tested diverse types of bacteria: *Pseudomonas, Bacillus, and Micrococcus* and *Proteus* spp., to observe the permeability changes by growth of bacteria. They observed that the bacterial growth clogged and reduced the permeability, but the presence of acids effectively restored the permeability of sandstone cores, which had been previously plugged by bacteria (Kalish et al., 1964).

3.4.3 Pressure Support by Biogenic Gas Producers
In theory, biogenic gases generated through fermentation can help elevating the depleted fluid pressure in an oil reservoir. However, this pressure support by biogenic gas is considered as not effective in EOR at a field scale (Sen, 2008). A mathematical simulation study suggests that the amount of gas produced by *C. acetobutylicum* is not sufficient to enhance oil recovery (Marsh et al.,

TABLE 3.4
Strategies Used to Improve the Biosurfactant Production Economics (Geetha et al., 2018).

Strategy	Biosurfactant	Microorganisms
USE OF CHEAPER RAW MATERIALS/SUBSTRATES		
Fatty acids from soybean oil refinery wastes	Rhamnolipids	*Pseudomonas aeruginosa* AT10
Glycerin from biodiesel production waste	Rhamnolipids	*P. aeruginosa* MSIC02
Sunflower oil refinery waste	Rhamnolipids	*P. aeruginosa* LBI
Soybean oil waste	Lipopeptide	*Bacillus pseudomycoides* BS6
Canola waste frying oil and corn steep liquor	Rhamnolipids	*Pseudomonas cepacia* CCT6659
Glycerol	Rhamnolipids	*P. aeruginosa* UCP0992
Clarified cashew apple juice	Lipopeptide	*Bacillus subtilis* LAMI005
Vinasse and waste frying oil	Lipopeptide	*Bacillus pumilus*
Cassava wastewater	Lipopeptide	*B. subtilis* LB5a
Soybean oil refinery residue and corn steep liquor	Sophorolipids	*Candida sphaerica* UCP0995
Groundnut oil refinery residue and corn steep liquor	Sophorolipids	*C. sphaerica* UCP0995
Animal fat and corn steep liquor	Sophorolipids	*Candida lipolytica* UCP0988
Vegetable fat	Sophorolipids	*C. glabrata* UCP1002
Waste frying oil	Rhamnolipids	*Candida tropicalis* UCP0996
OPTIMIZATION OF MEDIA COMPONENTS, ENVIRONMENTAL PARAMETERS, AND UPSTREAM AND DOWNSTREAM PROCESSES		
Statistical model–based optimization of media components	Lipopeptide	*Bacillus licheniformis* R2
Statistical model–based optimization of media components	Lipopeptide	*B. licheniformis* K51
Statistical model–based optimization of media components	Lipopeptide	*B. subtilis* SPB1
Statistical model–based optimization of media components	Lipopeptide	*Bacillus circulans*
Statistical model–based optimization of inoculum age and size	Surfactin	*B. subtilis*
Optimization study of batch recovery including concentration and purification using ultrafiltration	Surfactin	*B. subtilis*
Statistical model–based optimization of environmental conditions	Surfactin	*B. subtilis*
Statistical model–based optimization of medium composition and adequate control of aeration	Lipopeptide	*B. subtilis*
Optimization of Bench-Scale Production using %DO control	Lipopeptide	*B. licheniformis* R2
Hyperproducing strains: random or site-directed mutagenesis/genetic recombination	Lipopeptide	*B. subtilis*
HYPERPRODUCING STRAINS: RANDOM OR SITE-DIRECTED MUTAGENESIS/GENETIC RECOMBINATION		
Random mutagenesis with chemical mutagen	Lipopeptide	*B. licheniformis* KGL11
Replacement of native promoter of lichenysin biosynthesis operon and medium optimization	Lichenysin	*B. licheniformis*
Transposon Tn5-GM-induced mutation	Rhamnolipids	*P. aeruginosa*

Random mutagenesis with chemical mutagen	Rhamnolipids	*P. aeruginosa* PTCC 1637
Expression of cloned rhlAB genes in nonpathogenic strains	Rhamnolipids	*Pseudomonas putida* KT2442 and *Pseudomonas Fluorescens*
Insertion of *Escherichia coli* lacZY genes into the chromosomes of *P. aeruginosa* strains	Rhamnolipids	Recombinant *P. aeruginosa* strains
Selection of the mutant from media containing cationic detergent CTAB as selection pressure	Emulsan	*Acinetobacter calcoaceticus* RAG-1
Stable maintenance and expression of *Vitreoscilla* hemoglobin gene (vgb) for better aeration efficiency	Trehalose	*Gordonia amarae*
Random mutagenesis with ultraviolet radiation	Lipopeptides and glycolipids	*Rhodococcus erythropolis* SB-1A

1995). The oil recovery process is associated with various mechanisms, but the pressure buildup by microbially produced gases still needs further investigation.

3.5 PERMEABILITY AND POROSITY IN RESERVOIR ROCKS

The reservoir rocks are primarily characterized by the quantities of fluids trapped within the void space of these rocks and the ability of these fluids to flow through the rocks (Tiab and Donaldson, 2015). The amount of void space is defined as the *porosity* of the rock, and the ability of the rock to transmit fluids is referred to as the *permeability*. This section briefly introduces the porosity and permeability of reservoir rocks, which is further extended to the modification of these properties for MEOR.

3.5.1 Porosity

Porosity is a volumetric property of porous media, which indicates the volumetric ratio of the void space (pores, fractures, cracks) occupied in the unit volume of the porous medium. Therefore, the porosity of a reservoir is a measure of the ability of a reservoir to store fluid (oil, gas, and water). Porosity φ is determined, as follows:

$$\phi = \frac{V_{pore}}{V_{bulk}} = \frac{V_{bulk} - V_{solid}}{V_{bulk}} \qquad (3.3)$$

where V_{pore} is the volume of all pores, V_{bulk} is the bulk volume of the reservoir rock, and V_{solid} is the volume of the solid grains. The porosity of porous media could be any value from 0% to 100%. In general, the porosity of reservoir rocks ranges from 5% to 30%, and typical range of porosity is between 10% and 20% (Tiab and Donaldson, 2015).

The pore types are classified by both morphological and geological features (Selley, 1998; Hook, 2003). The morphological pore types are determined according to a throat passage connecting with another: catenary, cul-desac, and closed, as shown in Fig. 3.16. Total (absolute) porosity is the ratio of the total void space in the sample to the bulk volume of that sample, regardless of whether or not those void spaces are interconnected. By contrast, effective porosity is determined by the pores connected with others—catenary and cul-de-sac (or dead-end) pores, from which hydrocarbons can be produced. Hydrocarbons that fill the catenary pores can be flushed out by water drive. On the contrary, hydrocarbons residing in cul-de-sac (dead-end) pores are hardly affected by water drive but could be produced with some aids by chemicals, such as surfactants and/or solvents. Closed pores are unable to yield hydrocarbons as ineffective pores. Difference between total (absolute) porosity and effective porosity is the isolated or noneffective porosity, and the effective porosity directly influences the fluid transport and hence the hydrocarbon yields.

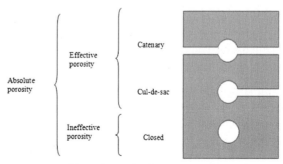

FIG. 3.16 The three basic types of pores and morphological classification of porosity. (Credit: Selley, R.C., 1998. Elements of Petroleum Geology. Gulf Professional Publishing.)

TABLE 3.5
Classification of Porosity in Sedimentary Rocks (Selley, 1998).

Time of Formation	Type	Origin
Primary or depositional	Intergranular, or interparticle Imtragranular, or intraparticle Intercrystalline Fenestral	Sedimentation Cementation
Secondary or postdepositional	Vuggy Moldic Fracture	Solution Tectonics, compaction Dehydration, diagenesis

The porosity types can be also classified according to their origins of formation (Selley, 1998; Tiab and Donaldson, 2015). Primary pores indicate the pores that are formed during the time of sedimentation, as shown in Table 3.5. Interparticle pores are initially present in all sediments, but their sizes are quickly reduced by compaction. These interparticle porosities are physically affected by gradation of grains, grain shape, degree of cementation, amount of compaction, and packing method. By contrast, intraparticle pores are generally found within the skeletal grains, mostly in carbonate sands, diatoms.

The secondary porosity is the result of geological processes, diagenesis, and catagenesis, after the deposition of sediment. The pores caused by solution are often called as the secondary pores. The secondary porosity includes the solution porosity, dolomitization, and fracture porosity (Tiab and Donaldson, 2015). The solution-induced porosity, which includes moldic and vuggy porosity, is more common in carbonate reservoirs than in sandstone reservoirs. Fracture porosity barely contributes to the total porosity increase; however, it is considered to be extremely important because it significantly enhances permeability of reservoirs. Fractures are at the much larger scale than the other types of porosity; the rock cores alone cannot capture the fractures. Thereby, the integrated approach should be taken to assess the fractures in a reservoir by combining borehole logging data, seismic data, and the production history of a well.

The fluid saturation is expressed as the fraction or percent of the total pore volume occupied by a fluid,

which can be water, oil, or gas. Thus, a fluid saturation S_f is defined as follows:

$$S_f = \frac{V_{fluid}}{V_{pore}}, \qquad (3.4)$$

where V_{pore} is the pore volume of the reservoir rock, and V_{fluid} is the volume of fluid in the reservoir rock. The saturation ranges from 0% to 100%. When a reservoir rock is fully saturated with water, the water saturation is 1% or 100%. If a rock is dry, then the water saturation would be zero. Normally, if the reservoir contains multiple fluid components, for an example, oil, gas, and water, the following relation is achieved for each fluid saturation:

$$S_o + S_g + S_w = 1, \qquad (3.5)$$

where the subscripts o, g, and w indicate oil, gas, and water, respectively.

3.5.2 Permeability
Permeability (or intrinsic permeability) is the ability of rock to conduct fluids. Assuming a laminar flow regime, Darcy's law is used as the standard formulation, as follows:

$$Q = \frac{K_a(\Delta P)A_{cs}}{\mu_f L_s}, \qquad (3.6)$$

where Q is the flow rate, A_{cs} is the cross-sectional area of the porous medium sample, L is the length of the sample, μ_f is the viscosity of the fluid, L_s is the length of the sample, ΔP is the pressure drop across the sample, and K_a is the permeability. The dimension of permeability K_a is [Length^2], and thus m^2 and µm^2 are typically used as the SI units. In petroleum engineering, Darcy (D) is often adopted as a unit of the permeability value, which is equal to 0.986,923 µm^2, and millidarcy (mD) is widely used. Permeability of petroleum reservoir rocks typically ranges from 0.1 to 1000 mD. The quality of a reservoir can be determined by its permeability value: poor if $K_a < 1$ mD, fair if 1 mD $< K_a < 10$ mD, moderate if 10 mD $< K_a < 50$ mD, good if 50 mD $< K_a < 250$ mD, and very good if $K_a > 250$ mD. The reservoirs with the permeability less than 1 mD are often referred to as "tight" reservoirs.

3.5.3 Porosity and Permeability Relation
The permeability of a rock primarily depends on its effective porosity, and it is consequently affected by the rock grain size, grain shape, grain size distribution, grain packing, and the degree of consolidation and

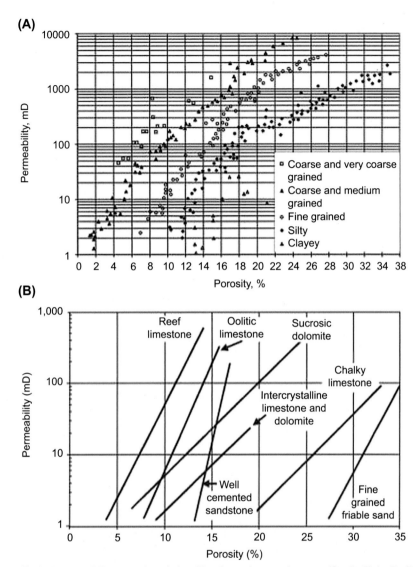

FIG. 3.17 Typical permeability-porosity relationships for various rock types. (Credit: Tiab, D., Donaldson, E.C., 2015. Petrophysics: Theory and Practice of Measuring Reservoir Rock and Fluid Transport Properties. Gulf Professional Publishing.)

cementation. Fig. 3.17 shows relations between permeability and porosity for various rock types. Such a porosity-permeability relationship plays an important role in modeling selective plugging caused by bacterial products. Herein, some of these relationships are introduced, which is further extended for modeling permeability reduction by biomasses in the later section of this chapter.

The porosity-permeability relations show a wide variation not only with rock types but also depending on the significance of secondary porosity and presence of fractures. Fig. 3.18 depicts the porosity-permeability relationships for different types of pore systems (Selley, 1998).

A family of Kozeny-Carman model is widely used to correlate permeability and porosity, and one type of formulation is expressed as follows (Kozeny, 1927; Carman, 1939; Tiab and Donaldson, 2015):

$$K_a = \frac{1}{2 \bullet \tau \bullet S_{sv}^2} \bullet \frac{\phi^3}{(1 - \phi)^2}, \qquad (3.7)$$

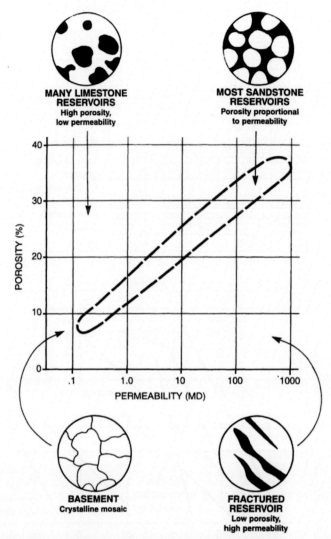

FIG. 3.18 Illustration of the porosity-permeability relationship for the different types of pore systems. (Credit: Selley, R.C., 1998. Elements of Petroleum Geology. Gulf Professional Publishing.)

where K_a is the permeability, ϕ is the porosity, τ is the tortuosity, and S_{sv} is the specific surface area per unit volume. The tortuosity is defined as follows:

$$\tau = \left(\frac{L_a}{L_c}\right)^2, \tag{3.8}$$

where L_a is the actual flow path length and L_c is the core length. One of the generalized Kozeny-Carman equations can be formulated, as follows (Costa, 2006):

$$K_a = \frac{1}{C_g \bullet \tau \bullet S_{sv}^2} \bullet \frac{\phi^{m_g}}{(1-\phi)^{m_g-1}}, \tag{3.9}$$

The coefficients C_g and m_g are introduced to consider some variations among rocks.

3.6 PRINCIPLES OF SELECTIVE PLUGGING IN HETEROGENEOUS RESERVOIRS

3.6.1 Heterogeneity in Reservoirs

The geologic formations in reservoirs show inherent heterogeneity with spatial variabilities in many petrophysical properties. These include from the basic physical indices, such as mineralogy, porosity, and permeability, to complex chemical characteristics, such as salinity, pH, hydrocarbon components, and

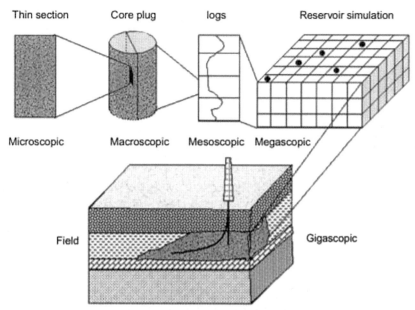

FIG. 3.19 Scales of reservoir heterogeneity. (Credit: Tiab, D., Donaldson, E.C., 2015. Petrophysics: Theory and Practice of Measuring Reservoir Rock and Fluid Transport Properties. Gulf Professional Publishing.)

microbial diversity. The petrophysical properties vary with the length scale from pore level to field level, as shown in Fig. 3.19. Table 3.6 summarizes the type of petrophysical information from each length scale.

3.6.2 Basic Mechanism of Selective Plugging

The selective plugging is one of the promising techniques of tertiary oil production to enhance oil production. The fundamental concept of the selective plugging is the permeability reduction and thus plugging of high-permeability channels in the targeted reservoir formation, which is also referred to as "thief zones." Thereby, the waterflooding can be redirected to the unswept, oil-rich channels and increases the oil recovery rate. During waterflooding, such selective plugging can be achieved chemically by using polymers or biologically by exploiting microbial reactions.

The biomasses, such as microorganisms, biofilms, and biopolymers, grow on the surfaces of rock minerals or/and accumulate in pore spaces of reservoir rocks, which decreases pore sizes and eventually leads to a reduction in permeability. Therefore, biological plugging, bioclogging, by microbial formation of biopolymers and biofilms, can be utilized as a means to selective plugging for MEOR. Fig. 3.20 shows pore-scale selective plugging and its effect on the residual or trapped oil production.

Biopolymers that have potentials to be used for MEOR include xanthan gum, pullulan, levan, curdlan, dextran, and scleroglucan produced by *Xanthomonas, Aureobasidium, Bacillus, Alcaligenes, Leuconostoc, Sclerotium, Brevibacterium*, and *Enterobacter*, as mentioned in Chapter 2.

The microbial selective plugging can be implemented by two different ways: (**A**) biostimulation, in which in situ indigenous bacteria that already exist within the reservoir are stimulated to produce biomasses, and (**B**) bioaugmentation, in which living bacterial cells are inoculated with the nutrients into targeted highly permeable zones for bacterial growth and the following biomass production. Once the growth of biopolymer-producing bacteria is confirmed, nutrients can be further fed to facilitate biomass production for selective plugging.

Several screening criteria are suggested for application of MEOR processes in oil fields (Petroleum et al., 1990; Bryant and Lindsey, 1996; Sheng, 2013). The important parameters for successful selective plugging are the reservoir temperature and permeability. The temperature should be appropriate for bacteria to thrive and for their enzyme activity, which governs biopolymer production. The targeted reservoir must have highly permeable channels which cause poor sweep efficiency for validity of this mechanism (Gray et al., 2008). One successful field-scale microbial

TABLE 3.6
Heterogeneity in Petrophysical Property at Different Length Scales (Tiab and Donaldson, 2015).

Scale	Property or Characteristics	Method
Microscopic Heterogeneity (Pore-scale)	(1) Grain size and shape (2) Pore size and shape (3) Distributions of grain size, pore size, and pore throat size (4) Fabric arrangement (5) Pore wall roughness	SEM, Petrographic image analysis (PIA), MRI, and NMR
Macroscopic Heterogeneity (Core-scale)	(1) Porosity (2) Permeability (3) Fluid saturation (4) Capillary pressure (5) wettability	Laboratory measurement
Mesoscopic Heterogeneity (Borehole-scale)	(1) Correlation and compatibility between the measured parameters (2) Downhole measurements with data from pore studies, core analysis, and geophysical surveys through interstate reconciliation (using upscaling functions) (3) Lithofacies (4) Petrophysical interpretation with geochemical, sediment logical, stratigraphic, and structural information (5) Different reservoir parameters such as porosity, permeability, net thickness, tops and bottoms, fluid saturation, and fluid contact	Well log
Megascopic Heterogeneity (Reservoir-scale)	(1) Lateral discontinuity of individual strata (2) Porosity pinch-out (3) Reservoir fluid contacts (4) Vertical and lateral permeability trends (5) Shale and sand intercalation (6) Reservoir compartmentalization	Reservoir simulation
Gigascopic Heterogeneity (Basin-scale)	(1) Division of reservoir into more than on producing zones or reservoirs (2) Position, size, shape, architecture, and connectivity of facies or reservoir units (3) Evaluation of the spatial distribution or lithological heterogeneity that comprises barriers, baffles, widespread sealing bed unconformities, and high-permeability zones (4) Large-scale structural features of folds and faults (5) The relationship of lithofacies to depositional environment and hydraulic flow units	Subsurface mapping

MRI, magnetic resonance imaging; *NMR*, nuclear magnetic resonance; *SEM*, scanning electron microscopy.

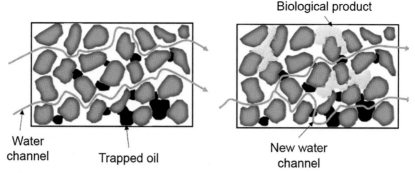

FIG. 3.20 Schematic diagram of selective clogging mechanism. (Credit: Courtesy of Dong-Hwa Noh, Taehyung Park and Tae-Hyuk Kwon.)

selective plugging has been reported, which was implemented in Fuyu oil field, China (Nagase et al., 2002). The temperature of the reservoir was 28°C, and the reservoir permeability was as high as 240 mD (Nagase et al., 2002). Indigenous microorganism, CJF-002, was isolated from the reservoir, and they were augmented with molasses to stimulate the production of insoluble biopolymers and induce selective plugging. Oil production rate was increased, almost doubled, and the water cut decreased upon this selective plugging treatment.

3.7 SELECTIVE PLUGGING MECHANISMS BY BACTERIAL EXTRACELLULAR POLYMERIC SUBSTANCES AND BIOPOLYMERS

This section presents the experimental observations on bioclogging and permeability reduction due to the bacterial biopolymer production by *Leuconostoc mesenteroides* in porous media. Previous studies conducted at the pore scale and at the core scale are described, and their results are summarized.

3.7.1 Pore-Scale Observations on Bioclogging

L. mesenteroides (NRRL B-523; ATCC 14935) is widely chosen as the model bacterium (e.g., Ham et al., 2018), because of its several advantages. (1) It is nonmotile and has a size of ∼400−500 nm, so it can be readily transported through pores by advective fluid flows, as shown in Fig. 3.1. (2) This bacterium is a facultative anaerobe that can grow under anoxic conditions and a high fluid pressure and low temperature, which broadens the applicability. (3) It is edible and harmless to humans (biosafety level 1) and is found in sugarcane and fermented foods. (4) Most importantly, *L. mesenteroides* produces dextran, which is a type of insoluble polysaccharide biopolymer, when

metabolizing sucrose (Lappan and Fogler, 1996). (5) It does not produce biopolymers when not feeding on nutrients based on glucose and fructose. And, lastly (6) the amount of dextran produced can be controlled with the sucrose concentration. Therefore, this unique metabolism feature enables a good control on biopolymer formation in various engineering practices.

In the experimental study by Ham et al. (2018), the growth medium contains yeast extract (YE, in short) to provide the necessary nutrients for bacterial growth. A potassium phosphate buffer of 0.1 M (both monobasic (KH_2PO_4) and dibasic (K_2HPO_4)) is used to control the pH of the medium at 7.0 and provide a consistent environment for bacterial growth (Ham et al., 2018). Table 3.7 shows the chemical composition of the growth medium, which was referenced from Lappan and Fogler (1996).

Figs. 3.21 and 3.22 show the morphology of the produced dextran and cultured cells imaged by using scanning electron microscopy (SEM) and optical microscopy. The insoluble dextran produced by

TABLE 3.7
Chemical Composition of the Growth Medium Used in Various Studies.[a]

Component	Concentration
Sucrose	15, 40, 80, 150, 300 g/L[b]
Yeast extract	10 g/L
1M Monobasic (KH_2PO_4)	41 mL/L
1M Dibasic (K_2HPO_4)	59 mL/L

[a] The growth media were used in studies by Noh et al. (2016), Ham et al. (2018), Ta et al. (2017), and Kim et al. (2019).
[b] Noh et al. (2016) and Ta et al. (2017) used 15 g/L sucrose, and Kim et al. (2019) used 40 g/L sucrose at a constant concentration, respectively. Ham et al. (2018) used various sucrose concentration from 40 to 300 g/L.

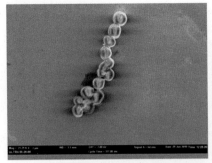
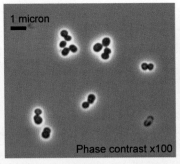
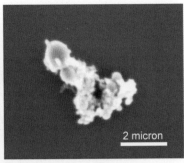

FIG. 3.21 The morphology of the cultured *Leuconostoc mesenteroides* cells and the produced dextran, imaged by using scanning electron microscopy and optical microscopy. (Credit: Courtesy of Tae-Hyuk Kwon.)

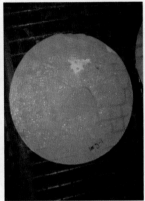

FIG. 3.22 Images of the produced dextran in a liquid culture and filtered dextran. (Credit: Courtesy of Dong-Hwa Noh, Yong-Min Kim and Tae-Hyuk Kwon.)

L. mesenteroides appears initially as small filaments between 10 and 100 μm in length; these filaments can then accumulate as porous deposits or layers in pore spaces. When *L. mesenteroides* are cultured in the sand-pack saturated with the growth medium above, they produce insoluble biopolymer dextran, as shown in Fig. 3.23. With an aid of SEM, a porous nature of insoluble biopolymer dextran is confirmed. Interestingly, an increase in the sucrose concentration stimulates the bacteria to produce more dextran (Fig. 3.22).

Kim and Fogler (2000) and Stewart and Fogler (2002) have conducted the pore-scale micromodel experiments on accumulation of biomass, particularly insoluble biopolymers and its effect on permeability. In their studies, *L. mesenteroides* (NRRL B-523; ATCC 14935) was used as the model bacteria. *L. mesenteroides* are known to produce insoluble biopolymer *dextran* when sucrose-based media are fed (Kim and Fogler, 2000; Stewart and Scott Fogler, 2002). More detailed descriptions on this biopolymer and bacteria can be found in Chapter 2.

L. mesenteroides are considered as the model bacteria for subsurface operations, such as MEOR, due to its rapid growth and ability to form exopolymers (Wang, 1991).

Growth medium composition. In the study by Kim and Fogler (2000), the inoculant cells are grown in a growth medium. The growth medium is composed of a phosphate-buffered saline solution with yeast extract (11.1 g/L), NaCl (0.07 M), NH_2Cl (0.06 M), CH_3COONa (0.06 M), ascorbic acid (0.5 g/L), trace minerals, and the carbon source (a combination of glucose and fructose, 0.1 M). The combination of glucose and fructose can maintain cell growth without inducing the production of exopolymers. For biopolymer production, the sucrose nutrient solution is prepared by replacing glucose and fructose with sucrose in the growth medium (Kim and Fogler, 2000). The trace minerals are prepared by individually dissolving $MgSO_4 \cdot 7H_2O$ (12.3 g/mL), $FeSO_4 \cdot 7H_2O$ (1.39 g/mL), and $MnSO_4$ (0.85 g/mL) in deionized (DI) water to make stock solutions. Each stock solution (2 mL) is

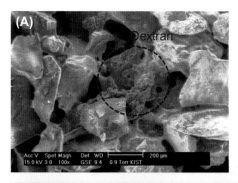
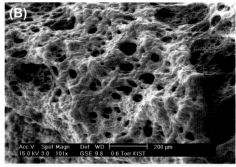

FIG. 3.23 **(A)** An environmental scanning electron microscopy (ESEM) image of dextran produced in silica sand, **(B)** an ESEM image of dextran itself, and **(C)** digital image of dextran produced by *Leuconostoc mesenteroides* aerobically cultured in different sucrose concentrations. (Credit: Ham, S.-M., Chang, I., Noh, D.-H., Kwon, T.-H., Muhunthan, B., 2018. Improvement of surface erosion resistance of sand by microbial biopolymer formation. Journal of Geotechnical and Geoenvironmental Engineering 144, 06018004.)

added to 1 L of phosphate-buffered saline solution. The final concentration of trace minerals in both the growth and the starvation medium is 12.7 mM for $MnSO_4$, 10 mM for $FeSO_4 \cdot 7H_2O$, and 100 mM for $MgSO_4 \cdot 7H_2O$. For starvation, the same phosphate-buffered saline solution with the same concentrations of trace minerals is used for the growth medium but without carbon sources. The experiments are performed at 25°C in a laminar hood (Kim and Fogler, 2000).

The experiment by Stewart and Fogler (2002) uses a growth medium with 7.9 g/L glucose, 7.9 g/L fructose, salts, trace minerals, and 10 g/L yeast extract for cell growth without dextran production. And, the sucrose medium with 15g/L sucrose, salts, trace minerals, and 10 g/L yeast extract, in which only glucose and fructose are basically replaced with sucrose, is used for dextran production (Stewart and Scott Fogler, 2002).

Main observations. The micromodel is a useful tool for pore-scale observation of emergent phenomena occurring in porous media, such as biomass accumulation and evolution. Micromodels, which is formed by etching on glass plates, are used to observe the biopolymer formation by *L. mesenteroides* in porous media. During the production of biomass including biofilms and biopolymer dextran in the micromodels, permeability decreases by up to more than two orders of magnitudes, which is equivalent to the relative permeability reduction K_a/K_o of ~0.01 (Stewart and Scott Fogler, 2002). Herein, the relative permeability reduction, K_a/K_o, is defined by the permeability K_a normalized by the baseline (initial) permeability K_o. Meanwhile, in the micromodels, biopolymer accumulates mostly in pore bodies, and it hardly coats the solid walls, as shown in Fig. 3.24.

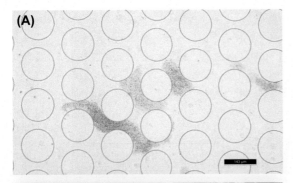

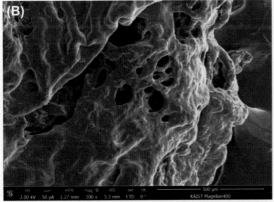

FIG. 3.24 **(A)** Accumulation of bacterial biopolymer in a micromodel, and **(B)** a scanning electron microscopy image of exopolymer (biopolymer) gel and bacterial cells growing on a solid surface. (Credit: Courtesy of Yong-Min Kim and Tae-Hyuk Kwon.)

Meanwhile, the biomasses are readily detachable upon the shear stress applied. When the applied shear stress exceeds the critical shear stress that biomass can withstand, biomass is detached, which is often referred to as "sloughing." By contrast, when the applied shear stress is less than the critical shear stress, the biomass stays in the porous media. As the produced biopolymers become denser and form a gel, their resistance against shear stress becomes greater, indicating the greater critical shear stress.

As a result, during the nutrient injection in the micromodel, fluid pressure oscillates by the growth and subsequent shearing of biomass from selected pore throats to form a breakthrough channel that is subsequently refilled. In detail, as the nutrient flows through the micromodel, the initial biomass plug occurs at the nutrient-inoculum interface. Soon, the first breakthrough channel in the first plug takes place when the shear stress exceeds the Bingham yield stress (or critical shear stress) of the biomass, resulting in its

being sheared from pore throats. After the first breakthrough, the channel begins refilling with biomass almost immediately to form the second plug. With biomass accumulation and fluid pressure gradient buildup, the shear stress by increased fluid velocity exceeds the critical shear stress, and a second breakthrough occurs, in turn causing the fluid pressure drop. Thereafter, such cycles of plugging and breakthrough follow. However, with subsequent plug development and breakthrough channel cycles, the pore area in the micromodel is filled with biomass and become less fluid-percolated.

The resiliency of bioplugs, whether or not the bioplug is stable, is heavily affected by the applied shear stress associated with flow velocity, the critical shear stress, and lastly the geometric characteristics of pores, including pore throat size.

3.7.2 Column Experiments at a Core Scale on Permeability Reduction

There are several studies which performed column experiments to explore the permeability reduction by biopolymer produced by *L. mesenteroides*. The first experimental study on bioclogging by *L. mesenteroides* was conducted by Lappan and Fogler (1992). Consolidated porous ceramic cores with two different pore size distributions are selected as an ideal medium to exclude such artifacts as fines migration, have a uniform surface chemistry composition, and have a specific and uniform pore-size distribution. One high-permeability core has a permeability ~ 15 Darcies with the mean pore size of 33.2 μm. The other low-permeability core has a permeability of 0.01 Darcies with the mean pore size of 3.27 μm. Therefore, the permeability differs by three orders of magnitude, and the pore size by one order of magnitude. The core samples have a diameter of 2.54 cm and a length of 5 cm (Lappan and Fogler, 1992).

The growth medium for *L. mesenteroides* contains yeast extract, tryptone, sodium acetate, sodium chloride, ascorbic acid, and some trace elements, as shown in Table 3.8. As the carbon source, a combination of glucose and fructose is used for bacterial growth without dextran production; by contrast, sucrose is used for bacterial growth and dextran production. Yeast extract and tryptone are fed to provide the sources of amino acids, purines, pyrimidines, and vitamins, for bacterial growth.

The core experiments consisted of two phases: the initial inoculation of bacteria into the porous medium, and followed by the nutrient feed into the medium to promote cell growth and biopolymer production.

TABLE 3.8
Chemical Composition of the Growth Medium
Used in Lappan and Fogler (1992).

Component	Concentration
Water	900 mL
Yeast extract	5 g
Tryptone	20 g
NaCl	4 g
Sodium acetate	1.5 g
Ascorbic acid	0.5 g
Trace elements	Ca, Mn, Fe, Mg
Carbohydrate solution	
Distilled water	100 ML
Sucrose	15 g
Fructose and glucose	7.9 g each

The inoculum cells grown in a glucose/fructose medium are first injected into the core samples. Then, a sucrose base medium is injected to monitor the pressure drop and permeability reduction. On the other hand, a control test is also conducted by continuously feeding a glucose/fructose solution instead of a sucrose solution.

Their results show that for both high- and low-permeability core samples, the permeability values are dropped by more than two orders of magnitude, close to three orders of magnitude, when sucrose solution is fed. This is equivalent to the permeability reduction by ~99%–99.9%. The sucrose-fed cores that are associated with both cell growth and biopolymer (dextran) production show more permeability reduction than those of cores fed glucose/fructose which have the cell growth only. This indicate that dextran production causes significant core plugging. The effect of such plugging (or clogging) is more pronounced in the high-permeability cores, where the permeability reduction by three orders of magnitude is observed. Note that the cell inoculation and growth into the high-permeability cores barely affect the permeability. By contrast, the injection of bacterial inoculum and cell growth in low-permeability cores causes a considerable reduction in permeability, by ~1.5 orders of magnitude in their study, even without biopolymer production. Nonetheless, in the low-permeability cores, the permeability reduction with dextran production is more severe by approximately 10 times than that without dextran production.

Meanwhile, this result has implications to MEOR implementation in low permeability formations. The bacterial growth itself can reduce the permeability near the injection well and cause damage when the reservoir permeability is fairly low. In such a condition, the injection (or inoculation) of cells should be carefully designed at low concentrations sufficient to transport to the target area while hindering plugging near wells. This is also valid when implementing MEOR strategy exploiting bacteria which produce biosurfactant, biogas, or solvents.

Recently, some experimental studies on permeability reductions in sand-pack columns have been reported (e.g., Kwon and Ajo-Franlkin, 2013; Noh et al., 2016; Ta et al., 2017; Kim et al., 2019). While the previous study by Lappan and Fogler (1992) uses the synthetic ceramic cores as a porous medium, these studies examine bioclogging in unconsolidated sand packs using clean silica sands. The growth media these studies used are summarized in Table 3.9.

The study by Kwon and Ajo-Franklin (2013) use a defined growth medium, which contains trace minerals, vitamins, 0.1 M phosphate buffers, 0.6 g/L ammonium citrate, 1 g/L sodium acetate, 0.5 g/L yeast extract, and 15 g/L sucrose (Table 3.9). This is used to stimulate the growth of the model bacteria *L. mesenteroides* and their biopolymer production. The pH of the growth medium is conditioned to be 6.8 using phosphate buffers. The control of pH using a buffer is essential for optimum dextran production as *L. mesenteroides* produces carbon dioxide during metabolism, thereby lowering the pH of the pore water (Kwon and Ajo-Franklin, 2013).

Noh et al. (2016), Ta et al. (2017), Kim et al. (2019), as well as Ham et al. (2018), basically design the growth media similar to each other, in which the growth media are composed of sucrose with various concentrations as a carbon source, 10 g/L yeast extract as a nitrogen source and vitamins, and 0.1 M phosphate buffer (or potassium phosphate monobasic, KH_2PO_4, 13.6 g/L; potassium phosphate dibasic, K_2HPO_4, 17.4 g/L), which is also shown in Table 3.7. The pH value of the fresh growth media is kept at ~7.0. Specifically, in the study by Noh et al. (2016) and Ta et al. (2018), the growth media contain 15 g/L sucrose, similar to Lappan and Fogler (1992) and Kwon and Ajo-Franklin (2013), whereas Kim et al. (2019) uses 40 g/L sucrose.

In these studies, a rigid-walled column, which is instrumented with one differential pressure transducer, is mostly used, as shown in Fig. 3.25. And, a transfer vessel is also connected to a syringe pump and used to inject growth media into the column as the syringe

TABLE 3.9
Composition of Defined growth Medium for
Leuconostoc mesenteroides (Kwon and Ajo-
Franklin, 2013).

Compound	Concentration	Molarity
CARBON/NITROGEN SOURCES		
Sucrose	15 g/L	43.82 mM
Na-acetate	1 g/L	12.2 mM
$(NH_4)_3$-citrate	0.6 g/L	2.47 mM
Yeast extract	0.5 g/L	-
TRACE MINERALS AND BUFFER		
$MnSO_4 \cdot H_2O$	1 g/L	0.0592 mM
$MgCl_2 \cdot 6H_2O$	20 g/L	0.984 mM
Monobasic—KH_2PO_4	13.609 g/L	100 mM
Dibasic—K_2HPO_4	17.418 g/L	100 mM
VITAMINS		
d-Biotin	0.02 mg/L	0.082 µM
Folic Acid	0.02 mg/L	0.045 µM
Pyridoxine hydrochloride	0.1 mg/L	0.49 µM
Thiamine hydrochloride	0.05 mg/L	0.15 µM
Riboflavin	0.05 mg/L	0.13 µM
Nicotinic Acid	0.05 mg/L	0.41 µM
Pantothenic Acid	0.05 mg/L	0.21 µM
p-Aminobenzoic Acid	0.05 mg/L	0.31 µM
Thioctic Acid	0.05 mg/L	0.24 µM
Choline Chloride	2.0 mg/L	14 µM
Vitamin B12	0.01 mg/L	0.0074 µM

pump provided water at a constant flow rate. The pore fluid pressure in the column is controlled by a back-pressure regulator to keep the sediment sample water saturated and avoid bubble formation. Typically, the fluid pressure (or back pressure) of 300—500 kPa is applied and kept constant throughout the experiments to dissolve microbially generated gas in aqueous solution and avoid any bubble formation in the column. These experiments are conducted in a pulsed-flow mode, and hence a shut-in period and a short-duration nutrient refilling injection period are repeated per cycle. In each cycle, a shut-in period of 2—3 days is allowed for the bacteria to grow and produce biopolymer without no mass flux under ambient room temperature ($\sim$20—25°C). After the shut-in

period, a fresh growth medium is refilled through a transfer vessel which is connected to a syringe pump at a constant flow rate (i.e., a refilling period). The refilling volume is typically $\sim$1.5—3 times of pore volume, but in particular, it is 5 times of pore volume in the study by Kim et al. (2019). During this refilling at a constant flow rate, differential pressure is recorded and used to compute the permeability.

Main results on permeability reduction and biopolymer accumulation. In the study by Kwon and Ajo-Franklin (2013), the initial (baseline) permeability is measured as 5.25×10^{-12} m^2 (or 5.3 D), and the permeability decreases to 2.25×10^{-12} m^2 (or 0.23 D) after 17 days of the first run, due to insoluble dextran and biomass formation. Herein, the final dextran pore saturation is estimated to be approximately 5.1%. In the second run, the permeability is reduced from 4.91×10^{-12} m^2 (or 5.0 D) to 1.09×10^{-12} m^2 (or 1.1 D), whereas the final dextran pore saturation is approximately 3.9% after 23 days, as shown in Fig. 3.26. Consequently, accumulated dextran with the pore saturation of $\sim$4%—5% reduces the permeability by more than one order of magnitude.

In the study by Noh et al. (2016), the permeability is reduced from 4.9×10^{-12} m^2 to 8.7×10^{-13} m^2 after 20 days of the first run, and the final dextran pore saturation is estimated to be $\sim$4.6%. In the repeated experiment (the second run), the permeability decreases from 4.1×10^{-12} to 1.8×10^{-13} m^2 after 38 days (Fig. 3.11). After 38 days of the second run, the dextran is accumulated to occupy $\sim$10% of pore spaces, which is the final dextran pore saturation of $\sim$10% (Fig. 3.26). Again, the permeability reduction appears to be by more than one order of magnitude, attributable to the accumulation of insoluble dextran produced by *L. mesenteroides* in the pores of the sand pack.

Herein, the pore saturation of biopolymer dextran (S_{BP}; defined as biopolymer volume divided by the pore volume) is estimated based on the mass of the carbon accumulating in the column which can be calculated on the basis of the net influx of carbon to the column (Kwon and Ajo-Franklin, 2013; Noh et al., 2016). The mass of organic carbon accumulating in the column is estimated by using the total inorganic and organic carbon analysis method, often referred to as total inorganic carbon/total organic carbon analysis. The assumption is as follows: (1) the mass of organic carbon accumulating in the column is completely converted into the mass of dextran using the mass ratio of carbon to dextran, and the average density of dextran is 1.5 g/cm^3. Note that the generic formula for dextran

(a)

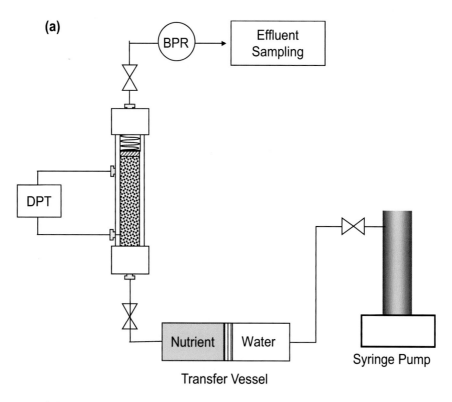

(b)

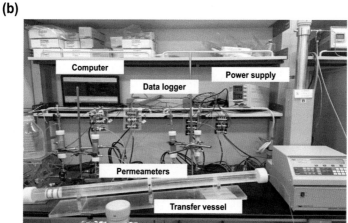

FIG. 3.25 A setup for column experiments: **(A)** schematic drawing, and **(B)** a digital picture of the column setup. This sketch is redrawn by referring the following references: Kwon and Ajo-Franlkin (2013), Noh et al. (2016), Ta et al. (2017), and Kim et al. (2019). (Credit: Courtesy of Yong-Min Kim and Tae-Hyuk Kwon.)

is $(C_6H_{10}O_5)_n$ OH_2). The resultant S_{BP} can be considered as an upper bound of S_{BP}.

In the study by Ta et al. (2017), the accumulation of insoluble biopolymer dextran produced by *L. mesenteroides* decreases the permeability from 9.3×10^{-12} m^2 to 3.2×10^{-13} m^2 after 41 days of the experiment. Again, the permeability reduction is as much as $\sim$96%–97%, which is more than one order of magnitude.

Kim et al. (2019) have performed the longest column experiment which had last 66 days, and thereby their result shows the most significant permeability

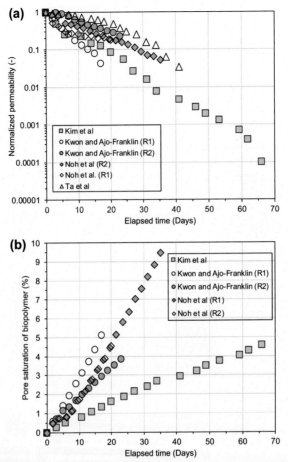

FIG. 3.26 **(A)** Permeability reduction versus time, and **(B)** biopolymer pore saturation versus time. The data are gathered from Kwon and Ajo-Franlkin (2013), Noh et al. (2016), Ta et al. (2017), and Kim et al. (2019). (Credit: Courtesy of Tae-Hyuk Kwon.)

reduction (Fig. 3.26). The permeability is remarkably reduced from $\sim 2 \times 10^{-10}$ to $\sim 2 \times 10^{-14}\, \mathrm{m^2}$ after 66 days. The permeability reduction by $\sim 99.99\%$, in other words, by approximately four orders of magnitudes is achieved, whereas the pore saturation of the produced dextran is estimated to be $\sim 5\%$. Note that the dextran estimation method in Kim et al. (2019) is different from the other previous studies (Kwon and Ajo-Franklin, 2013; Noh et al., 2016). The method in the former study is based on filtering of insoluble dextran grown in batch cultures, which gives a lower bound, whereas the method in the later studies estimates an upper bound because it assumes 100% conversion of accumulated carbon into dextran. Therefore, the dextran pore saturation value obtained in Kim et al. (2019) considers only the volume of solid (insoluble biopolymer), but the biopolymers also present in an aqueous phase (Jeon et al., 2017). It is

important to note that the apparent bulk volume of dextran can be much greater than its net solid volume. The data are summarized in Table 3.10.

The production of dextran is clearly observed when *L. mesenteroides* is fed with sucrose-rich media, as shown in Fig. 3.27. Thin layers of sluggish biopolymer aggregates on the top of the sand pack are observed a few days elapsed upon the commencement of experiments. In addition, when the sand packs are dismantled after the completion of experiments, the optical microscopy and SEM images of the sand samples show a notable amount of dextran accumulated in the pore space as well as coating the sand grains, as shown in Fig. 3.28.

In reservoir modeling, a family of permeability-porosity relationships is widely adopted to describe permeability modification, in which the porosity change due to chemical or biological reactions is correlated to permeability of reservoir rocks. Such porosity

TABLE 3.10
Summary of Bioclogging Experiments Using *Leuconostoc mesenteroides.*

Reference	Experiment Condition	PHYSICAL PROPERTIES OF MEDIUM		PERMEABILITY ($\times 10^{-12}$ m²)		Final Dextran Saturation
				Initial	Final	
Lappan and Fogler (1992)	Continuous flow mode at Q = 0.1 mL/min 15 g/L sucrose	High K Ceramic core	Porosity = 0.555 Mean pore size = 33.2 μm	14.7	0.015	—
		Low K Ceramic core	Porosity = 0.536 Mean pore size = 3.27 μm	0.098	~0.00008	—
Kwon and Ajo-Franklin (2013)	Pulsed mode at Q = 2 mL/min 15 g/L sucrose Refilling every 2 d	Fine quartz sand Ottawa F110; uniform grain size; mean particle diameter = 120 μm		5.25 4.91	0.225 1.09	~5.1% ~3.9%
Noh et al. (2016)	Pulsed mode at Q = 2 mL/min 15 g/L sucrose Refilling every 2 d	Fine quartz sand Ottawa F110; mean particle size = 110 μm		4.89 4.06	0.873 0.215	~4.6% ~9.5%
Ta et al. (2018)	Pulsed mode at Q = 2 mL/min 15 g/L sucrose Refilling every 2 d	Fine quartz sand Ottawa F110; uniform grain size; mean particle diameter = 120 μm		9.3	0.32	—
Kim et al. (2019)	Pulsed mode at Q = 1 −4 mL/min 40 g/L sucrose Refilling every 2−3 d	Coarse quartz sand Ottawa 20/30; uniform grain size; mean particle diameter = 840 μm		227	0.023	~4.6%

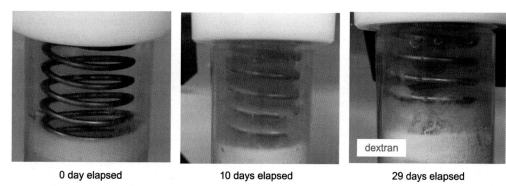

| 0 day elapsed | 10 days elapsed | 29 days elapsed |

FIG. 3.27 Deposited dextran of *Leuconostoc mesenteroides* on the top of the sand pack. (Credit: Courtesy of Yongmin-Kim and Tae-Hyuk Kwon.)

change due to insoluble biopolymer production is related to the amount of biopolymer produced in pore spaces, such that the pore saturation of biopolymer (or pore fraction of biopolymer) can be used. Thereby, correlation of permeability reduction ratio to biopolymer saturation plays a critical role in reservoir modeling.

Fig. 3.29 shows the relations between the permeability reduction and the pore saturation of biopolymer (dextran). In general, the biopolymer saturation less than 10% can readily cause the permeability reduction by more than one order of magnitude and up to four orders of magnitude at maximum. Such significant permeability reduction associated with this low level

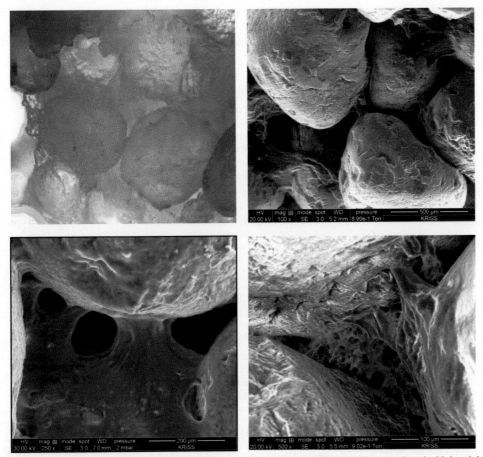

FIG. 3.28 Optical microscopy and scanning electron microscopy images of sands associated with insoluble biopolymer dextran by *Leuconostoc mesenteroides*. (Credit: Courtesy of Yongmin-Kim and Tae-Hyuk Kwon.)

of biopolymer saturation is possibly caused by combined features of the pore-throat clogging behavior and the large apparent volume of the twined stringlike biopolymer having complex internal structures and large specific surface area. These experimental results can be extended to establish permeability-biopolymer saturation relations.

3.7.3 Analytical Pore-Scale Models for Correlation Between Biopolymer Saturation and Permeability

Pore-scale clogging mechanisms caused by bacterial biopolymer accumulation affect the permeability of porous media. In this section, let us explore analytical pore-scale permeability models associated with biopolymer accumulation. Then, the analytical models are compared with previously complied experimental

data to correlate the permeability reduction to the biopolymer saturation in porous media.

3.7.3.1 Variations in capillary tube models and Kozeny grain models with pore-scale habits of biopolymer accumulation

Simple capillary tube and Kozeny grain models for permeability reduction caused by biopolymer saturation S_{BP} are described. These models are originally derived for hydrate-bearing sediments (Kleinberg et al., 2003), and they are further applied to biopolymer-containing sediments (Noh et al., 2016).

A. Parallel capillary tube model. The flow rate through a unit cross-sectional area containing a bundle of parallel cylindrical capillary tubes having the inner radius a_i and length L_s is as follows (Scheidegger, 1958):

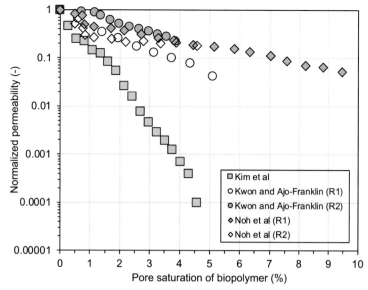

FIG. 3.29 Permeability reduction versus biopolymer pore saturation. The data are gathered from Kwon and Ajo-Franklin (2013), Noh et al. (2016), Ta et al. (2017), and Kim et al. (2019). (Credit: Courtesy of Tae-Hyuk Kwon.)

$$Q = \frac{N_t \pi a_i^4}{8\mu_f} \frac{\Delta P_a}{L_s}, \tag{3.10}$$

where N_t is the number of capillary tubes per unit cross-sectional area, μ_f is the dynamic viscosity, and ΔP_a is the differential pressure. Fig. 3.30 depicts the parallel capillary tube model.

Thus, Darcy's law defines the permeability K as

$$Q = \frac{K_a}{\mu_f} \frac{\Delta P_a}{L_s}, \tag{3.11}$$

and the porosity is related to the number of capillaries per unit cross-sectional area N_a,

$$\phi = N_a \pi a_i^2. \tag{3.12}$$

Thus, the baseline permeability K_o before biopolymer production is

$$K_0 = \frac{\phi a_i^2}{8}. \tag{3.13}$$

A-1. Grain-coating capillary model. When an impermeable layer of insoluble biopolymer layer uniformly coats the capillary walls, as shown in Fig. 3.30B, the radius of the water-filled pore space can be assumed to decrease to a_r. Then, Eq. (3.10) becomes

$$Q = \frac{N_a \pi a_r^4}{8\mu_f} \frac{\Delta P_a}{L_s}. \tag{3.14}$$

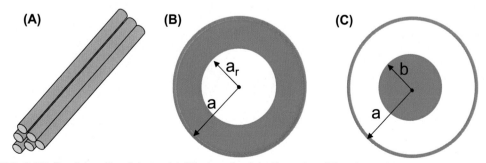

FIG. 3.30 Parallel capillary tube model: **(A)** a bun,dle of capillary tubes, **(B)** grain-coating capillary model, and **(C)** pore-filling capillary model. Gray color denotes solid grains and blue color indicates biopolymer. (Credit: Courtesy of Tae-Hyuk Kwon.)

Because the number of capillaries per unit cross-section remains $N = \varphi/\pi a^2$, the permeability to water is reduced to

$$K(S_{BP}) = \frac{\phi a_r^4}{8a_{ini}^2}. \tag{3.15}$$

S_{BP} can be related to the initial radius a and the reduced radius a_r as

$$S_{BP} = 1 - \left(\frac{a_r}{a_{ini}}\right)^2, \tag{3.16}$$

thus, the reduced permeability is

$$K(S_{BP}) = \frac{\phi a_{ini}^2 (1 - S_{BP})^2}{8}. \tag{3.17}$$

The normalized permeability or the (relative) permeability reduction ratio $K_r = K_a/K_0$ of the grain-coating capillary model is defined as

$$K_r = \frac{K_a(S_{BP})}{K_0} = (1 - S_{BP})^2. \tag{3.18}$$

A-2. Pore-filling capillary model. When impermeable biopolymer aggregates occupy the centers of cylindrical pores, leaving an annular flow path for water, as shown in Fig. 3.30C, the fluid flux of a single tube is (Lamb, 1945).

$$Q = \frac{\pi}{8\mu_f}\frac{\Delta P_a}{L_s}\left[a_p^4 - b_b^4 - \frac{\left(a_p^2 - b_b^2\right)^2}{\log\left(\frac{a_p}{b_b}\right)}\right], \tag{3.19}$$

where a_p is the radius of the pore, and b_b is the radius of the biopolymer core. The S_{BP} is

$$S_{BP} = \left(\frac{b_b}{a_p}\right)^2. \tag{3.20}$$

Accordingly, the permeability of the pore-filling capillary model is

$$K(S_{BP}) = \frac{\phi a_p^2}{8}\left[1 - S_{BP}^2 - \frac{(1 - S_{BP})^2}{\log\left(\frac{1}{S_{BP}^{0.5}}\right)}\right]. \tag{3.21}$$

Then, the normalized permeability (or relative permeability reduction ratio K_r) is

$$K_r = \frac{K_a(S_{BP})}{K_0} = 1 - S_{BP}^2 + \frac{2(1 - S_{BP})^2}{\log(S_{BP})} \tag{3.22}$$

B. Kozeny grain models. Kozeny-type permeability equations are widely used because of their simplicity, in which the ratio of the pore surface area to the pore volume is considered as

$$K_a = \frac{\theta_{sf}}{\theta_{sf}\tau(A_{cs}/V_{pore})^2}, \tag{3.23}$$

where θ is a shape factor, and τ is the tortuosity. Fig. 3.31 depicts the Kozeny grain model.

The tortuosity is

$$\tau = \left(\frac{L_a}{L_{sl}}\right)^2, \tag{3.24}$$

where L_a is the path length for flow, which is longer than the straight-line distance L_{sl} associated with the pressure drop. The tortuosity can be correlated with the electrical formation factor F_e and the porosity as follows (Hearst et al., 2000):

$$\tau = F_e\phi. \tag{3.25}$$

Combining Eqs. (3.23) and (3.25), the normalized permeability K_N is

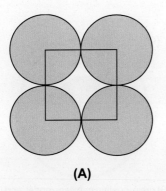
(A)

(B)

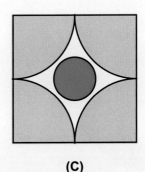
(C)

FIG. 3.31 Kozeny grain model: **(A)** a grain-pack, **(B)** grain-coating Kozeny model (GCKM), and **(C)** pore-filling Kozeny model (PFKM). Gray color denotes solid grains, and blue color indicates biopolymer. (Credit: Courtesy of Tae-Hyuk Kwon.)

$$K_N = \frac{K_a(S_{BP})}{K_0} = \frac{F_e}{F(S_{BP})}\left(\frac{A_0}{A_{cs}(S_{BP})}\right)^2\left(\frac{V_{pore}(S_{BP})}{V_0}\right)^2. \quad (3.26)$$

Thus far, no model is available to show how the electrical formation factor and the surface-to-volume ratio change in the presence of biopolymers. Herein, despite the differences between the hydrate and the biopolymer, the relationship developed for a hydrate-containing grain pack is adopted (Spangenberg, 2001), because both insoluble biopolymers are semisolid similar to gas hydrates. The relationship between the formation factor in a biopolymer-saturated medium $F(S_{BP})$ and in a fully water-saturated rock F_0 is

$$\frac{F(S_{BP})}{F_0} = (1 - S_{BP})^{-n_A}, \quad (3.27)$$

where n is the Archie saturation exponent. Hence, the pore water volume ratio $V(S_{BP})/V_0 = 1 - S_{BP}$ is

$$K_N = \frac{K_a(S_{BP})}{K_0} = (1 - S_{BP})^{n_A+2}\left(\frac{A_0}{A_{cs}(S_{BP})}\right)^2. \quad (3.28)$$

B-1. Grain-coating Kozeny model.
The surface area of the water-filled pore volume decreases as the grain surfaces are coated by the biopolymer, increasing the S_{BP}. If we assume the cylindrical pore model, the pore radius with no biopolymer is a_{wo} and the pore radius with S_{BP} is a_w. Then, the surface area ratio is

$$\frac{A_0}{A_{cs}(S_{BP})} = \frac{a_{wo}}{a_w}, \quad (3.29)$$

and recalling Eq. (3.16) $S_{BP} = 1 - (a_w/a_{wo})^2$,

$$\frac{A_0}{A_{cs}(S_{BP})} = (1 - S_{BP})^{-0.5}. \quad (3.30)$$

Hence, Eq. (3.28) becomes

$$K_N = \frac{K_a(S_{BP})}{K_0} = (1 - S_{BP})^{n_A+1} \quad (3.31)$$

B-2. Pore-filling Kozeny model.
When the biopolymer occupies the pore center, the surface area of the water-filled pore volume increases. If we assume cylindrical pores, combining Eq. (3.20), the surface area ratio is derived as

$$\frac{A_{cs}(S_{BP})}{A_0} = 1 + S_{BP}^{0.5}; \quad (3.32)$$

thus,

$$K_N = \frac{K_a(S_{BP})}{K_0} = \frac{(1 - S_{BP})^{n+2}}{(1 + S_{BP}^{0.5})^2}. \quad (3.33)$$

C. Generalized model for K_r-S_{BP} correlation.
The analytical models for K_r-S_{BP} correlation derived earlier include the upper-bound and lower-bound parallel capillary tube models: the capillary grain-coating model (CGCM; $K_a/K_0 = (1 - S_{cc})^2$) and the capillary pore-filling model ($K_a/K_0 = 1 - (S_{BP}^2 + 2(1 - S_{BP})^2/\ln(S_{BP}))$); and two types of Kozeny grain models: the grain-coating Kozeny model (GCKM; $K_a/K_0 = (1 - S_{BP})^{n+1}$) and the pore-filling Kozeny model ($K_a/K_0 = (1 - S_{BP})^{n+2}/(1 + S_{BP}^{0.5})^2$). Herein, the Archie saturation exponent n is adopted as an empirical parameter, which indicates the extent of pore clogging and the level of permeability reduction. The GCKM with the exponent $n = 1$ becomes identical to the CGCM. The pore volume fraction of biopolymer is also correlated to the porosity normalized to the baseline value, i.e., $\varphi/\varphi_o = 1 - S_{BP}$.

By using the empirical power exponent, a more generalized power law type equation of the GCKM can be proposed for permeability reduction associated with biogeochemical reactions (Hommel et al., 2018), as follows:

$$K_a/K_0 = (\phi/\phi_0)^n = (1 - S_{BP})^n. \quad (3.34)$$

Herein, the greater value for exponent n indicates greater reduction in permeability and, hence, more significant pore-filling or even throat-occluding behaviors (e.g., $n > 10$). Similarly, this type of equation has been also proposed for hydrate-bearing sediments, called as the University of Tokyo model, i.e., $K_a/K_0 = (1 - S_{hydrate})^n$ (Masuda, 1997).

3.7.3.2 Correlations between permeability reduction K_r and biopolymer saturation S_{BP}
Reservoir-scale modeling of MEOR requires a mathematical model that correlates the normalized permeability ($K_r = K_a/K_0$) with the pore volume fraction of biopolymer (S_{BP}). The pore volume fraction of biopolymer is also correlated to the porosity normalized to the baseline value i.e., $\phi/\phi_o = 1 - S_{BP}$. Herein, the experiment results complied are compared with several analytical models, including the capillary tube models and Kozeny-type grain models.

Fig. 3.32 shows the variations in permeability with increasing biopolymer saturation superimposed with

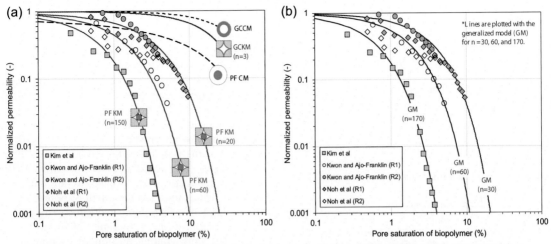

FIG. 3.32 Relative permeability reduction ratio (or normalized permeability, K/Ko) versus the biopolymer saturation: **(A)** comparison with grain-coating capillary model (GCCM), pore-filling capillary model (PFCM), grain-coating Kozeny model (GCKM), and pore-filling Kozeny model (PFKM), and **(B)** comparison with the generalized model (GM). Herein, several analytical models are superimposed for comparison: capillary grain-coating model (CGCM), capillary pore-filling model (CPFM), GCKM, PFKM, and GM. (Credit: Courtesy of Tae-Hyuk Kwon.)

several analytical models, which is further extended from the plot drawn in Noh et al. (2016) with more complied and rich datasets. Again, note the permeability is substantially reduced as the bacterial biopolymer accumulated in porous media though the rates differ due to the different culture and experiment conditions.

The experimental data appear to follow the pore-filling models, particularly the pore-filling Kozeny model (PFKM) with a remarkably large exponent value n larger than 20, rather than the grain-coating models. The exponent value for PFKM ranges approximately $n = 20-150$. And, the range of the exponent value for the generalized model (GM) is n = $\sim 30-170$. This is consistent with the result in previous study by Noh et al. (2016). The remarkably high exponent n greater than 20 indicates that the bacterial biopolymers are likely to accumulate at the pore throats and effectively occlude the flow paths in porous media.

On the other hand, the produced biopolymer, dextran, is reported to have complex internal structures with high specific surface areas, e.g., a ball of dextran thread in Kwon and Ajo-Franklin (2013); porous dextran in Jeon et al. (2017). Therefore, such significant reductions in permeability with a low biopolymer saturation are possibly attributable to the large apparent volume caused by the twined stringlike biopolymer, which is significantly larger than its actual volume.

The production and accumulation of such twined stringlike biopolymers in the sand pack is expected to enlarge the contact area between fluid and solid, which leads to the largely increased, specific surface area of the biopolymer-associated sediments.

Conclusively, the permeability reduction K_a/K_o can be correlated to the biopolymer saturation S_{BP} (or porosity reduction $\phi/\phi_o = 1 - S_{BP}$) with a simple power relation, as shown in Eq. (3.34). Biopolymer, in particular dextran produced by *L. mesenteroides*, is an effective permeability reducer which exhibits a pore-throat clogging behavior by its large apparent volume and high specific surface area. This consistent mechanistic model for describing permeability alteration as a function of dextran saturation provides a path forward for more accurate reservoir reactive transport models of the profile control process.

REFERENCES

Abdel-Mawgoud, A.M., Aboulwafa, M.M., Hassouna, N.A.-H., 2008. Optimization of surfactin production by *Bacillus subtilis* isolate BS5. Applied Biochemistry and Biotechnology 150, 305–325.

Al-Bahry, S., Al-Wahaibi, Y., Elshafie, A., Al-Bemani, A., Joshi, S., Al-Makhmari, H., Al-Sulaimani, H., 2013. Biosurfactant production by *Bacillus subtilis* B20 using date molasses and its possible application in enhanced oil recovery. International Biodeterioration & Biodegradation 81, 141–146.

Al-Wahaibi, Y., Joshi, S., Al-Bahry, S., Elshafie, A., Al-Bemani, A., Shibulal, B., 2014. Biosurfactant production by *Bacillus subtilis* B30 and its application in enhancing oil recovery. Colloids and Surfaces B: Biointerfaces 114, 324–333.

Amani, H., Müller, M.M., Syldatk, C., Hausmann, R., 2013. Production of microbial rhamnolipid by *Pseudomonas aeruginosa* MM1011 for ex situ enhanced oil recovery. Applied Biochemistry and Biotechnology 170, 1080–1093.

Amani, H., Sarrafzadeh, M.H., Haghighi, M., Mehrnia, M.R., 2010. Comparative study of biosurfactant producing bacteria in MEOR applications. Journal of Petroleum Science and Engineering 75, 209–214.

Arashiro, E.Y., Demarquette, N.R., 1999. Use of the pendant drop method to measure interfacial tension between molten polymers. Materials Research 2, 23–32.

Behlulgil, K., Mehmetoglu, T., Donmez, S., 1992. Application of microbial enhanced oil recovery technique to a Turkish heavy oil. Applied Microbiology and Biotechnology 36, 833–835.

Bendure, R.L., 1971. Dynamic surface tension determination with the maximum bubble pressure method. Journal of Colloid and Interface Science 35, 238–248.

Berry, J.D., Neeson, M.J., Dagastine, R.R., Chan, D.Y., Tabor, R.F., 2015. Measurement of surface and interfacial tension using pendant drop tensiometry. Journal of Colloid and Interface Science 454, 226–237.

Biswas, S., Dubreil, L., Marion, D., 2001. Interfacial behavior of wheat puroindolines: study of adsorption at the air–water interface from surface tension measurement using Wilhelmy plate method. Journal of Colloid and Interface Science 244, 245–253.

Bryant, R.S., Lindsey, R.P., 1996. World-wide applications of microbial technology for improving oil recovery. In: SPE/DOE Improved Oil Recovery Symposium. Society of Petroleum Engineers.

Carman, P.C., 1939. Permeability of saturated sands, soils and clays. The Journal of Agricultural Science 29, 262–273.

Castorena-Cortés, G., Zapata-Peñasco, I., Roldán-Carrillo, T., Reyes-Avila, J., Mayol-Castillo, M., Román-Vargas, S., Olguín-Lora, P., 2012. Evaluation of indigenous anaerobic microorganisms from Mexican carbonate reservoirs with potential MEOR application. Journal of Petroleum Science and Engineering 81, 86–93.

Cayias, J., Schechter, R., Wade, W., 1975. The measurement of low interfacial tension via the spinning drop technique. Adsorption at interfaces 8, 234–247.

Chong, H., Li, Q., 2017. Microbial production of rhamnolipids: opportunities, challenges and strategies. Microbial Cell Factories 16, 137.

Cooper, D., Macdonald, C., Duff, S., Kosaric, N., 1981. Enhanced production of surfactin from *Bacillus subtilis* by continuous product removal and metal cation additions. Applied and Environmental Microbiology 42, 408–412.

Costa, A., 2006. Permeability-porosity relationship: a reexamination of the Kozeny-Carman equation based on a fractal pore-space geometry assumption. Geophysical Research Letters 33.

Da Cunha, C.D., Rosado, A.S., Sebastián, G.V., Seldin, L., Von der Weid, I., 2006. Oil biodegradation by *Bacillus strains* isolated from the rock of an oil reservoir located in a deep-water production basin in Brazil. Applied Microbiology and Biotechnology 73, 949–959.

Das, M.D., 2018. Application of biosurfactant produced by an adaptive strain of *C. tropicalis* MTCC230 in microbial enhanced oil recovery (MEOR) and removal of motor oil from contaminated sand and water. Journal of Petroleum Science and Engineering 170, 40–48.

Davis, D., Lynch, H., Varley, J.J.E., Technology, M., 1999. The production of surfactin in batch culture by *Bacillus subtilis* ATCC 21332 is strongly influenced by the conditions of nitrogen metabolism, 25, 322–329.

Dong, H., Xia, W., Dong, H., She, Y., Zhu, P., Liang, K., Zhang, Z., Liang, C., Song, Z., Sun, S., 2016. Rhamnolipids produced by indigenous *Acinetobacter junii* from petroleum reservoir and its potential in enhanced oil recovery. Frontiers in Microbiology 7, 1710.

Drelich, J., Fang, C., White, C., 2002. Measurement of interfacial tension in fluid-fluid systems. Encyclopedia of Surface and Colloid Science 3, 3158–3163.

du Noüy, P.L., 1925. An interfacial tensiometer for universal use. The Journal of General Physiology 7, 625.

Dubeau, D., Déziel, E., Woods, D.E., Lépine, F., 2009. *Burkholderia thailandensis* harbors two identical rhl gene clusters responsible for the biosynthesis of rhamnolipids. BMC Microbiology 9, 263.

Eastoe, J., Tabor, R.F., 2014. Surfactants and Nanoscience. Colloidal Foundations of Nanoscience. Elsevier, pp. 135–157.

Fernandes, P., Rodrigues, E., Paiva, F., Ayupe, B., McInerney, M., Tótola, M., 2016. Biosurfactant, solvents and polymer production by *Bacillus subtilis* RI4914 and their application for enhanced oil recovery. Fuel 180, 551–557.

Geetha, S., Banat, I.M., Joshi, S.J., 2018. Biosurfactants: production and potential applications in microbial enhanced oil recovery (MEOR). Biocatalysis and Agricultural Biotechnology 14, 23–32.

Gray, M., Yeung, A., Foght, J., Yarranton, H.W., 2008. Potential microbial enhanced oil recovery processes: a critical analysis. In: SPE Annual Technical Conference and Exhibition. Society of Petroleum Engineers.

Gudiña, E.J., Rodrigues, A.I., Alves, E., Domingues, M.R., Teixeira, J.A., Rodrigues, L.R., 2015. Bioconversion of agro-industrial by-products in rhamnolipids toward applications in enhanced oil recovery and bioremediation. Bioresource Technology 177, 87–93.

Ham, S.-M., Chang, I., Noh, D.-H., Kwon, T.-H., Muhunthan, B., 2018. Improvement of surface erosion resistance of sand by microbial biopolymer formation. Journal of Geotechnical and Geoenvironmental Engineering 144, 06018004.

Harkins, W.D., Brown, F., 1919. The determination of surface tension (free surface energy), and the weight of falling drops: the surface tension of water and benzene by the capillary height method. Journal of the American Chemical Society 41, 499–524.

Hearst, J., Nelson, P., Paillet, F., 2000. Well Logging for Physical Properties, Chapter 15. McGraw-Hill, New York.

Ta, H.X., Balasingam, M., Somayeh, R., Nehal, A.-L., Tae-Hyuk, K., 2017. Effects of bacterial dextran on soil geophysical properties. Environmental Geotechnics 5 (2), 114–122.

Hommel, J., Coltman, E., Class, H., 2018. Porosity–permeability relations for evolving pore space: a review with a focus on (bio-) geochemically altered porous media. Transport in Porous Media 124, 589–629.

Hook, J.R., 2003. An introduction to porosity. Petrophysics 44.

Hui, M.-H., Blunt, M.J., 2000. Effects of wettability on three-phase flow in porous media. The Journal of Physical Chemistry B 104, 3833–3845.

Ingham, D.B., Pop, I., 1998. Transport Phenomena in Porous Media. Elsevier.

Jeon, M.-K., Kwon, T.-H., Park, J.-S., Shin, J.H., 2017. In situ viscoelastic properties of insoluble and porous polysaccharide biopolymer dextran produced by *Leuconostoc mesenteroides* using particle-tracking microrheology. Geomechanics and Engineering 12, 849–862.

Johnson Jr., R.E., Dettre, R.H., 1964. Contact angle hysteresis. III. Study of an idealized heterogeneous surface. The Journal of Physical Chemistry 68, 1744–1750.

Jung, J.-W., Wan, J., 2012. Supercritical CO₂ and ionic strength effects on wettability of silica surfaces: equilibrium contact angle measurements. Energy & Fuels 26, 6053–6059.

Kalish, P., Stewart, J., Rogers, W., Bennett, E., 1964. The effect of bacteria on sandstone permeability. Journal of Petroleum Technology 16, 805–814.

Kato, T., Haruki, M., Imanaka, T., Morikawa, M., Kanaya, S., 2001. Isolation and characterization of long-chain-alkane degrading *Bacillus thermoleovorans* from deep subterranean petroleum reservoirs. Journal of Bioscience and Bioengineering 91, 64–70.

Khademolhosseini, R., Jafari, A., Mousavi, S.M., Hajfarajollah, H., Noghabi, K.A., Manteghian, M., 2019. Physicochemical characterization and optimization of glycolipid biosurfactant production by a native strain of *Pseudomonas aeruginosa* HAK01 and its performance evaluation for the MEOR process. RSC Advances 9, 7932–7947.

Kim, D.S., Fogler, H.S., 2000. Biomass evolution in porous media and its effects on permeability under starvation conditions. Biotechnology and Bioengineering 69, 47–56.

Kim, H.-S., Yoon, B.-D., Lee, C.-H., Suh, H.-H., Oh, H.-M., Katsuragi, T., Tani, Y., 1997. Production and properties of a lipopeptide biosurfactant from *Bacillus subtilis* C9. Journal of Fermentation and Bioengineering 84, 41–46.

Kim, Y.-M., Park, T., Kwon, T.-H., 2019. Engineered bioclogging in coarse sands by using fermentation-based bacterial biopolymer formation. Geomachanics and Engineering 17, 485–496.

Kleinberg, R., Flaum, C., Griffin, D., Brewer, P., Malby, G., Peltzer, E., Yesinowski, J., 2003. Deep sea NMR: methane hydrate growth habit in porous media and its relationship to hydraulic permeability, deposit accumulation, and submarine slope stability. Journal of Geophysical Research: Solid Earth 108.

Kozeny, J., 1927. Uber kapillare leitung der wasser in boden. Royal Academy of Science, Vienna, Proc. Class I 136, 271–306.

Kwon, T.-H., Ajo-Franklin, J.B., 2013. High-frequency seismic response during permeability reduction due to biopolymer clogging in unconsolidated porous media. Geophysics 78, EN117–EN127.

Lamb, H., 1945. Hydrodynamics, sixth ed. Dover, New York. NY.

Lappan, R., Fogler, H.S., 1992. Effect of bacterial polysaccharide production on formation damage. SPE Production Engineering 7, 167–171.

Lappan, R.E., Fogler, H.S., 1996. Reduction of porous media permeability from in situ *Leuconostoc mesenteroides* growth and dextran production. Biotechnology and Bioengineering 50, 6–15.

Law, K.-Y., Zhao, H., 2016a. Surface Wetting. Springer International Publishing, Switzerland. ISBN 978-3-319-25212-4.

Law, K.-Y., Zhao, H., 2016b. Surface Wetting: Characterization, Contact Angle, and Fundamentals. Springer International Publishing.

Liu, Q., Lin, J., Wang, W., Huang, H., Li, S., 2015. Production of surfactin isoforms by Bacillus subtilis BS-37 and its applicability to enhanced oil recovery under laboratory conditions. Biochemical Engineering Journal 93, 31–37.

Marsh, T., Zhang, X., Knapp, R., McInerney, M., Sharma, P., Jackson, B., 1995. Mechanisms of Microbial Oil Recovery by *Clostridium Acetobutylicum* and *Bacillus* Strain JF-2. BDM Oklahoma, Inc., Bartlesville, OK (United States).

Masuda, Y., 1997. Numerical calculation of gas production performance from reservoirs containing natural gas hydrates. In: Annual Technical Conference, Soc. of Petrol. Eng., San Antonio, Tex., Oct. 1997.

Melrose, J.C., 1965. Wettability as related to capillary action in porous media. Society of Petroleum Engineers Journal 5, 259–271.

Morrow, N.R., 1990. Wettability and its effect on oil recovery. Journal of Petroleum Technology 42 (1), 476-471, 484.

Nagase, K., Zhang, S., Asami, H., Yazawa, N., Fujiwara, K., Enomoto, H., Hong, C., Liang, C., 2002. A successful field test of microbial EOR process in Fuyu Oilfield, China. In: SPE/DOE Improved Oil Recovery Symposium. Society of Petroleum Engineers.

Nicholson, W.L., Munakata, N., Horneck, G., Melosh, H.J., Setlow, P., 2000. Resistance of *Bacillus* endospores to extreme terrestrial and extraterrestrial environments. Microbiology and Molecular Biology Reviews 64, 548–572.

Noh, D.H., Ajo-Franklin, J.B., Kwon, T.H., Muhunthan, B., 2016. P and S wave responses of bacterial biopolymer formation in unconsolidated porous media. Journal of Geophysical Research: Biogeosciences 121, 1158–1177.

Ohno, A., Ano, T., Shoda, M., 1995. Effect of temperature on production of lipopeptide antibiotics, iturin A and surfactin by a dual producer, *Bacillus subtilis* RB14, in solid-state fermentation. Journal of Fermentation and Bioengineering 80, 517–519.

Olajire, A.A., 2014. Review of ASP EOR (alkaline surfactant polymer enhanced oil recovery) technology in the petroleum industry: prospects and challenges. Energy 77, 963–982.

Park, T., Jeon, M.-K., Yoon, S., Lee, K.S., Kwon, T.-H., 2019. Modification of interfacial tension and wettability in oil-brine-quartz system by in situ bacterial biosurfactant production at reservoir conditions: implications to microbial enhanced oil recovery. Energy & Fuels 33, 4909–4920.

Peet, K.C., Freedman, A.J., Hernandez, H.H., Britto, V., Boreham, C., Ajo-Franklin, J.B., Thompson, J.R., 2015. Microbial growth under supercritical CO_2. Applied and Environmental Microbiology 81, 2881–2892.

Pereira, J.F., Gudiña, E.J., Costa, R., Vitorino, R., Teixeira, J.A., Coutinho, J.A., Rodrigues, L.R., 2013. Optimization and characterization of biosurfactant production by Bacillus subtilis isolates towards microbial enhanced oil recovery applications. Fuel 111, 259–268.

Perfumo, A., Banat, I.M., Canganella, F., Marchant, R., 2006. Rhamnolipid production by a novel thermophilic hydrocarbon-degrading *Pseudomonas aeruginosa* AP02-1. Applied Microbiology and Biotechnology 72, 132.

Petroleum, N.I.f., Research, E., Bryant, R.S., 1990. Screening Criteria for Microbial EOR Processes.

Pornsunthorntawee, O., Arttaweeporn, N., Paisanjit, S., Somboonthanate, P., Abe, M., Rujiravanit, R., Chavadej, S., 2008. Isolation and comparison of biosurfactants produced by Bacillus subtilis PT2 and Pseudomonas aeruginosa SP4 for microbial surfactant-enhanced oil recovery. Biochemical Engineering Journal 42, 172–179.

Qazi, M.A., Subhan, M., Fatima, N., Ali, M.I., Ahmed, S., 2013. Role of biosurfactant produced by *Fusarium* sp. BS-8 in enhanced oil recovery (EOR) through sand pack column. International Journal of Bioscience, Biochemistry and Bioinformatics 3, 598.

Rocha, C., San-Blas, F., San-Blas, G., Vierma, L., 1992. Biosurfactant production by two isolates of *Pseudomonas aeruginosa*. World Journal of Microbiology and Biotechnology 8, 125–128.

Rosen, M.J., Kunjappu, J.T., 2012. Surfactants and Interfacial Phenomena. John Wiley & Sons.

Sarafzadeh, P., Niazi, A., Oboodi, V., Ravanbakhsh, M., Hezave, A.Z., Ayatollahi, S.S., Raeissi, S., 2014. Investigating the efficiency of MEOR processes using Enterobacter cloacae and Bacillus stearothermophilus SUCPM# 14 (biosurfactant-producing strains) in carbonated reservoirs. Journal of Petroleum Science and Engineering 113, 46–53.

Scheidegger, A.E., 1958. The physics of flow through porous media. Soil Science 86, 355.

Selley, R.C., 1998. Elements of Petroleum Geology. Gulf Professional Publishing.

Sen, R., 2008. Biotechnology in petroleum recovery: the microbial EOR. Progress in Energy and Combustion Science 34, 714–724.

Shavandi, M., Mohebali, G., Haddadi, A., Shakarami, H., Nuhi, A., 2011. Emulsification potential of a newly isolated biosurfactant-producing bacterium, *Rhodococcus* sp. strain TA6. Colloids and Surfaces B: Biointerfaces 82, 477–482.

Sheng, J.J., 2013. Enhanced Oil Recovery Field Case Studies. Gulf Professional Publishing.

Simpson, D.R., Natraj, N.R., McInerney, M.J., Duncan, K.E., 2011. Biosurfactant-producing *Bacillus* are present in produced brines from Oklahoma oil reservoirs with a wide range of salinities. Applied Microbiology and Biotechnology 91, 1083.

Spangenberg, E., 2001. Modeling of the influence of gas hydrate content on the electrical properties of porous sediments. Journal of Geophysical Research: Solid Earth 106, 6535–6548.

Stalder, A., Kulik, G., Sage, D., Barbieri, L., Hoffmann, P., 2006. A snake-based approach to accurate determination of both contact points and contact angles. Colloids and Surfaces A: Physicochemical and Engineering Aspects 286, 92–103.

Stewart, T.L., Scott Fogler, H., 2002. Pore-scale investigation of biomass plug development and propagation in porous media. Biotechnology and Bioengineering 77, 577–588.

Syldatk, C., Lang, S., Wagner, F., Wray, V., Witte, L., 1985. Chemical and physical characterization of four interfacial-active rhamnolipids from *Pseudomonas* spec. DSM 2874 grown on n-alkanes. Zeitschrift für Naturforschung C 40, 51–60.

Ta, H.X., Muhunthan, B., Ramezanian, S., Abu-Lail, N., Kwon, T.-H., 2017. Effects of bacterial dextran on soil geophysical properties. Environmental Geotechnics 5, 114–122.

Tiab, D., Donaldson, E.C., 2015. Petrophysics: Theory and Practice of Measuring Reservoir Rock and Fluid Transport Properties. Gulf Professional Publishing.

Varjani, S.J., Upasani, V.N., 2016. Core flood study for enhanced oil recovery through ex-situ bioaugmentation with thermo-and halo-tolerant rhamnolipid produced by *Pseudomonas aeruginosa* NCIM 5514. Bioresource Technology 220, 175–182.

Veshareh, M.J., Azad, E.G., Deihimi, T., Niazi, A., Ayatollahi, S., 2019. Isolation and screening of *Bacillus subtilis* MJ01 for MEOR application: biosurfactant characterization, production optimization and wetting effect on carbonate surfaces. Journal of Petroleum Exploration and Production Technology 9, 233–245.

Wang, Q., Fang, X., Bai, B., Liang, X., Shuler, P.J., Goddard III, W.A., Tang, Y., 2007. Engineering bacteria for production of rhamnolipid as an agent for enhanced oil recovery. Biotechnology and Bioengineering 98, 842–853.

Wang, X.-Y., 1991. Ch. F-9 Advances in Research, Production and Application of Biopolymers Used for Eor in China. Developments in Petroleum Science: Elsevier 467–481.

Wei, Y.-H., Chou, C.-L., Chang, J.-S., 2005. Rhamnolipid production by indigenous *Pseudomonas aeruginosa* J4 originating from petrochemical wastewater. Biochemical Engineering Journal 27, 146–154.

Willenbacher, J., Rau, J.-T., Rogalla, J., Syldatk, C., Hausmann, R., 2015. Foam-free production of Surfactin via anaerobic fermentation of *Bacillus subtilis* DSM 10 T. AMB Express 5, 21.

Yakimov, M.M., Timmis, K.N., Wray, V., Fredrickson, H.L., 1995. Characterization of a new lipopeptide surfactant produced by thermotolerant and halotolerant subsurface *Bacillus licheniformis* BAS50. Applied and Environmental Microbiology 61, 1706–1713.

Zhao, F., Shi, R., Zhao, J., Li, G., Bai, X., Han, S., Zhang, Y., 2015. Heterologous production of *Pseudomonas aeruginosa* rhamnolipid under anaerobic conditions for microbial enhanced oil recovery. Journal of Applied Microbiology 118, 379–389.

Zhao, F., Zhou, J.-D., Ma, F., Shi, R.-J., Han, S.-Q., Zhang, J., Zhang, Y., 2016. Simultaneous inhibition of sulfate-reducing bacteria, removal of H_2S and production of rhamnolipid by recombinant *Pseudomonas stutzeri* Rhl: applications for microbial enhanced oil recovery. Bioresource Technology 207, 24–30.

Zheng, C., Yu, L., Huang, L., Xiu, J., Huang, Z., 2012. Investigation of a hydrocarbon-degrading strain, *Rhodococcus ruber* Z25, for the potential of microbial enhanced oil recovery. Journal of Petroleum Science and Engineering 81, 49–56.

CHAPTER 4

Modeling and Simulation

4.1 BIOLOGICAL GROWTH AND METABOLISM KINETICS

4.1.1 Basic Concept for Growth Rate

In the microbial process model, mass balances between biomass and primary substrates that inhibit the rate of biomass growth should always be observed. In most cases, the rate-limiting substrate acts as an electron donor. Since this assumption applies equally here, from now on the substrate means the primary electron donor substrate. To complete the mass balance equations, the rates of biomass growth and substrate consumption must be expressed (Rittmann and McCarty, 2001).

The most commonly used relationship to describe the growth kinetics of microorganisms is the Monod equation, which was developed in 1949 by Jacques Monod. The specific growth rate of active biomass is related to the concentration of rate-limiting substrate as follows:

$$\mu_{syn} = \left(\frac{1}{X_a}\frac{dX_a}{dt}\right)_{syn} = \hat{\mu}\frac{S}{K+S} \tag{4.1}$$

where μ_{syn} is the specific growth rate for cell synthesis, X_a is the concentration of active biomass, t is the time, S is the concentration of the rate-limiting substrate, $\hat{\mu}$ is the maximum specific growth rate, and K is the concentration giving one-half the maximum rate (Fig. 4.1). This equation is a very convenient method to express for smooth transition from first-order relation at low substrate concentration to zero-order relations at high substrate concentration (Rittmann and McCarty, 2001).

The maintenance of microbes requires the energy consumption for cell functions such as motility, repair and resynthesis, osmotic regulation, transport, and heat loss, which slows down the growth rate. The energy and electrons flow to meet the maintenance needs is defined as the endogenous decay. That is, the bacterial cells oxidize themselves to meet the maintenance-energy needs. The rate of endogenous decay is represented as follows:

$$\mu_{dec} = \left(\frac{1}{X_a}\frac{dX_a}{dt}\right)_{dec} = -b \tag{4.2}$$

where μ_{dec} is the specific growth rate due to decay, and b is the endogenous decay coefficient. This equation shows that the decay of active biomass is a first-order function. However, not all of the active biomass lost by decay is oxidized to generate the maintenance energy. Although most decayed biomass is oxidized, some of them accumulate as inert biomass. The oxidation rate is defined as follows:

$$\left(\frac{1}{X_a}\frac{dX_a}{dt}\right)_{ox} = -f_d b \tag{4.3}$$

where f_d is the fraction of active biomass, which is biodegradable. The rate at which the active biomass is converted to the inert biomass is represented as the difference between the overall decay rate and the oxidation decay rate as follows:

$$-\frac{1}{X_a}\frac{dX_i}{dt} = \left(\frac{1}{X_a}\frac{dX_a}{dt}\right)_{inert} = -(1-f_d)b \tag{4.4}$$

where X_i is the inert biomass concentration.

As a result, the net specific growth rate (μ) and net growth rate (r_{net}) for bacteria can be expressed as the sum of cell synthesis and decay processes as follows:

$$\mu = \frac{1}{X_a}\frac{dX_a}{dt} = \mu_{syn} + \mu_{dec} = \hat{\mu}\frac{S}{K+S} - b \tag{4.5}$$

and

$$r_{net} = \hat{\mu}\frac{S}{K+S}X_a - bX_a \tag{4.6}$$

Since the growth of biomass is fueled by substrate utilization, Eq. (4.6) can also be expressed as:

$$r_{net} = Y\frac{\hat{q}S}{K+S}X_a - bX_a \tag{4.7}$$

where $\hat{q}$ is the maximum specific rate of substrate utilization, and Y is the true yield for cell synthesis.

Mitchell et al. (2004) reported the various kinetic equations for bacterial growth in solid-state fermentation systems, including linear, exponential, logistic, and fast acceleration/slow deceleration. The typical trends of these curves are represented in Fig. 4.2. The empirical equations as followed are simply fitted to experimental results by nonlinear regression.

Theory and Practice in Microbial Enhanced Oil Recovery. https://doi.org/10.1016/B978-0-12-819983-1.00004-1

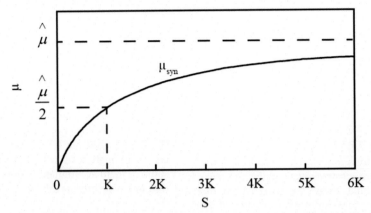

FIG. 4.1 Schematic of the specific growth rate depending on substrate concentration. (Credit: Self-editing)

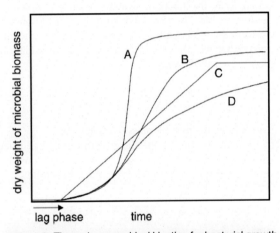

FIG. 4.2 The various empirical kinetics for bacterial growth in solid-state fermentation systems: **(A)** exponential, **(B)** logistic, **(C)** linear, and **(D)** fast acceleration/slow deceleration. (Credit: from Mitchell, D.A., von Meien, O.F., Krieger, N., Dalsenter, F.D.H., 2004. A review of recent developments in modeling of microbial growth kinetics and intraparticle phenomena in solid-state fermentation. Biochemical Engineering Journal 17 (1), 15–26.)

The linear form of bacterial growth is shown as follows:

$$\frac{dX}{dt} = K \tag{4.8}$$

and

$$X = \alpha t + X_0 \tag{4.9}$$

The exponential form is represented as follows:

$$\frac{dX}{dt} = \mu X \tag{4.10}$$

and

$$X = X_0 e^{\mu t} \tag{4.11}$$

The logistic form is expressed as:

$$\frac{dX}{dt} = \mu X \left(1 - \frac{X}{X_{max}}\right) \tag{4.12}$$

and

$$X = \frac{X_{max}}{1 + \left(\dfrac{X_{max}}{X_0} - 1\right)e^{-\mu t}} \tag{4.13}$$

where α is the linear growth rate, X_0 is the initial biomass, and X_{max} is the maximum possible microbial biomass. The fast acceleration and slow deceleration model is proposed as:

$$\frac{dX}{dt} = \mu X, \ t < t_a \tag{4.14a}$$

$$\frac{dX}{dt} = \mu L e^{-k(t-t_a)} X, \ t \geq t_a \tag{4.14b}$$

and

$$X = X_0 e^{\mu t}, \ t < t_a \tag{4.15a}$$

$$X = X_A \exp\left[\frac{\mu L}{k}\left(1 - e^{-k(t-t_a)}\right)\right], \ t \geq t_a \tag{4.15b}$$

In this model, an exponential stage is followed by a deceleration stage. The t_a is the time at the transition from exponential to the deceleration stage and X_A is the bacterial concentration at time t_a. The L is the ratio of the specific growth rate at the beginning of the deceleration state to previous exponential stage. The quantity of $\mu L e^{-k(t-t_a)}$ in Eq. (4.14b) represents the specific growth rate during the deceleration stage. This equation assumes that a sudden deceleration occurs at time t_a, and further deceleration is described as exponential decay by the first-order rate constant, k, in the exponential term. The difficulty in applying this equation is that the exponential period is very short and the biomass concentration is very low during this period, resulting in fewer experimental points and more error. These equations do not include the effect of substrate concentration on growth and are simply fitted to experimental growth results by nonlinear regression (Mitchell et al., 2004).

The Monod model is very useful for expressing microbial growth and substrate utilization. However, other models could be used in special cases. One famous alternative is the Contois model as follows:

$$r_{ut} = -\frac{\hat{q}S}{BX_a + S}X_a \qquad (4.16)$$

where r_{ut} is the rate of substrate utilization, and B is a constant. In this equation, the specific reaction rate is affected by the concentration of the active biomass. In high biomass concentration, the reaction rate decreases, reaching a first-order reaction being dependent on S:

$$r_{ut} = -\frac{\hat{q}}{B}S, \ X_a \to \infty \qquad (4.17)$$

The Contois model is useful for describing hydrolysis of suspended particulate organic matter in primary and waste-activated sludge (Henze et al., 1995).

The following Moser and Tessier equations are also two other alternatives, respectively:

$$r_{ut} = -\frac{\hat{q}S}{K + S^{-\gamma}}X_a \qquad (4.18)$$

$$r_{ut} = -\hat{q}\left(1 - e^{\frac{S}{K}}\right)X_a \qquad (4.19)$$

where γ is a constant. When $-\gamma = 1$ for Moser equation or $BX_a = K$ for Contois equation, these relationships are reverted to Monod equation. However, the Tessier equation shows quite a different response from that of Monod equation when the same K and $\hat{q}$ are used in both equations. The Tessier model reaches its maximum rate faster than the Monod model.

One last alternative model applies to the dual limitation situation where more than one substrate acts as rate limiting. The most commonly used method to describe this dual limitation is the multiplicative Monod approach as follows:

$$r_{ut} = -\hat{q}\frac{S}{K + S}\frac{S'}{K_A + S'}X_a \qquad (4.20)$$

where S' is the concentration of the second substrate, and K_A is the half-maximum rate concentration of the second substrate.

4.1.2 Effects of Environmental Parameters

As noted in Chapter 1, environmental factors have a critical effect on microbial growth. Too high or too low temperature causes irreversible morphology change of the biomass, greatly affecting microbial activity. Salinity and pH are also known to affect enzyme activity. Moreover, high pressures accelerate the decay of biomass. Therefore, many researchers have developed models to simulate microbial growth that reflect these environmental factors.

The effect of temperature on the maximum specific growth rate is similar to that observed in enzyme activity. The rate of microbial growth continues to increase until a certain temperature at which protein denatures and rapidly decreases above that temperature. The maximum specific growth rate at the temperature below the protein denaturation proceeds is expressed by the Arrhenius equation like the normal chemical reaction rate as follows:

$$\hat{\mu} = F \exp\left(-\frac{E_a}{RT}\right) \qquad (4.21)$$

where F is a constant, E_a is activation energy, R is gas constant, and T is temperature. Assuming that protein denaturation by temperature occurs by a reversible chemical reaction involving a free energy change (ΔG) and that these denatured proteins are inactive, $\hat{\mu}$ (Roels, 1983) being closely related to the Hougen-Watson equation for catalyst activity can be described as follows (Villadsen et al., 2011):

$$\hat{\mu} = \frac{F_1 \exp\left(-\frac{E_a}{RT}\right)}{1 + F_2 \exp\left(-\frac{\Delta G}{RT}\right)} \qquad (4.22)$$

where F_1 and F_2 are constants.

Hong et al. (2019a,b) used the modified exponential term in the Arrhenius equation (Eq. 4.21) to show the effect of growth temperature as follows:

$$f(T) = \exp\left(A_j - \frac{E_{a,j}}{RT_j}\right) \qquad (4.23)$$

and

$$A_j - \frac{E_{a,j}}{RT_{j+1}} = A_{j+1} - \frac{E_{a,j+1}}{RT_{j+1}} \qquad (4.24)$$

where A_j and A_{j+1} are constants. It only remains to fix A_j for one interval since all the other A_j values can be calculated from it. It would be convenient to specify a reference interval (CMG, 2018).

Additionally, temperature-dependent microbial growth models have been suggested in several researches.

Ratkowsky et al. (1982) proposed the square root method as follows:

$$\sqrt{r} = \omega(T - T_0) \tag{4.25}$$

where r is the growth rate, ω is a constant, and T_0 is a conceptual temperature of no metabolic significance.

Rosso et al. (1995) used a cardinal temperature model defined as follows:

$$\hat{\mu} = \mu_{opt}\gamma(T) \tag{4.26}$$

and

$$\gamma(T) = \frac{(T - T_{max})(T - T_{min})^2}{(T_{opt} - T_{min})[(T_{opt} - T_{min})(T - T_{opt}) - (T_{opt} - T_{max})(T_{opt} + T_{min} - 2T)]} \tag{4.27}$$

where μ_{opt} is the specific growth rate at the optimum conditions, T_{opt} is the optimum temperature for bacterial growth, T_{max} is the maximum temperature for microbial growth, and T_{min} is the minimum temperature for microbial growth.

They also suggested the cardinal pH model as follows:

$$\hat{\mu} = \mu_{opt}\gamma(pH) \tag{4.28}$$

and

$$\gamma(pH) = \frac{(pH - pH_{min})(pH - pH_{max})}{(pH - pH_{min})(pH - pH_{max}) - (pH - pH_{opt})^2} \tag{4.29}$$

where pH_{opt} is the optimum pH for bacterial growth, pH_{max} is the maximum pH for microbial growth, and pH_{min} is the minimum pH for microbial growth. The equation for describing the two effects in combination is as follows:

$$\hat{\mu} = \mu_{opt}\gamma(T)\gamma(pH) \tag{4.30}$$

The descriptions of growth models that reflect temperature and pH effects at the same time have been existed before. Wijtes et al. (1993) proposed a model with temperature and pH as control factors as follows:

$$\hat{\mu} = b_1\left\{(pH - pH_{min})\left[1 - e^{c_1(pH - pH_{max})}\right](T - T_{min})\right\}^2 \tag{4.31}$$

where b_1 and c_1 are constants (biologically meaningless parameters).

Zwietering et al. (1993) also showed the combined model as follows:

$$\hat{\mu} = \mu_{opt}\tau(pH)\tau(T) \tag{4.32}$$

with

$$\tau(pH) = \left\{\frac{(pH - pH_{min})\left[1 - e^{c_2(pH - pH_{max})}\right]}{(pH_{opt} - pH_{min})\left[1 - e^{c_2(pH_{opt} - pH_{max})}\right]}\right\}^2 \tag{4.33}$$

and

$$\tau(T) = \left\{\frac{(T - T_{min})\left[1 - e^{c_3(T - T_{max})}\right]}{(T_{opt} - T_{min})\left[1 - e^{c_3(T_{opt} - T_{max})}\right]}\right\}^2 \tag{4.34}$$

where c_2 and c_3 are regression coefficients which are verified by the following two equations:

$$1 - (c_2 pH_{opt} - c_2 pH_{min} + 1)e^{c_2(pH_{opt} - pH_{max})} = 0 \tag{4.35}$$

and

$$1 - (c_3 T_{opt} - c_3 T_{min} + 1)e^{c_3(T_{opt} - T_{max})} = 0 \tag{4.36}$$

The changes in bacterial activity due to pH can also be determined by the type of enzyme present (Villadsen et al., 2011). The enzyme can exist in three forms such as:

$$E \leftrightarrow E^- + H^+ \leftrightarrow E^{2-} + 2H^+ \tag{4.37}$$

where E^- is an active form of the enzyme, whereas the two other forms are assumed to be completely inactive. The fraction of active enzyme is calculated as follows:

$$\frac{E^-}{E_{tot}} = \frac{1}{1 + \frac{[H^+]}{k_1} + \frac{k_2}{[H^+]}} \tag{4.38}$$

where E_{tot} is total number of enzymes, k_1 and k_2 are the dissociation constants for E and E^-, respectively. If the cell activity is determined by the enzyme activity, the maximum specific growth rate can be given as:

$$\hat{\mu} = \frac{aE_{tot}}{1 + \frac{[H^+]}{k_1} + \frac{k_2}{[H^+]}} \tag{4.39}$$

where a is the activity of the active enzyme.

Microbial activity by enzyme activity is also affected by salinity. Park and Raines (2001) described a dramatic effect of salt concentration on enzymatic catalysis. The profile of enzyme E, substrate S, and product E by chemical reaction in the presence of $[Na^+]$ is as follows:

$$E + S \leftrightarrow ES \rightarrow EP \tag{4.40}$$

The overall reaction (r) is described with second-order rate constant as follows:

$$r = \frac{k_3 k_5}{k_4 + k_5} \tag{4.41}$$

in which k_3 is rate constant for association $(E + S \rightarrow ES)$, k_4 is rate constant for dissociation $(E + S \leftarrow ES)$, and k_5 is rate constant for chemistry step $(ES \rightarrow EP)$. The association and dissociation constants can be written as a function of $[Na^+]$:

$$k_3 = k_3^\theta [Na^+]^{-\beta n} \tag{4.42}$$

and

$$k_4 = k_4^\theta [Na^+]^{(1-\beta)n} \tag{4.43}$$

where k_3^θ and k_4^θ are rate constants when $[Na^+] = 1M$ (with the θ superscript denoting the state in which $[Na^+] = 1M$), β is a constant of proportionality between thermodynamic changes and kinetic changes due to $[Na^+]$, and n is a constant. By using Eqs. (4.42) and (4.43), r can be expressed as a function of $[Na^+]$ as follows:

$$r = k_3^\theta [Na^+]^{-\beta n} \left(\frac{C_f}{1 + C_f} \right) \tag{4.44}$$

where

$$C_f = C_f^\theta [Na^+]^{(\beta-1)n} \tag{4.45}$$

and

$$C_f = \frac{k_5}{k_4} \tag{4.46}$$

From Eqs. (4.44) and (4.45), the dependence of r on $[Na^+]$ can be calculated as follows:

$$\frac{\partial(\log r)}{\partial(\log[Na^+])} = -n \left(\frac{1 + \beta C_f}{1 + C_f} \right) \approx -n' \tag{4.47}$$

where n' is a constant. By integrating Eq. (4.47), r can be written as:

$$r = r^\theta [Na^+]^{-n'} \tag{4.48}$$

Although the Coulombic forces are stronger at low Na^+ concentration, the r value is limited by the collision of the enzyme and substrate. The value of r at this physical limit is r_{lim}. Thus, the r is limited by the smaller of r_{lim} and $r^\theta[Na^+]^{-n'}$ as follows:

$$\frac{1}{r} = \frac{1}{r_{lim}} + \frac{1}{r^\theta[Na^+]^{-n'}} \tag{4.49}$$

Eq. (4.49) can be simplified as:

$$r = \frac{r_{lim}}{1 + \left(\frac{[Na^+]}{k_{Na^+}} \right)^{n'}} \tag{4.50}$$

where

$$k_{Na^+} = \left(\frac{r^\theta}{r_{lim}} \right)^{\frac{1}{n'}} \tag{4.51}$$

Park and Marchand (2006) proposed another salinity model by modifying the Monod equation to reflect the salinity inhibition effect. The salinity inhibition parameter reduces specific growth rate in high concentrations of NaCl. The modified specific growth rate is as follows:

$$\mu = \frac{(\hat{\mu} - I_S)S}{K + S + \frac{S^2}{K_i}} \tag{4.52}$$

where I_S is the salinity inhibition constant, and K_i is the substrate inhibition constant. Since the I_S should be zero for a nonsaline environment, it can be expressed as:

$$I_S = \frac{\%NaCl}{0.01 + \%NaCl} I_S^* \tag{4.53}$$

where I_S^* is a constant depending on the cultures.

Leroi et al. (2012) also developed a salinity effect model using a cardinal model as follows:

$$\gamma(c_{NaCl}) = \frac{(c_{NaCl} - 2c_{NaCl_{opt}} + c_{NaCl_{max}})(c_{NaCl_{max}} - c_{NaCl})}{(c_{NaCl_{max}} - c_{NaCl_{opt}})^2} \tag{4.54}$$

where c_{NaCl} is NaCl concentration, $c_{NaCl_{opt}}$ is the optimum NaCl concentration for bacterial metabolism, and $c_{NaCl_{max}}$ is the maximum NaCl concentration for bacterial metabolism.

Additionally, a microbial decay model with pressure effect was represented by Basak et al. (2002). They showed the destruction kinetics of microorganisms at high pressure. During the pressure-hold time, the destruction of cells follows a traditional first-order kinetics as follows:

$$\ln \left(\frac{N}{N_o} \right) = -\alpha_d t \tag{4.55}$$

where N is the number of surviving microbes after pressure treatment, N_o is the initial number of microbes, α is the destruction rate constant, and t is the pressure treatment time. The decimal reduction time, D or D-value, is defined as the time which takes for the number of microbes to reduce to one-tenth at a certain pressure. They showed the relationship between D-value and pressure as follows:

$$\log \left(\frac{D_i}{D_{i+1}} \right) = \frac{p_{i+1} - p_i}{z_p} \tag{4.56}$$

where p_i is pressure at i-th step, D_i is decimal time at i-th pressure, and z_p is negative reciprocal slope of the $logD$ versus p.

4.1.3 Model for Metabolites Generation

Microorganisms consume substrates to produce new biomass, as well as soluble microbial products (SMPs). The SMPs consist of cellular components that come out of the cell lysis process, remain in the cell synthesis process, or are released for some purpose. They have moderate formula weights and are biodegradable. SMPs are important because they exist in all cases and form the majority of the chemical oxygen demand (COD) and biochemical oxygen demand (BOD) of effluents in many cases (Rittmann and McCarty, 2001).

The SMPs can be subdivided in to two categories such as substrate-utilization-associated products (UAP) and biomass-associated products (BAP). The UAPs are generated directly during substrate utilization. The production kinetics can be described as follows:

$$r_{UAP} = -\alpha_{UAP} r_{ut} \quad (4.57)$$

where r_{UAP} is the production rate of UAP, and α_{UAP} is the UAP formation coefficient. The BAPs are generated directly from biomass and rate expression is as follows:

$$r_{BAP} = \alpha_{BAP} X_a \quad (4.58)$$

in which r_{BAP} is the production rate of BAP, and α_{BAP} is the BAP formation coefficient.

Although SMPs are generally known to be biodegradable, the details of various degradation kinetics for highly heterogeneous mixtures are not fully resolved. However, UAP and BAP are known to have distinct degradation kinetics, which can be described as separate Monod degradation equations as follows:

$$r_{deg-UAP} = \frac{-\hat{q}_{UAP} c_{UAP}}{K_{UAP} + c_{UAP}} X_a \quad (4.59a)$$

and

$$r_{deg-BAP} = \frac{-\hat{q}_{BAP} c_{BAP}}{K_{BAP} + c_{BAP}} X_a \quad (4.59b)$$

where $r_{deg-UAP}$ and $r_{deg-BAP}$ are the degradation rates of UAP and BAP, respectively; $\hat{q}_{UAP}$ and $\hat{q}_{BAP}$ are the maximum specific rates of UAP and BAP degradations, respectively; c_{UAP} and c_{BAP} are UAP and BAP concentrations, respectively; and K_{UAP} and K_{BAP} are half-maximum rate concentrations for UAP and BAP, respectively.

The steady-state mass balances on UAP and BAP are as follows:

$$0 = -\alpha_{UAP} r_{ut} V - \frac{\hat{q}_{UAP} c_{UAP}}{K_{UAP} + c_{UAP}} X_a V - Q c_{UAP} \quad (4.60a)$$

and

$$0 = \alpha_{BAP} X_a V - \frac{\hat{q}_{BAP} c_{BAP}}{K_{BAP} + c_{BAP}} X_a V - Q c_{BAP} \quad (4.60b)$$

where V is the volume of chemostat, and Q is the flow rate for influent and effluent. These equations can be calculated for c_{UAP} and c_{BAP} as follows:

$$c_{UAP} = -\frac{\hat{q}_{UAP} X_a t_d + K_{UAP} + \alpha_{UAP} r_{ut} t_d}{2}$$
$$+ \frac{\sqrt{\left(\hat{q}_{UAP} X_a t_d + K_{UAP} + \alpha_{UAP} r_{ut} t_d\right)^2 - 4 K_{UAP} \alpha_{UAP} r_{ut} t_d}}{2}$$

$$(4.61a)$$

and

$$c_{BAP} = \frac{-K_{BAP} - (\hat{q}_{BAP} - \alpha_{BAP}) X_a t_d}{2}$$
$$+ \frac{\sqrt{\left[K_{BAP} + (\hat{q}_{BAP} - \alpha_{BAP}) X_a t_d\right]^2 + 4 K_{BAP} \alpha_{BAP} X_a t_d}}{2}$$

$$(4.61b)$$

where t_d is hydraulic detention time. Rittmann and McCarty (2001) provided a detailed example.

The production of metabolites is based on enzymatic reactions. The rate of production can be described by a very well-known relationship called Michaelis–Menten kinetics as follows:

$$r_p = r_{max} \frac{S}{K_M + S} \quad (4.62)$$

where r_p is the production rate, r_{max} is the maximum specific rate of metabolite production, and K_M is the Michaelis–Menten constant (Villadsen et al., 2011; Jeong et al., 2019a).

The empirical model to describe the production rate of the metabolic products is proposed by Bajpai and Reuss (1982) as follows:

$$r_p = \hat{\mu}_p \frac{S - S*}{K_{p/s} + S - S*} X_a \quad (4.63)$$

in which $\hat{\mu}_p$ is the maximum specific production rate, $K_{p/s}$ is the saturation constant for metabolite to consumption of substrate S, and $S*$ is the critical concentration of the substrate for metabolic production. Lappan (1994) also suggested a model as follows:

$$r_p = r_{max} \frac{S}{K_M + S} Y \quad (4.64)$$

where Y is the yield coefficient.

4.2 SIMULATION OF MICROBIAL ENHANCED OIL RECOVERY MECHANISMS

4.2.1 Permeability Alteration

There are two general numerical approaches to simulate the permeability alterations (Patel et al., 2015). The approach using the equilibrium relationship between flowing and adsorption bioproducts (such as bacteria cell or biopolymer) mainly utilizes the Langmuir isotherm equation to describe the adsorption as follows:

$$ad = \frac{(A + Bc_{\text{NaCl}})X_b}{1 + CX_b} \qquad (4.65)$$

where ad is adsorbed moles of bioproducts, c_{NaCl} is salinity, X_b is the concentration of bioproducts, and A, B, and C are constants. This adsorption causes a blockage effect that impedes the flow of fluid, which affects effective permeability as follows:

$$k_e = 1 + (rrft - 1)\frac{ad}{ad_{\max}} \qquad (4.66)$$

in which k_e is effective permeability, $rrft$ is constant for residual resistance factor, and $ad_{\max}$ is maximum adsorption capacity.

The other method is to determine the permeability resulting from the porosity change. The most widely used method is to use the Carman-Kozeny equation as follows:

$$k_f = k_o \left(\frac{\varphi_f}{\varphi_o}\right)^{CK} \left(\frac{1 - \varphi_o}{1 - \varphi_f}\right)^2 \qquad (4.67)$$

where k_f is changed permeability, k_o is original permeability, φ_f is changed porosity, φ_o is original porosity, and CK is Carman-Kozeny constant. Commercial simulators also use other relationships between porosity and permeability as follows:

$$\frac{\varphi_f}{\varphi_o} = \frac{k_f}{k_o} \qquad (4.68)$$

$$\frac{\varphi_f}{\varphi_o} = \ln\left(\frac{k_f}{k_o}\right) \qquad (4.69)$$

$$k_f = k_o \exp\left(k_{mul}\frac{\varphi_f - \varphi_o}{1 - \varphi_o}\right) \qquad (4.70)$$

where k_{mul} is permeability multiplier factor.

The change in porosity is usually calculated by the amount of the adsorbed bioproducts. If the products are insoluble, it can also be calculated using the products' volume generated in pore space as follows:

$$\varphi_f = \varphi_o\left(1 - \frac{c_s}{\rho_s}\right) \qquad (4.71)$$

where c_s is the concentration of the insoluble products, and ρ_s is the density of insoluble products.

4.2.2 Interfacial Tension Reduction

The main mechanism of the biosurfactant is to reduce the interfacial tension (IFT) between the oil and water interfaces, just like a typical chemical surfactant. In numerical simulations, the direct input of experimental results and then interpolation (or extrapolation) is the easiest and most common method to calculate the IFT changes. However, this section introduces some of the relationships between IFT and surfactant solubilization, and the capillary number associated with displacement efficiency.

Healy and Reed (1997a,b) suggested the correlation by fitting experimental results as follows:

$$\log \sigma_{ws} = H_{wo} + \frac{H_{ww}}{1 + H_{ws}R_{ws}} \qquad (4.72a)$$

and

$$\log \sigma_{os} = H_{oo} + \frac{H_{ow}}{1 + H_{os}R_{os}} \qquad (4.72b)$$

where σ is the IFT, H is the fitting parameter, and R_{ws} and R_{os} are the solubilization ratios ($\frac{c_{wsurf}}{c_{surf}}$ and $\frac{c_{osurf}}{c_{surf}}$, respectively). The subscript w means water phase, o means oil phase, and $surf$ means surfactant. However, this method has a problem that it is difficult to determine fitting parameters without proper experimental data.

Huh (1979) proposed a theoretical relationship between the IFT and solubilization parameter for microemulsion as follows:

$$\sigma_{mw} = \frac{C_H \cos\left(\frac{\pi}{2}\phi_w\right)}{\left(\frac{V_{wm}}{V_{sm}}\right)^2} \qquad (4.73a)$$

and

$$\sigma_{mo} = \frac{C_H \cos\left(\frac{\pi}{2}\phi_o\right)}{\left(\frac{V_{om}}{V_{sm}}\right)^2} \qquad (4.73b)$$

where V_{sm} is the surfactant volume in the microemulsion phase, C_H is an empirical constant, ϕ_w is $\frac{V_{wm}}{V_{wm}+V_{om}}$, and ϕ_o is $\frac{V_{om}}{V_{wm}+V_{om}}$. The subscript m means microemulsion phase.

Reduction of residual oil saturation by biosurfactant effect is closely related to capillary number. The capillary number is the ratio of viscous to capillary force (Moore and Slobod, 1955) defined as follows:

$$N_c = \frac{F_v}{F_c} = \frac{\nu\mu_{\text{displacing}}}{\sigma\cos\theta} \qquad (4.74)$$

where N_c is the capillary number, F_v is the viscous force, F_c is the capillary force, v is the pore flow velocity of the displacing fluid, $\mu_{displacing}$ is the displacing fluid viscosity, and σ is the interfacial tension between the displacing and displaced fluids. This capillary number calculated by (Eq. 4.74) is a semi-empirical parameter.

Another type of capillary number, developed by Foster (1973) and Green and Willhite (1998), uses interstitial velocity and ignores the $\cos\theta$ term as follows:

$$N_c = \frac{u\mu_{displacing}}{\varphi\sigma} \quad (4.75)$$

where u is the Darcy velocity of the displacing fluid, φ is the porosity, and $\frac{u}{\varphi}$ is defined as interstitial velocity. In a simpler form, the Darcy velocity is used as follows:

$$N_c = \frac{u\mu_{displacing}}{\sigma} \quad (4.76)$$

The capillary number at which residual oil starts to move is defined as the critical capillary number, $(N_c)_c$. To extract this residual oil, the value of N_c must be greater than $(N_c)_c$. In general waterflooding case, the injection velocity is 1 ft/day (3.528×10^{-6} m/s), the water viscosity is 1 mPa·s, and the interfacial tension is 30 mN/m. The corresponding capillary number is calculated to be approximately 10^{-7} by (Eq. 4.76).

The typical relationship between residual saturation of aqueous or non-aqueous phase and local capillary number is represented by capillary desaturation curve (CDC). The normalized residual saturation ($\overline{S}_{pr}$) can be calculated as follows:

$$\overline{S}_{pr} \equiv \frac{S_{pr} - S_{pr}^{(N_c)_{max}}}{S_{pr}^{(N_c)_c} - S_{pr}^{(N_c)_{max}}} \quad (4.77)$$

where S_{pr} is residual saturation of phase p. In Fig. 4.3, there is no clear-cut for the $(N_c)_c$ from laboratory data, but there are gradual change regimes near $(N_c)_c$. As illustrated in Fig. 4.3, $(N_c)_c$ is larger than N_c of normal waterflooding condition in typical oil reservoir. At this situation, IFT decrease by biosurfactant increases the N_c value of waterflooding, thus enabling additional oil production.

4.2.3 Relative Permeability Change

A general relative permeability curve for water and oil is shown in Fig. 4.4. The relative permeability of a particular phase is a function of the saturation of its own saturation only. Delshad et al. (1987) suggested the relative permeability function for each phase as follows:

$$k_{rp} = k_{rp}^e (\overline{S}_p)^{n_p} \quad (4.78)$$

and

$$\overline{S}_p = \frac{S_p - S_{pr}}{1 - \sum S_{pr}} \quad (4.79)$$

where k_{rp} is the relative permeability of phase p, S_p is the saturation of phase p, k_{rp}^e is the end-point relative permeability of phase p at its maximum saturation, n_p is the exponent of phase p, $\overline{S}_p$ is the normalized saturation, and S_{pr} is the irreducible or residual saturation of phase p. The subscript p can be w (aqueous phase), o (oleic phase), or m (microemulsion phase). These parameters are highly dependent on the capillary number. If the capillary number changes, these values must also be recalculated by interpolation or extrapolation.

In the case of two-phase flow system, (Eq. 4.79) becomes:

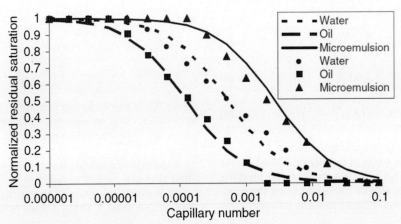

FIG. 4.3 Normalized capillary desaturation curve. The lines are the results calculated by the CDC model and the points are the experimental results. (Credit: from Sheng, J.J., 2011. Modern Chemical Enhanced Oil Recovery: Theory and Practice. Gulf Professional Publishing, Houston, Texas, USA.)

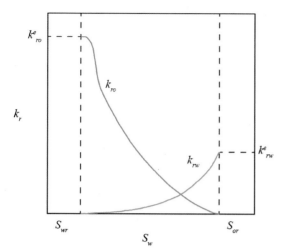

FIG. 4.4 Relative permeability curve for water and oil. (Credit: self editing)

$$\overline{S}_p = \frac{S_p - S_{pr}}{1 - S_{pr} - S_{p'r}} \qquad (4.80)$$

where $S_{p'r}$ is the residual saturation of the other phase. In this equation, S_{pr} and $S_{p'r}$ are dependent on N_c and can be estimated by:

$$S_{pr} = S_{pr}^{(N_c)_{max}} + \frac{S_{pr}^{(N_c)_c} - S_{pr}^{(N_c)_{max}}}{1 + C_p N_c} \qquad (4.81)$$

where $(N_c)_{max}$ is the maximum desaturation capillary number, above which the residual saturation do not decrease in practical conditions even if the capillary number is increased, and C_p is the parameter used to fit the laboratory measurements.

The end-point relative permeability is assumed that it depends on the residual saturation of the other phase. The k_{rp}^e at any N_c can be interpolated between those at the $(N_c)_c$ and $(N_c)_{max}$ as follows:

$$k_{rp}^e = \left(k_{rp}^e\right)^{(N_c)_c}$$
$$+ \frac{(S_{p'r})^{(N_c)_c} - S_{p'r}}{(S_{p'r})^{(N_c)_c} - (S_{p'r})^{(N_c)_{max}}} \left[\left(k_{rp}^e\right)^{(N_c)_{max}} - \left(k_{rp}^e\right)^{(N_c)_c} \right] \quad (4.82)$$

Similarly, the exponent of phase p (n_p) in (Eq. 4.78) is estimated as follows:

$$n_p = n_p^{(N_c)_c} + \frac{(S_{p'r})^{(N_c)_c} - S_{p'r}}{(S_{p'r})^{(N_c)_c} - (S_{p'r})^{(N_c)_{max}}} \left[n_p^{(N_c)_{max}} - n_p^{(N_c)_c} \right] \quad (4.83)$$

These relationships assume that k_{rp}^e and n_p depends on the residual saturation of conjugate phase through

linear interpolation, although the assumption may not be exactly correct (Fulcher et al., 1985; Anderson, 1987; Masalmeh, 2002; Tang and Firoozabadi, 2002).

Amaefule and Handy (1982) showed two empirical equations to describe water and oil relative permeabilities at decreased IFT. In their equations, water relative permeability is a function of its own saturation and reduced residual saturation at low IFT. However, the oil relative permeability is a function of the saturations and reduced residual saturations of water and oil. The models are expressed as follows:

$$k_{rw}(S_w, \sigma) = B S_{wr(\sigma)} \left(\frac{S_w - S_{wr(\sigma)}}{1 - S_{wr(\sigma)}} \right)^\delta + \left(1 - B S_{wr(\sigma)}\right) \frac{S_w - S_{wr(\sigma)}}{1 - S_{wr(\sigma)}}$$

$$(4.84)$$

and

$$k_{ro}(S_w, \sigma) = A S_{or(\sigma)} \left(\frac{1 - S_{or(\sigma)} - S_w}{1 - S_{or(\sigma)} - S_{wr(\sigma)}} \right)^\varepsilon$$
$$+ \left(1 - A S_{or(\sigma)}\right) \frac{1 - S_{or(\sigma)} - S_w}{1 - S_{or(\sigma)} - S_{wr(\sigma)}} \qquad (4.85)$$

where A, B, δ, and ε are constants.

The last model is proposed by Bang and Caudle (1984) which describes the relative permeability curve depending on IFT as follows:

$$k_{rp}(S_p, \sigma) = k_{rp}(S_p, \sigma_{max}) + [k_{rp}(S_p, \sigma_{min}) - k_{rp}(S_p, \sigma_{max})]$$
$$\times] \frac{\sigma_{max} - \sigma(c_{surf})}{\sigma_{max} - \sigma_{min}}$$

$$(4.86)$$

where σ_{min} and σ_{max} are minimum and maximum IFT from experimental measurements, respectively, and $\sigma(c_{surf})$ is the IFT at a given surfactant concentration.

4.3 APPLICATIONS OF NUMERICAL SIMULATION

4.3.1 Permeability Alteration

Islam (1990) used the flow equations that were solved in a three-phase and three-dimension for MEOR processes. The equations were calculated by IMPES (implicit pressure, explicit saturation) procedure. The bacterial growth was described by Monod equation. The bacteria transport was solved by implicit method, whereas the nutrient transport was solved by explicit method. This procedure was repeated until the convergence is obtained for each time step. The solution time became longer when there were permeability alterations.

The first run of numerical simulations was implemented using single-phase fluid. The bacterial plugging was investigated in three-dimensional case. The

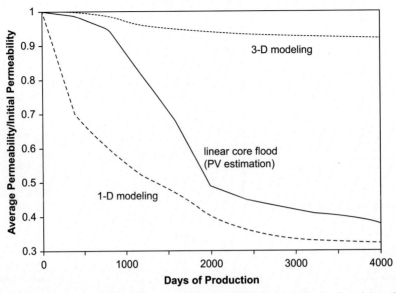

FIG. 4.5 Comparison of permeability reduction due to bacteria injection. (Credit: from Islam, M.R., 1990. Mathematical modeling of microbial enhanced oil recovery. Proceedings of the SPE Annual Technical Conference and Exhibition, New Orleans, LA, USA, 23–26 September, 1990.)

plugging occurred by two different ways such as shear multiplication of bacteria (Islam and Gianetto, 1993) or generation of in situ polymer. The modeling results of the first method were simulated in both three- and one-dimensional cases and compared with the linear core results by Islam and Gianetto (1993). All these simulations assumed that bacteria, nutrients, and chase water are injected in sequence. In one-dimensional analysis, bacteria plugging occurred rapidly in the first block and continued only in the first few blocks. It decreased the overall permeability drastically. In three-dimensional case, the local plugging problem was overcome, and overall permeability reduction was much lower than that of one-dimensional model. Also, correct paths of nutrients and bacteria could be tracked in three-dimensional simulation. When comparing these results with the result of linear core flood, the importance of time scaling in a laboratory study could be shown. Because laboratory experiments took a very short time, the tendency of time-dependent variables, such as bacterial growth, was different. Compared with the one-dimensional model, the residence time was shorter, and the plugging was not intense in the core model. The results for permeability reduction were shown in Fig. 4.5.

The simulation of in situ polymer generation was performed at the maximum growth rate of 6.5 per hour. In this case, the permeability reduction was

apparent compared with the model in which no polymer was generated. Permeability was reduced more intensively near the wellbore. The low permeability near wellbore was likely to cause injectivity problems in field situation (Fig. 4.6).

Chang et al. (1991) developed a three-dimensional, three-phase, multiple-component numerical model to describe the microbial transport in porous media. The models for bacterial transport were built in two steps. The first step predicted the propagation and distribution of bacteria and nutrients. This transport model included diffusion, adsorption, growth and decay of bacteria, and consumption of nutrients. The growth of microbes was assumed to follow the Monod equation, and the porosity and permeability alterations were occurred due to the cell deposition. This deposition was defined by clogging and declogging process. The clogging rate was the function of microbial concentration. The declogging rate was a function of adsorption saturation on rock surface (Corapcioglu and Haridas, 1984). The second step was to incorporate the transport model into the three-dimensional, three-phase (oil, water, and gas) black oil simulator. The pressure distribution and fluid saturations in porous media were calculated by continuity equation. The Darcy flow velocity was used in the calculation of bacteria and nutrients transport.

Fluid flow for Darcy velocity was solved using IMPES procedure. As no-flow boundary of the reservoir was

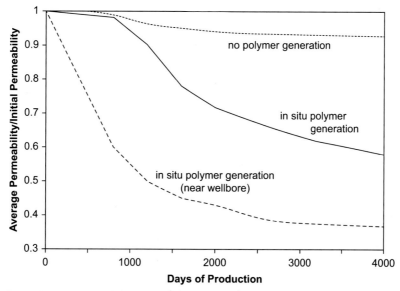

FIG. 4.6 Comparison of permeability reduction due to bacteria injection and in situ polymer generation. (Credit: from Islam, M.R., 1990. Mathematical modeling of microbial enhanced oil recovery. Proceedings of the SPE Annual Technical Conference and Exhibition, New Orleans, LA, USA, 23–26 September, 1990.)

assumed, the pressure gradient at boundary interfaces was set to zero. The concentrations of microbes and nutrients were calculated implicitly. The microbial deposition was then solved, and the permeability was changed for pressure calculations in the next time step.

Certain parameters such as diffusion and clogging coefficients of bacteria were required to apply the simulator. These parameters obtained from the experiments and simulation matches. From the microbial growth and nutrient consumption test, the maximum growth rate was 8.4 per day, and decay rate was 0.22 per day. The yield coefficient was 0.5 calculated by the nutrient consumption rate. The simulation parameters for diffusion, clogging, and declogging reactions were matched to core-flood experiments. As a result, the diffusion coefficient was 8.93×10^{-5} cm^2/s (or 8.3×10^{-3} ft^2/day). From the microbial effluent matching results, the clogging and declogging rates were determined as 25 and 37 per day, respectively. The pressure profile of core-flood experiment was matched for permeability reduction. The simulation results showed the dramatic increase of pressure at the early time due to the microbial plugging. As the waterflooding was implemented, declogging of the microbes occurred, and the reduced permeability was a little recovered.

Field-scale simulations were implemented to analyze the microbial transport. The microbial clogging and chemotaxis in transports were simulated in a one-dimensional reservoir model. The application of

microbial system for various injection profiles was conducted in a two-dimensional cross section model. From the simulation results, the clogging or adsorption of microbes to the rock surface had critical effect on the transport of microorganisms in the reservoir (Fig. 4.7). To confirm the chemotaxis effect, the results according to the chemotaxis coefficient were shown in Fig. 4.8. Due to the high nutrient concentration near the injector, the microbial concentration was higher near the wellbore with high chemotaxis coefficient (k_{ch}). Based on the microbial transport models, the effects of selective plugging on oil productivity were analyzed. Simulations were carried out on a two-layer cross section model. The top layer had a permeability of 1000 md and thickness of 20 ft. The bottom layer had a permeability of 100 md and thickness of 10 ft. The bacteria and nutrients were injected after water breakthrough from the high-permeability layer. The microbial plugging near the injector in high-permeability layer diverted the injected water into the low-permeability layer (Fig. 4.9). Compared with the case without microbial treatment, the water oil ratio (WOR) was reduced (Fig. 4.10). Therefore, oil recovery during the waterflooding was increased. The fraction of water injection into the low-permeability layer decreased during the later stage of waterflooding due to microbial declogging and the stop of nutrient injection.

Zhang et al. (1992) developed a mathematical models including convection-dispersion equations for

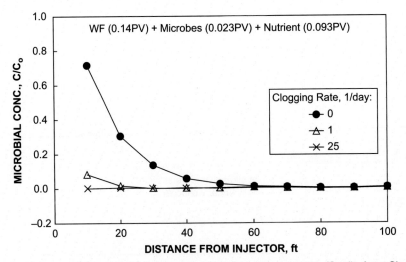

FIG. 4.7 Microbial profiles in reservoir with various clogging rate constants. (Credit: from Chang, M.M., Bryant, R.S., Chung, T.H., Gao, H.W., Burchfield, T.E., 1991. Modeling and laboratory investigations of microbial transport phenomena in porous media. Proceedings of the SPE Annual Technical Conference and Exhibition, Dallas, Texas, USA, 6–9 October, 1991.)

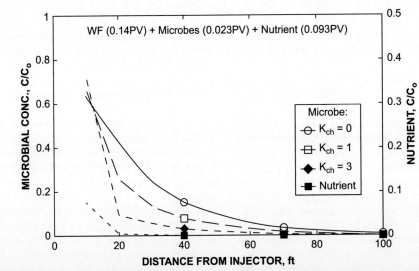

FIG. 4.8 Effect of chemotaxis coefficient on microbial transport. (Credit: from Chang, M.M., Bryant, R.S., Chung, T.H., Gao, H.W., Burchfield, T.E., 1991. Modeling and laboratory investigations of microbial transport phenomena in porous media. Proceedings of the SPE Annual Technical Conference and Exhibition, Dallas, Texas, USA, 6–9 October, 1991.)

transport, kinetics for microbial growth and metabolism, empirical formula for permeability alteration, and continuity equations for solving the pressure and saturations. The model had some assumptions. They assumed one-dimension horizontal linear flow. The model was homogeneous, isotropic, and incompressible porous media. Three-phase Newtonian fluids such as water, oil, and gas were considered. Both glucose and ammonium nitrate were growth-limiting factors in Monod equation. The microbes were partitioned into two phases such as planktonic phase flowing in aqueous phase and sessile phase retained on rock surfaces. Finally, they assumed that anaerobic bacterial growth occurred in both the planktonic and sessile phases.

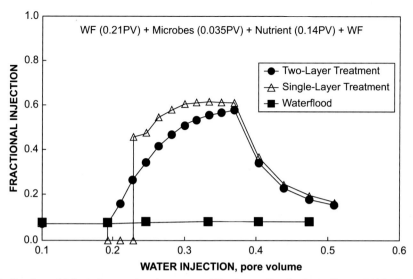

FIG. 4.9 Fraction of injected water into low-permeability layer. (Credit: from Chang, M.M., Bryant, R.S., Chung, T.H., Gao, H.W., Burchfield, T.E., 1991. Modeling and laboratory investigations of microbial transport phenomena in porous media. Proceedings of the SPE Annual Technical Conference and Exhibition, Dallas, Texas, USA, 6–9 October, 1991.)

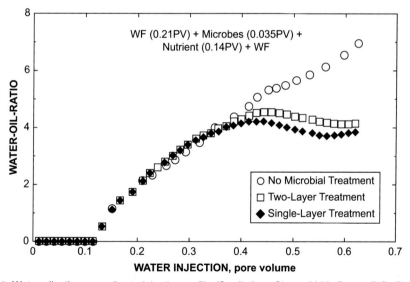

FIG. 4.10 Water-oil ratios according to injection profile. (Credit: from Chang, M.M., Bryant, R.S., Chung, T.H., Gao, H.W., Burchfield, T.E., 1991. Modeling and laboratory investigations of microbial transport phenomena in porous media. Proceedings of the SPE Annual Technical Conference and Exhibition, Dallas, Texas, USA, 6–9 October, 1991.)

The developed model used experimental data from laboratory results of bacterial growth in static sand pack. The maximum specific growth rate was 0.602 per hour, and yield coefficients for microbial metabolisms were calculated from the chemical balances for stoichiometric equations. Fig. 4.11 showed the comparison results of simulation with experiments.

The mathematical model also tested in the MEOR core flooding. The results for different types of permeability reduction were shown in Fig. 4.8. This model

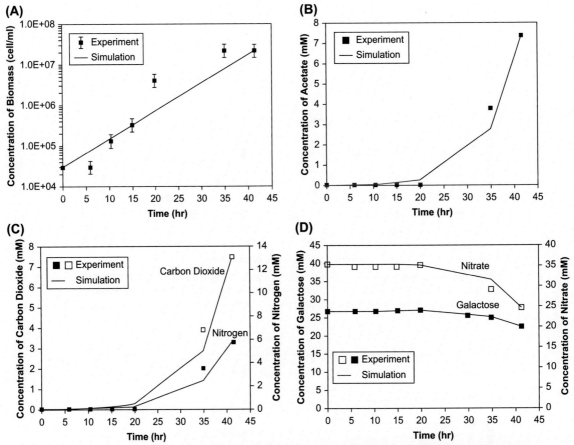

FIG. 4.11 Comparisons between simulation and experiment: **(A)** biomass growth, **(B)** acetate production, **(C)** gas production, **(D)** nutrient consumption. (Credit: from Zhang, X., Knapp, R.M., McInerney, M.J., 1992. A mathematical model for microbially enhanced oil recovery process. Proceedings of the SPE/DOE Enhanced Oil Recovery Symposium, Tulsa, Oklahoma, USA, 22−24 April, 1992.)

considered pore-surface retention and/or pore-throat plugging. To describe the pore-throat plugging effect, a flow-efficiency coefficient (f) was introduced into the porosity-permeability correlation as follows:

$$\frac{k_f}{k_o} = f\left(\frac{\varphi_f}{\varphi_o}\right)^3 \tag{4.87}$$

In Fig. 4.12, top solid line ($f=1$) represented the permeability reduction due to only pore-surface retention. The middle line ($f<1$, $C=0$) considered both pore-surface retention and pore-throat plugging, whereas the sessile phase plugging on pore-throat was ignored. The bottom line ($f<1$, $C>0$) included all the type of plugging. The comparison results indicated that the consideration of both pore-surface retention and pore-throat plugging due to the planktonic phase

and in situ bacterial growth was a more effective approach to build the modified permeability model.

The microbial models have been coupled with UTCHEM, a three-dimensional chemical flooding simulator developed by the University of Texas at Austin. The simulator includes the microbial processes such as the destruction of substrate, consumption of electron acceptors, and bacterial growth. The mass transfer models considering multiple substrates, electron acceptors, and biological species are used. There are important general assumptions for the microbial model. Firstly, the microbial reactions only occur in the aqueous phase. The microbes are fully penetrated, which means that there is no internal resistance to mass-transfer within the attached biomass. The initial biomass is uniformly distributed in the porous

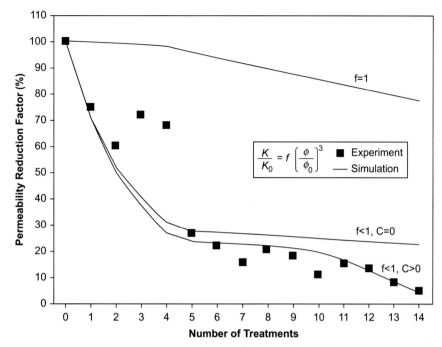

FIG. 4.12 The permeability reduction due to the plugging types. (Credit: from Zhang, X., Knapp, R.M., McInerney, M.J., 1992. A mathematical model for microbially enhanced oil recovery process. Proceedings of the SPE/DOE Enhanced Oil Recovery Symposium, Tulsa, Oklahoma, USA, 22–24 April, 1992.)

medium. The biomass is prevented from being destroyed below the lower limit by metabolism unless the concentration of active biomass is lowered below the natural state by the cometabolic reaction. The area available for oleic component transport into the attached biomass is directly proportional to the biomass quantity. The number of cells per microcolony, density of biomass, and volume of microcolony are constant. The microbial reactions are mutually independent unless they are related to competition or inhibition terms. The adsorption of biomass to the rock surface can be described by equilibrium partitioning. Finally, the chemical species of the attached biomass do not adsorb to aquifer solids.

This simulator can select the biokinetics such as Monod, first-order, or instantaneous model. It follows the first-order abiotic decay reactions and includes the option for external mass transfer resistances to microcolonies. It can also describe the enzyme competition between multiple substrates and the inhibition effect by electron acceptors and/or toxic substrates. The porosity is changed by the growth and retention of bacteria, and the permeability alteration is recalculated by the Carman-Kozeny equation.

Delshad et al. (2002) validated this MEOR model. To test the model, they compared the simulation results with the experimental results of Silfanus (1990). The core was saturated with crude oil to residual water saturation and then followed by brine flooding to residual oil saturation. After these steps, the core was treated with one or several incubations of bacteria. The experimental results included cumulative oil recovery, brine flow rates, and the pressure drop to define the effective brine permeability. The effective permeability reduction factor (PRF, $\frac{k_f}{k_o}$) was the ratio of the brine permeability for each microbial treatment to that at residual oil saturation. The PRF and oil recovery comparisons were represented in Fig. 4.13. These results indicated that UTCHEM could qualitatively simulate the permeability reduction mechanism, which occur during the MEOR process.

Stewart and Kim (2004) developed biomass accumulation and evolution model in porous media using a combination of biofilm evolution and removal model. These models were used to describe the plugging and removal of biomass. The biomass evolution model calculated the bacterial growth and exopolymer generation. The evolution model based on experimental results of microbial distribution in

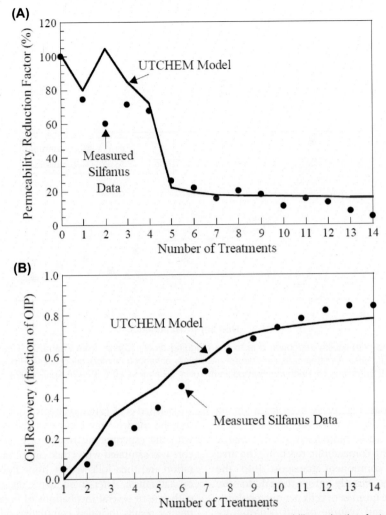

FIG. 4.13 Comparisons of measured and simulated results: **(A)** permeability reduction factor and **(B)** oil recovery. (Credit: from Delshad, M., Asakawa, K., Pope, G.A., Sepehrnoori, K., 2002. Simulations of chemical and microbial enhanced oil recovery methods. Proceedings of the SPE/DOE Improved Oil Recovery Symposium, 13–17 April, 2002.)

micromodel, which showed the development and propagation of microbial plugging (Fig. 4.14). The initial cell attachment in this model was followed by cell growth and biopolymer production. This attachment was a reversible and irreversible reaction, which occurred as the microbes in the bulk phase transported to the pore-throat surfaces. To simulate the microbial kinetics, mass balances and rate laws from Lappan and Fogler (1994) were used.

Fig. 4.15 showed the process of evolution and removal model of biofilm. To calculate the biofilm evolution process in pore throat, the cell and associated

biopolymers were assumed to be a homogeneous solid, which occupied a volume approximated by combining the biomass and biopolymer volumes. The volume of biofilm was calculated at the end of each time step as follows:

$$V_{\text{biofilm}} = \left(\frac{c_{att} m_{\text{cell}} + c_{\text{poly}}}{\rho_{\text{biofilm}}} \right) \frac{\pi D_0^2 l}{4} \qquad (4.88)$$

where V_{biofilm} is the volume of biofilm, c_{att} is the attached cell concentration, m_{cell} is the mass of bacterial cell, c_{poly} is the biopolymer concentration, ρ_{biofilm} is biofilm density, D_0 is the initial diameter of pore throat, and l is

(A)

(B)

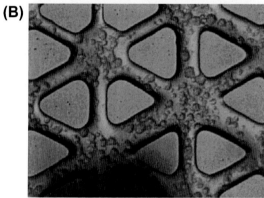

FIG. 4.14 Biomass distribution: **(A)** during exopolymer-induction phase, **(B)** during transition from induction to plugging phase. (Credit: from Stewart, T.L., Fogler, H.S., 2001. Biomass plug development and propagation in porous media. Biotechnology and Bioengineering 72 (3), 353–363.)

the length of pore throat. The changed diameter (d_1) at next time step (t_1 stage in Fig. 4.15) was solved as follows:

$$V_{\text{biofilm}} = \frac{\pi (D_0^2 - D_1^2) l}{4} \qquad (4.89a)$$

and

$$D_1 = \sqrt{D_0^2 - \frac{4V_{\text{biofilm}}}{\pi l}} \qquad (4.89b)$$

In flowing condition, microbial cells are continuously detached by shear stress and decay due to biofilm formation. This detachment has been simulated by the gradual reduction of biofilm layer thickness method (Rittmann, 1982), focusing on sloughing of biofilms. The biomass removal model solved the plugging propagation and channel breakthrough using the Bingham

yield stress model, which described the biofilm stability against shear stress as follows:

$$\tau_s = \frac{\Delta p_{\max} A_{ave}}{A_{cum}} = \frac{\Delta p_{\max} D_{\text{bond}}}{4 l_{\text{network}}} = \tau_B \qquad (4.90)$$

where τ_s is the shear stress, τ_B is the Bingham yield stress, $\Delta p_{\max}$ is the maximum pressure drop in the pore throat, A_{ave} is the average cross-sectional area along a breakthrough channel, A_{cum} is the cumulative wall surface area along a breakthrough channel, D_{bond} is the diameter of network bond, and l_{network} is the length of network. Therefore, neither individual cell detachment nor gradual layer removal was not considered in this model. The accumulation of biofilm continued until biofilm could withstand shear stress, and the diameter of the pore throat continued to decrease ($\tau_s < \tau_B$). When the shear stress exceeded the biofilm yield stress ($\tau_s \geq \tau_B$), the entire biofilm in the pore throat was detached and sloughed (t_2 stage in Fig. 4.15).

They chose a network model over an empirical model for two reasons in this study. First, the network model was easy to simulate microbial experiments by describing physical heterogeneities such as pore size distribution of porous media. Second, it could easily describe the biomass growth and removal in the pore throat and the subsequent change in diameter of the pore throat by treating the bacterial cells as discrete units. The network model was constructed of pore body and pore throat (pore bond), and their size was determined by the pore size distribution of the ceramic cores. The simulation results showed that the developed model could predict the permeability reduction by microbial kinetics (Fig. 4.16). The pressure drop indicated the process of biomass accumulation and channel breakthrough. They expected to extend the application areas of the model such as MEOR, in situ bioremediation, and biobarrier techniques.

Various reactive transport models (such as Geochemist's Workbench (GWB), Hydrogeochem, Bioslurp, CrunchFlow, etc.) can be used to describe microbial and geochemical reactive transport in porous media. Vilcáez et al. (2013) simulated the selective plugging in carbonate formation using CrunchFlow, which is advantageous for processing multicomponent geochemical and biological reactions in heterogeneous system (Steefel, 2009). The main functions of the CrunchFlow code include the complex calculation of advective, dispersive and diffusive transport, and chemical reaction processes in three dimensions. Simulations are possible to take into account microbial reactions with Monod equations,

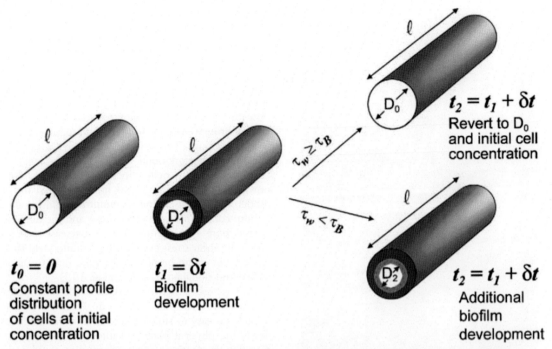

FIG. 4.15 Biofilm evolution and removal model. (Credit: from Stewart, T.L., Kim, D.S., 2004. Modeling of biomass-plug development and propagation in porous media. Biochemical Engineering Journal 17 (2), 107–119.)

aqueous and surface complexation reactions, sorption and desorption, and mineral precipitation and dissolution. This code is based on the standard EQ3/EQ6 thermodynamic database (Wolery et al., 1990).

The reaction network in their selective plugging model included biogeochemical reactions, mineral dissolution and precipitation reactions, and aqueous speciation reactions. The bacterial growth was formulated following the Rittmann and McCarty (2001) method, which was based on bacteria energetics and experiments. In formulating the chemical reactions by microorganisms in stoichiometric equations, it is very important to consider the mass and electron equivalents between the fermentation of nutrients by bacteria and the resulting end products. In their study, the stoichiometric equation of growth reaction was based on the sucrose fermentation equation determined by Dols et al. (1997). The bacteria growth was represented as follows:

$$0.2083C_{12}H_{22}O_{11} + 0.2NH_4^- + 0.341HCO_3^- \rightarrow$$

$$0.2C_5H_7O_2N + 0.14076C_6H_{12}O_6 + 0.0705CH_3CHOHCOO^- +$$

$$0.0705CH_3COO^- + 0.01408C_6H_{14}O_6 + 0.0705CH_3CH_2OH +$$

$$0.4184CO_2 + 0.7259H_2O$$

$$(4.91)$$

where $C_{12}H_{22}O_{11}$ is sucrose, $C_5H_7O_2N$ is bacteria (*Leuconostoc mesenteroides*), $C_6H_{12}O_6$ is fructose, $CH_3CHOHCOO^-$ is lactate, CH_3COO^- is acetate, $C_6H_{14}O_6$ is mannitol, and CH_3CH_2OH is ethanol. The Michaelis–Menten reaction model was used to determine the stoichiometry for biopolymer (dextran) production. It was mainly used in the form suggested by Santos et al. (2000) as follows:

$$nC_{12}H_{22}O_{11} \rightarrow (C_6H_{10}O_5)_n + nC_6H_{12}O_6 \quad (4.92)$$

where $(C_6H_{10}O_5)_n$ is biopolymer (dextran) and the value of n is determined by molecular weight of generated biopolymer.

The simulations were implemented to investigate the effect of selective plugging in high-permeability zones in a carbonate reservoir. The results showed that biopolymer generation and plugging effectiveness could be significantly affected by injection rates of bacteria and nutrient (Fig. 4.17). The selective plugging of high-permeability zones occurred when the injection rates were higher than the biopolymer generation rates. Otherwise, plugging only occurred near the injection wells. The chemistry of injection and formation water was also important factor due to the pH-dependent enzyme

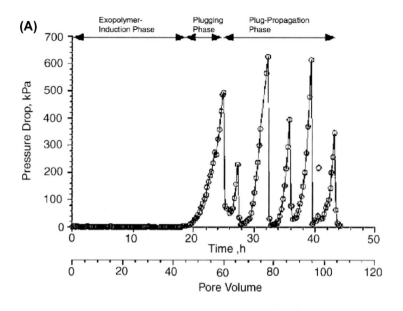

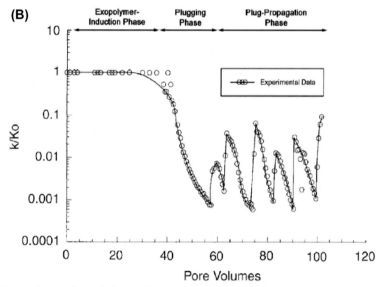

FIG. 4.16 Comparisons of simulation and experimental results: **(A)** pressure oscillation, **(B)** permeability reduction. (Credit: from Stewart, T.L., Kim, D.S., 2004. Modeling of biomass-plug development and propagation in porous media. Biochemical Engineering Journal 17 (2), 107–119.)

activity and the intrinsic reactivity of carbonate rocks. Injection at the optimum pH of 5.2 for dextran production increased the pH level due to calcite dissolution. This increase did not inhibit plugging with dextran because lactic acid and CO_2 generation during microbial growth buffered the water pH levels between 5.2 and 7.0 for continued biopolymer production (Fig. 4.18). At carbonate concentrations found in common oil reservoirs,

calcite precipitation induced at neutral and basic pH levels did not cause significant permeability alterations.

As mentioned earlier, reactive transport models are a powerful tool for understanding the flow, transport, and reactions of bacteria, the physical properties of the reservoir, and the effectiveness of the plugging process. Therefore, Surasani et al. (2013) followed Vicáez et al. (2013) to simulate the microbial plugging with CrunchFlow.

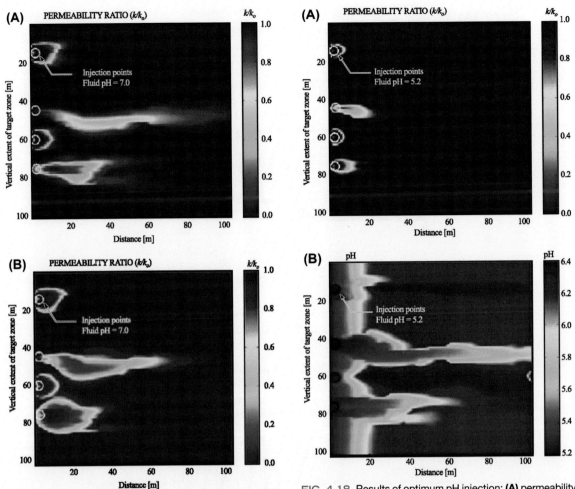

FIG. 4.17 Permeability alterations after 250 days of flooding with 1 g/L of bacteria and 25 g/L of nutrient injection under different injection rates: **(A)** 50 bbl/day, **(B)** 100 bbl/day. (Credit: from Vilcáez, J., Li, L., Wu, D., Hubbard, S.S., 2013. Reactive transport modeling of induced selective plugging by Leuconostoc mesenteroides in carbonate formations. Geomicrobiology Journal 30 (9), 813–828.)

FIG. 4.18 Results of optimum pH injection: **(A)** permeability profile, **(B)** pH profile. (Credit: from Vilcáez, J., Li, L., Wu, D., Hubbard, S.S., 2013. Reactive transport modeling of induced selective plugging by Leuconostoc mesenteroides in carbonate formations. Geomicrobiology Journal 30 (9), 813–828.)

They modeled a reservoir based on the King Island gas field in California (Fig. 4.19). The distributions of porosity and permeability were obtained by open-hole logging data. The reaction network, biopolymer (dextran) generation kinetics, and petrophysical relationships were validated using column-scale experiments (Wu et al., 2014). They analyzed microbial growth variables such as sucrose mass injection rate and attachment rates.

The reaction network was constructed identically to the method used by Vilcáez et al. (2013). The microbial cells can attach to the rock surface and therefore can

exist in both aqueous and solid phases. Dextran is generally present in solid form because of its relatively low solubility. In flow conditions, however, the immobilized dextran may detach and be transported in insoluble form. Therefore, in this study, bacteria and dextran were assumed as both aqueous and solid phases to describe the attachment and detachment.

Simulation results showed that there is an optimal injection rate for microbial plugging (Fig. 4.20). The injection rates needed to be fast enough to suppress microbial growth and plugging at the vicinity of the wellbore. They should also be able to ensure sufficient

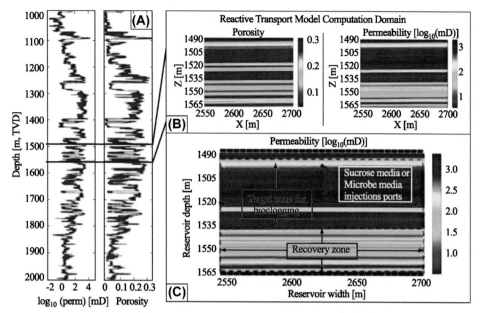

FIG. 4.19 Distribution of porosity and permeability from logging dat in King Island gas field. (Credit: from Surasani, V.K., Li, L., Ajo-Franklin, J.B., Hubbard, C., Hubbard, S.S., Wu, Y., 2013. Bioclogging and permeability alteration by L. mesenteroides in a sandstone reservoir: a reactive transport modeling study. Energy & Fuels 27 (11), 6538–6551.)

residence time for dextran production in reservoir. The dextran production and associated permeability alterations showed the water diversion to low-permeability zone. When the total mass of the sucrose to be injected was same, the efficiency of selective plugging was better when increasing the flow rate than increasing the sucrose concentration (Fig. 4.21).

MEOR modeling using commercial programs has also been performed. Jeong et al. (2019b) reported a mechanistic modeling approach for applying the in situ microbial selective plugging reservoir and optimizing the injection strategies for successful MEOR implementations in a pilot-scale reservoir. They carried out these numerical processes with a commercial reservoir simulator, CMG STARS. The microbial growth formulas were described using stoichiometric equations, and growth kinetics were constructed based on Arrhenius equation. The kinetic parameters for microbial activities were obtained from batch experiment data where the bacteria were cultured. The bacterial growth and biopolymer production were modeled using the parameters (Fig. 4.22). Thereafter, the permeability reduction simulation was verified with experimental test results (Noh et al., 2016) (Fig. 4.23).

Based on laboratory scale presimulation, the MEOR process was carried out in a pilot-scale system, which

had a high-permeability zone in low-permeability reservoir. The synthetic reservoir was two-dimensional layered system (Fig. 4.24). The reservoir was constructed with two permeability zone; the fifth and sixth layers were high-permeability area with 1 D (1000 md), and the other layers were low-permeability area with 0.05 D (50 md). This type of reservoir can cause early water breakthrough problems, which are serious conformance problems. The entire production was implemented for 2 years. First, the initial waterflooding was carried out for 1 year. The nutrient injection for MEOR treatment was then applied for 4 weeks. Thereafter, the waterflooding was reimplemented during the remaining period. In this MEOR treatment, the injection rate was 8 m³/day, and nutrient concentration was 0.005 mol fraction. The MEOR design improved oil recovery factor by 61.5% compared with waterflooding (Fig. 4.25). By setting various nutrient concentration, injection rate, and injection period as injection parameters, they confirmed the increase in oil recovery for each scenario (Fig. 4.26). The parametric simulations showed that the recovery efficiency was affected by the amount of biopolymer, as well as the distributions, and thus, the injection strategy was critical for determining the biopolymer distribution (Fig. 4.27). Additionally, they performed sensitivity analysis and

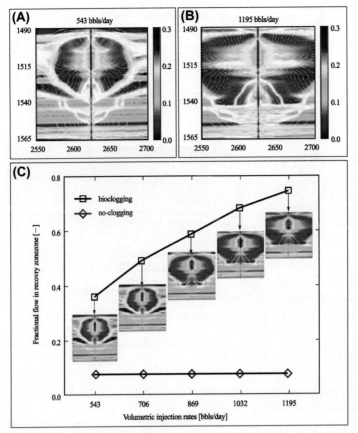

FIG. 4.20 Hydrodynamic alteration and the effectiveness of the microbial plugging under different injection conditions. (Credit: from Surasani, V.K., Li, L., Ajo-Franklin, J.B., Hubbard, C., Hubbard, S.S., Wu, Y., 2013. Bioclogging and permeability alteration by L. mesenteroides in a sandstone reservoir: a reactive transport modeling study. Energy & Fuels 27 (11), 6538–6551.)

optimization for maximizing the oil recovery. The multiple injection parameters such as nutrient concentration, flow rate, and injection period were considered to maximize the MEOR efficiency. The results showed a significant improvement of oil recovery by the optimization (Fig. 4.28).

A novel concept for microbial plugging process under nonisothermal conditions was suggested by Hong et al. (2019b). The model set the temperature-triggered biopolymer generated by microbes. The effects of environmental temperature on bacterial metabolisms were validated and calibrated by the cardinal temperature model (Fig. 4.29). They also used STARS program for overall numerical modeling. Batch and sand-pack simulations were performed in the same way as Jeong et al. (2019b). Each parameter for the nonisothermal parameters was gained by history matching

method against experimental results. With this process, quantification of the temperature effect on bacterial activities was established (Fig. 4.30).

A two-dimensional model was used to examine the effectiveness of thermally active MEOR. The multilayered system included a high permeability zone with 1500 md (Fig. 4.31). The rock heat capacity and the conductivity of surrounding formations were applied to heat transfer across the boundaries (Robertson, 1988). Recovery process was conducted in three steps to investigate thermally active MEOR for 2000 days. The first waterflooding was applied to not only increase the oil recovery but also organize the proper temperature environment for bacterial growth. The nutrient was then introduced for 60 days. Finally, the second waterflooding was conducted for 940 days. The injection rate was 15 m^3/day. They analyzed the

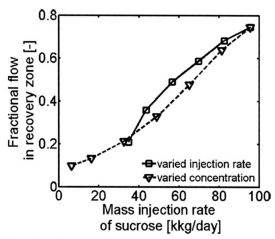

FIG. 4.21 Fractional flow in recovery zone as a function of flow rates and sucrose concentration. (Credit: from Surasani, V.K., Li, L., Ajo-Franklin, J.B., Hubbard, C., Hubbard, S.S., Wu, Y., 2013. Bioclogging and permeability alteration by L. mesenteroides in a sandstone reservoir: a reactive transport modeling study. Energy & Fuels 27 (11), 6538–6551.)

effects of injection parameters such as nutrient concentration, flow rate, and temperature on oil recovery. Higher concentration of nutrient induced the increasing bio-polymer production, thereby increasing oil recovery. In the other hand, injection rate and temperature affected biopolymer distributions. Higher flow rates led to lower polymer concentration with wider distributions. The injection temperature controlled the trapped biopolymer location in porous media. Through the optimization process, the optimal injection strategy was derived (Fig. 4.32), and the oil recovery was found to increase considerably. Furthermore, they conducted the thermally active MEOR in King Island field model (Surasani et al., 2013). The result showed that injection with low-temperature in high-temperature reservoir increased the oil recovery efficiency (Fig. 4.33).

4.3.2 Interfacial Tension Reduction

Modeling of microbial behavior and reactions in bio-surfactant simulation is similar as selective plugging simulation. The only difference is that the change in IFT is considered in the model. Islam (1990) assumed the IFT as a function of microbial concentration to simulate the bacteria-generated surfactant flooding (Fig. 4.34). The relative permeability curves were formed based on Bang and Caudle method. In this method, the relative permeability curves of water and oil were assumed to be a straight line when IFT was zero. The reservoir model was shown in Fig. 4.35.

Permeability of x- and y-direction was 5 D, and that of z-direction was 1D. In this model, IFT between oil and water was 35 dyne/cm. The surfactant-generating bacteria were introduced after waterflooding when the oil cut was less than 5%. To minimize the microbial plugging effect, the deposition rate was assumed to be very low. The injection and production rate was constant of 50 m^3/day. After 20% of pore volume of bacteria solution was injected, nutrient was then injected. The results demonstrated that surfactant-generating bacteria were effective in increasing oil recovery (Fig. 4.36). The technique was also applied to the Huff and Puff process. Only one well was used for this process. Bacteria was injected at 50 m^3/day for 10 days, and then nutrient was injected at 50 m^3/day for 10 days. Following this, the well was shut-in for 20 days to cultivate. Production then proceeded for 40 days. The result showed that Huff and Puff process had much quicker recovery response but smaller recovery amount than the drive process (Fig. 4.37).

Desouky et al. (1996) developed a one-dimensional model to simulate the microbial surfactant effect on enhanced oil recovery. The model included five components (such as water, oil, bacteria, nutrient, and metabolites) with adsorption, diffusion, chemotaxis, growth and decay of microbes, nutrient consumption, and changes of porosity and permeability. Several correlations between experimental data and numerical model were used in this simulator. The experiments were conducted with three bacterial strains such as *Streptococcus* (O_{6a}), *Staphylococcus* (O_9), and *Bacillus* (O_{12}), which were isolated from Saudi crude oil and formation water. In addition, two kinds of nutrients, glucose and molasses, were used to establish six cases: (1) O_{12} in glucose, (2) O_{12} in molasses, (3) O_9 in molasses, (4) O_{6a} in molasses, (5) indigenous bacteria with glucose, and (6) indigenous bacteria with molasses. The IFTs between bacterial culture and oil were represented in Fig. 4.38.

The accuracy of the developed simulator was confirmed by comparison with the experimental results. They investigated the effects of indigenous bacteria-to-injected bacteria ratio, slug size of nutrient, incubation time, residual oil saturation, absolute permeability, and injection rate on oil production. The effect of indigenous bacteria and injected bacteria ratio was described in Fig. 4.39, which showed that oil recovery increased with increasing the ratio. This ensured that the role of indigenous bacteria in oil recovery could not be ignored. Fig. 4.40 represented the effect of nutrient slug size on oil recovery. The result showed that the increase in oil recovery slowed down as the slug size increased. Additionally, oil recovery improved with

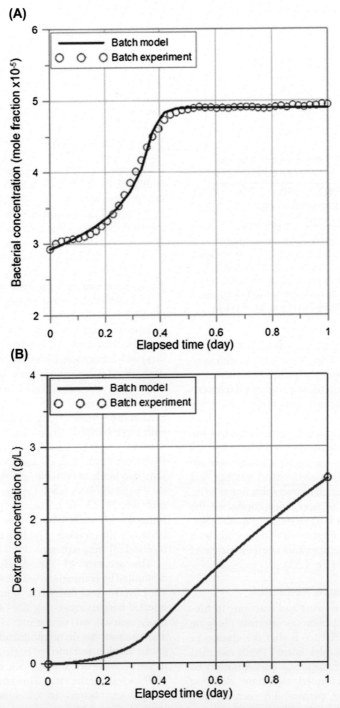

FIG. 4.22 Batch model simulation results: **(A)** bacterial growth, **(B)** biopolymer generation. (Credit: from Jeong, M.S., Noh, D., Hong, E., Lee, K.S., Kwon, T., 2019b. Systematic modeling approach to selective plugging using in situ bacterial biopolymer production and its potential for microbial-enhanced oil recovery. Geomicrobiology Journal 36 (5), 468–481.)

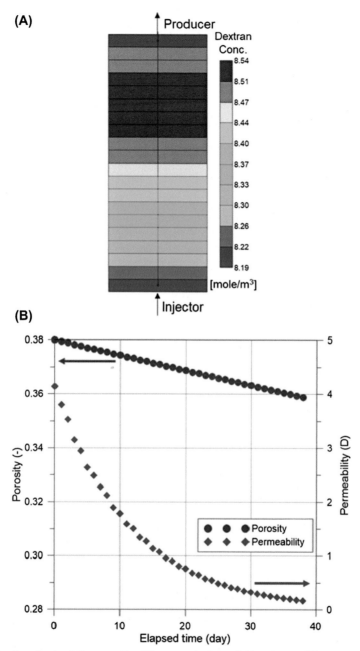

FIG. 4.23 Sand-pack simulation results: **(A)** concentration of biopolymer, **(B)** average porosity and permeability alterations. (Credit: from Jeong, M.S., Noh, D., Hong, E., Lee, K.S., Kwon, T., 2019b. Systematic modeling approach to selective plugging using in situ bacterial biopolymer production and its potential for microbial-enhanced oil recovery. Geomicrobiology Journal 36 (5), 468–481.)

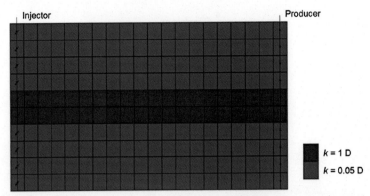

FIG. 4.24 Permeability distribution in the synthetic reservoir. (Credit: from Jeong, M.S., Noh, D., Hong, E., Lee, K.S., Kwon, T., 2019b. Systematic modeling approach to selective plugging using in situ bacterial biopolymer production and its potential for microbial-enhanced oil recovery. Geomicrobiology Journal 36 (5), 468–481.)

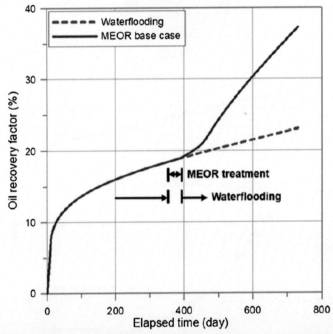

FIG. 4.25 Oil recovery factors of waterflooding and MEOR. *MEOR*, microbial enhanced oil recovery. (Credit: from Jeong, M.S., Noh, D., Hong, E., Lee, K.S., Kwon, T., 2019b. Systematic modeling approach to selective plugging using in situ bacterial biopolymer production and its potential for microbial-enhanced oil recovery. Geomicrobiology Journal36 (5), 468–481.)

increasing incubation time (Fig. 4.41). On the other hand, oil recovery and residual oil saturation seemed to be inversely proportional (Fig. 4.42). It was because the increase in residual oil saturation led to reduction of displacing fluid volume at constant nutrient slug size, and consequently, reduction of sweep efficiency. Fig. 4.43 showed the effects of absolute permeability and injection rate on oil recovery. According to their findings, oil recovery was affected by indigenous and injected bacteria ratio, nutrient slug size, incubation time, and residual oil saturation, but absolute permeability and injection rate had no significant effect on microbial surfactant flooding efficiency.

Behesht et al. (2008) suggested a numerical model for MEOR process. There were several assumptions for model development. The fluid flow was laminar in

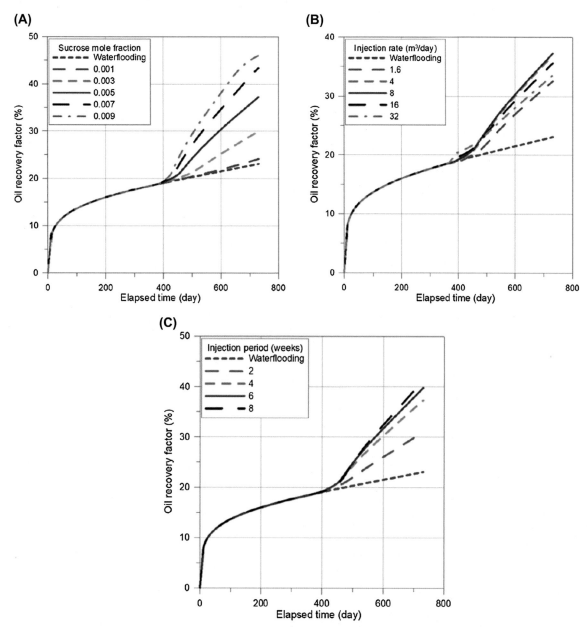

FIG. 4.26 Oil recovery factors according to **(A)** nutrient concentration, **(B)** injection rate, and **(C)** injection period. (Credit: from Jeong, M.S., Noh, D., Hong, E., Lee, K.S., Kwon, T., 2019b. Systematic modeling approach to selective plugging using in situ bacterial biopolymer production and its potential for microbial-enhanced oil recovery. Geomicrobiology Journal 36 (5), 468–481.)

three dimensions. The formation was homogenous, isotropic, and incompressible. The model was composed of only two phases such as water and oil. The MEOR process activated the indigenous anaerobic bacteria by injected nutrients. The bacteria, substrate, and metabolites did not enter the nonaqueous phase. Temperature and volume changes during the MEOR process were ignored. The microbial reactions only

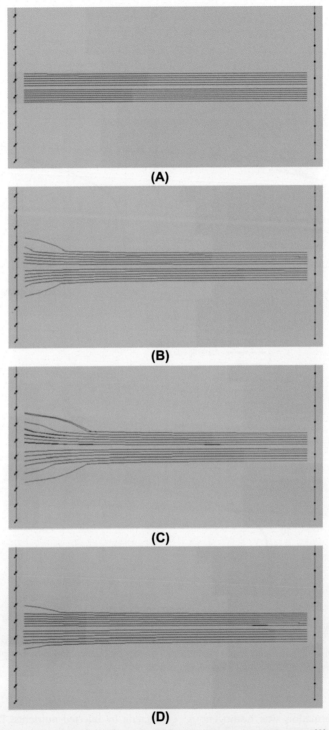

FIG. 4.27 Changes in water diversion effect of MEOR with different injection rates: **(A)** waterflooding, **(B)** 1.6 m³/day, **(C)** 8 m³/day, **(D)** 32 m³/day. MWOR, microbial enhanced oil recovery. (Credit: from Jeong, M.S., Noh, D., Hong, E., Lee, K.S., Kwon, T., 2019b. Systematic modeling approach to selective plugging using in situ bacterial biopolymer production and its potential for microbial-enhanced oil recovery. Geomicrobiology Journal 36 (5), 468–481.)

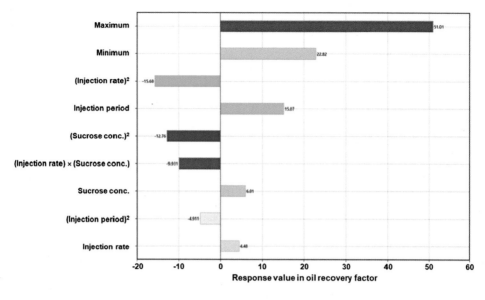

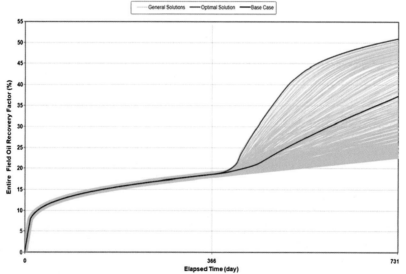

FIG. 4.28 Results of **(A)** sensitivity analysis and **(B)** optimization. (Credit: from Jeong, M.S., Noh, D., Hong, E., Lee, K.S., Kwon, T., 2019b. Systematic modeling approach to selective plugging using in situ bacterial biopolymer production and its potential for microbial-enhanced oil recovery. Geomicrobiology Journal 36 (5), 468–481.)

occurred in the aqueous phase. The microbial surfactant was the key metabolite that caused an increasing oil recovery. Wettability alteration and IFT reduction were induced by the surfactant.

The bacterial growth was described by Monod equation with single substrate. They used a Langmuir isotherm type to express the biosurfactant adsorption, which considered the salinity, surfactant concentration, and rock permeability. This adsorption was reversible with salinity and irreversible with concentration. The IFT model was based on exponential correlation between IFT and surfactant concentration. The IFT

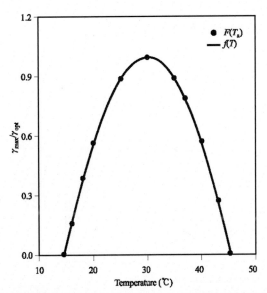

FIG. 4.29 Temperature effect on microbial biokinetics: $F(T_a)$ described the results calculated by Arrhenius equation, and $f(T)$ described the results from cardinal temperature model. (Credit: from Hong, E., Jeong, M.S., Lee, K.S., 2019b. Optimization of nonisothermal selective plugging with a thermally active biopolymer. Journal of Petroleum Science and Engineering 173, 434–446.)

between oil and water also depended on the salinity. The relative permeability curves for IFT alteration were calculated by standard Corey equations. Finally, the wettability alterations were calculated with relative permeability and capillary pressure changes.

Biosurfactant-based MEOR simulations have also been implemented through commercial software. Bültemeier et al. (2014) simulated the microbial surfactant mechanism with CMG STARS. In the same way as Jeong et al. (2019b) and Hong et al. (2019b), chemical reactions were described by stoichiometric equations, and their reaction rates were calculated by the Arrhenius equation. The oil-water relative permeability curves were calculated by Corey's approach.

In this simulation, one pore volume of bacterial media consisting of the bacteria and nutrient was injected into the sand pack after the first waterflooding. Fig. 4.44 showed the concentrations of bacteria and surfactant as well as the IFT changes. The surfactant concentration increased with increasing bacterial concentration, while at the same time the IFT decreased rapidly. As a result, it was indicated that the oil recovery factors were always

higher in the MEOR cases than in the sterile nutrient injection cases.

Hosseininoosheri et al. (2016) simulated and characterized the in situ biosurfactant generation in MEOR process with UTCHEM. To simulate the IFT reduction, they used biological model to generate biosurfactant and existing IFT model in the simulator to decrease the IFT between water and oil. The IFT model used relative permeability alteration method. Once the microbial surfactant was generated in a grid block, it reduced the IFT, and consequently, the trapping number increased. This trapping number changed the relative permeability.

The modeling of MEOR in this study coupled the kinetics transport with local equilibrium transport in the existence of surfactant phase behavior model. The surfactant-oil-water phase behavior can be described as a function of salinity after the binodal curves, and tie lines are depicted by Hand's rule (Hand, 1939). In the Hand's rule based on empirical observation, the concentration ratios of equilibrium phase show the straight line on log-log scale. When C_{ij} represents the volume fraction of component i in phase j, and i and j represent water (w), oil (o), and microemulsion (m), the binodal curves by the Hand's rule are calculated as follows:

$$\frac{C_{mj}}{C_{oj}} = A_H \left(\frac{C_{mj}}{C_{wj}}\right)^{B_H}, \quad j = w, \ o, \text{ or } m \qquad (4.93)$$

where A_H and B_H are empirical parameters.

As the salinity is increased, the type of microemulsion is altered from type II($-$) to type III to type II($+$). Conventionally, surfactant-brine-oil phase behavior is described on a ternary diagram as shown in Fig. 4.45. In this graph, the tie lines had negative slopes in low-salinity environment and positive slopes in high-salinity conditions. Therefore, type II($-$) means that there are two phases in the system with negative tie lines. Type III and II($+$) are also determined in the similar way (Nelson and Pope, 1978).

This work used first-order Monod kinetic equations as a function of temperature, salinity, and pH. Hosseininoosheri et al. (2016) investigated the effects of nutrient concentration, bacterial growth rate, temperature, salinity, and biosurfactant adsorption. A three-dimensional model was used (Fig. 4.46). Water was injected at 3000 ft³/day for 1000 days, and then substrate, electron acceptor, and microorganisms were injected at 3000 ft³/day for another 1000 days. The

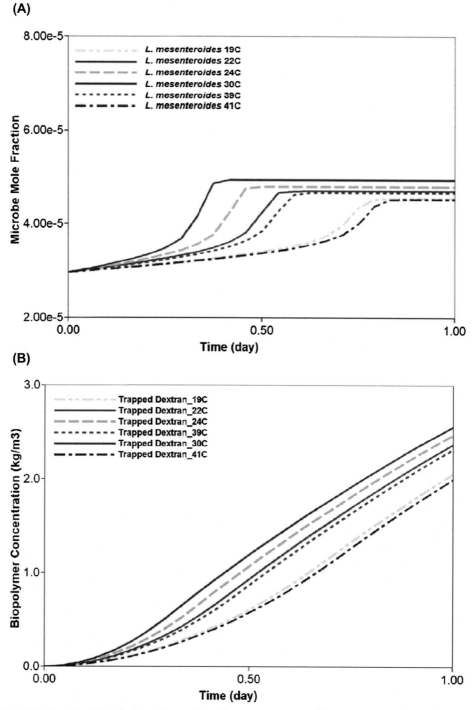

FIG. 4.30 Results of batch simulations with various temperatures: **(A)** bacterial growth, **(B)** biopolymer generation. (Credit: from Hong, E., Jeong, M.S., Lee, K.S., 2019b. Optimization of nonisothermal selective plugging with a thermally active biopolymer. Journal of Petroleum Science and Engineering 173, 434–446.)

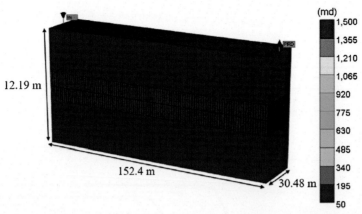

FIG. 4.31 Permeability distribution of two-dimensional hypothetical reservoir model. (Credit: from Hong, E., Jeong, M.S., Lee, K.S., 2019b. Optimization of nonisothermal selective plugging with a thermally active biopolymer. Journal of Petroleum Science and Engineering 173, 434–446.)

effect of nutrient concentration was represented in Fig. 4.47. The result showed that the recovery of MEOR was highly dependent on the nutrient concentration. The increase in the nutrient concentration increased the oil recovery. The results of the sensitivity analysis of the maximum growth rate for oil recovery were represented in Fig. 4.48. According to the results, the magnitude of the maximum growth rate should be increased by at least two orders to affect oil recovery. They used a Langmuir type isotherm model to simulate the surfactant adsorption. In this case, it could be seen that the oil productivity decreased when adsorption was taken into account (Fig. 4.49). Additionally, they showed that the salinity and temperature were the most significant parameters for increasing the oil recovery and that there are optimum range of these parameters for maximizing oil recovery (Fig. 4.50).

Hong et al. (2019a) also investigated the effect of temperature on biosurfactant-based MEOR efficiency. This cold-water MEOR was suggested to overcome the technical limitation associated with application temperature. The temperature effect on reaction rate was calculated with a multiactivation energy. Stoichiometric equations and Arrhenius reaction rate were obtained via history matching with experimental results. They assumed that the adsorption of any component was ignored to analyze the only effect of IFT reduction. Additionally, surfactant partitioning was not considered to simplify the simulation. Wettability alteration was designed based on the relative permeability changes according to IFT reduction.

The cold-water MEOR was performed in a high-temperature reservoir. This reservoir was a two-dimensional hypothetical homogenous model (Fig. 4.51). The rock heat capacity and conductivity were assumed to be general sandstone properties (Robertson, 1988). The initial temperature, porosity, and permeability were 71°C, 0.24, and 30 md, respectively. Fig. 4.52 showed the IFT change with biosurfactant, which was based on experimental data (Xiao et al., 2013). A total of 1 pore volume injection was performed in three stages. The first waterflooding of 0.48 pore volume was injected to generate a proper temperature environment for microbial growth, followed by 0.02 pore volume microbial solution. During this microbial treatment, biosurfactant was produced by bacteria and injected nutrients. The second waterflooding of 0.5 pore volume was then introduced to propagate the surfactant into the deeper reservoir.

The nutrient concentration, injection rate, and injection temperature were design parameters for the MEOR parametric analysis. The effect of each parameter on oil recovery was analyzed independently (Fig. 4.53). The oil recovery continued to increase as the nutrient concentration increased, whereas injection rate and temperature had the appropriate range to maximize oil recovery. Fig. 4.54 depicted the temperature and IFT distributions in the reservoir, depending on the injection rate. When water at lower temperature than the reservoir was injected at high flow rate, the temperature distribution of the reservoir was more favorable for microbial growth. However, the IFT reduction was less

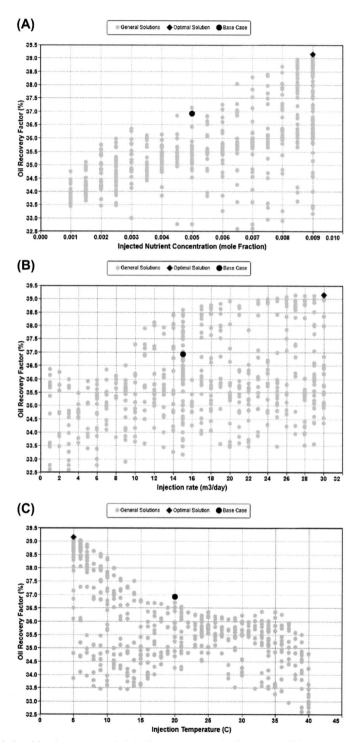

FIG. 4.32 Relationships between each injection strategy and oil recovery: **(A)** nutrient concentration, **(B)** injection rate, **(C)** injection temperature. (Credit: from Hong, E., Jeong, M.S., Lee, K.S., 2019b. Optimization of nonisothermal selective plugging with a thermally active biopolymer. Journal of Petroleum Science and Engineering 173, 434–446.)

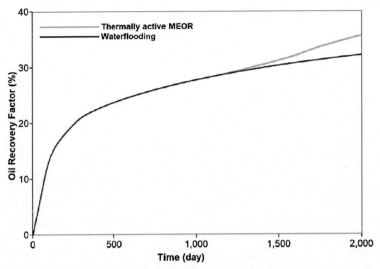

FIG. 4.33 Oil recovery factors obtained from realistic model application. (Credit: from Hong, E., Jeong, M.S., Lee, K.S., 2019b. Optimization of nonisothermal selective plugging with a thermally active biopolymer. Journal of Petroleum Science and Engineering 173, 434–446.)

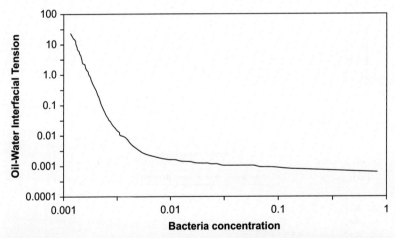

FIG. 4.34 Relationship between bacteria concentration and oil-water interfacial tension. (Credit: from Islam, M.R., 1990. Mathematical modeling of microbial enhanced oil recovery. Proceedings of the SPE Annual Technical Conference and Exhibition, New Orleans, LA, USA, 23–26 September, 1990.)

because the microbes did not have sufficient time to produce the biosurfactant. The temperature and IFT distributions in the reservoir with injection temperature were shown in Fig. 4.55. It could be seen that when injected at a sufficiently low temperature, an IFT environment was favored for oil production. Finally, injection design optimization to maximize oil recovery was performed to show the improved production results (Fig. 4.56).

4.3.3 Other Simulation Studies

Due to the limitations of experimental and modeling techniques, MEOR simulation studies have been carried out primarily with microbial plugging and IFT

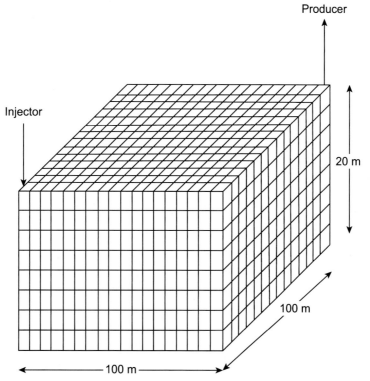

FIG. 4.35 Grid blocks and well locations of reservoir model. (Credit: from Islam, M.R., 1990. Mathematical modeling of microbial enhanced oil recovery. Proceedings of the SPE Annual Technical Conference and Exhibition, New Orleans, LA, USA, 23–26 September, 1990.)

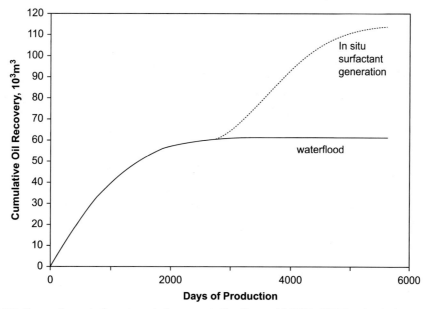

FIG. 4.36 Comparison of oil recovery between waterflooding and MEOR. *MEOR*, microbial enhanced oil recovery. (Credit: from Islam, M.R., 1990. Mathematical modeling of microbial enhanced oil recovery. Proceedings of the SPE Annual Technical Conference and Exhibition, New Orleans, LA, USA, 23–26 September, 1990.)

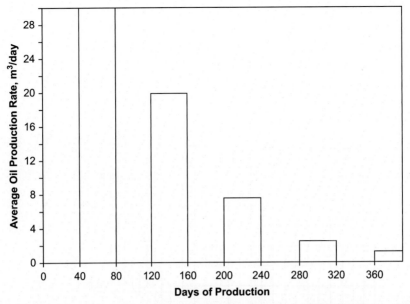

FIG. 4.37 Oil rates of Huff and Puff process with surfactant-generating bacteria. (Credit: from Islam, M.R., 1990. Mathematical modeling of microbial enhanced oil recovery. Proceedings of the SPE Annual Technical Conference and Exhibition, New Orleans, LA, USA, 23–26 September, 1990.)

reduction. However, simulation researches continue to involve various mechanisms, including biogas generation, oil viscosity reduction, hydrocarbon degradation, and increased sweep efficiency by soluble biopolymer.

Islam (1990) simulated the effect of viscosity-reducing bacteria on oil recovery. The viscosity of oil was 50 mPa·s which could be decreased considerably by the bacteria existence. The graph of viscosity reduction according to the concentration of bacteria was presented in Fig. 4.57. It was inferred from solvent flooding because there was no proper experimental result. Fig. 4.58 showed the improvement of oil recovery by the viscosity-reducing bacteria.

Islam (1990) also modeled the performance of CO_2-generating bacteria. Three-phase relative permeabilities were generated by using Stone's method, and gas-oil relative permeability was obtained from Islam et al. (1992). The correlation between CO_2 and bacteria concentration was assumed (Fig. 4.59). The oil viscosity decrease and IFT reduction due to the CO_2 presence were considered following Meszaros et al. (1990). The change in oil recovery due to CO_2 production was

shown in Fig. 4.60, and the result of CO_2 production was not very encouraging. This was because the generated CO_2 had the advantage of increasing the pressure in the reservoir but had the effect of disturbing the flow of oil.

Sitnikov and Eremin (1994) developed a mathematical model to describe MEOR process in fractured reservoir. The reservoir model was simplified and adapted from the Romashkino field in Russia. To make the work easier, only biosurfactant and carbonic gas productions were considered as a result of microbial activities. The injection was implemented for 10 days and then stopped for 20 days. The generations of metabolites and the properties changes were represented in Figs. 4.61 and 4.62. The result indicated that the imbibition sweep efficiency of microbial treatment was better than that of general waterflooding (Fig. 4.63).

There is a case in which MEOR simulation is performed on real field (Ibrahim et al., 2004). A total of eight wells in Bokor field, Sarawak, Malaysia, were applied for microbial treatment with Huff and Puff

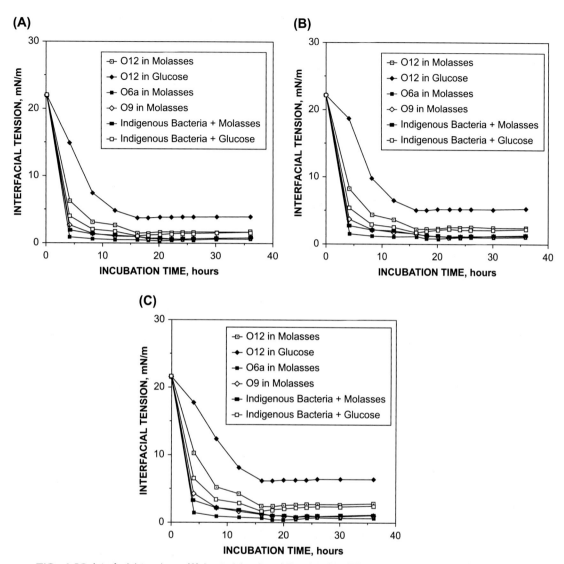

FIG. 4.38 Interfacial tensions: **(A)** bacterial culture/oil ratio of 1, **(B)** bacterial culture/oil ratio of 0.8, **(C)** bacterial culture/oil ratio of 0.6. (Credit: from Desouky, S.M., Abdel-Daim, M.M., Sayyouh, M.H., Dahab, A.S., 1996. Modelling and laboratory investigation of microbial enhanced oil recovery. Journal of Petroleum Science and Engineering 15 (2–4), 309–320.)

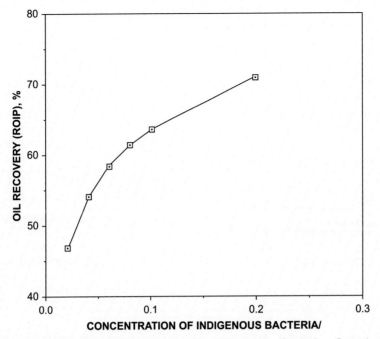

FIG. 4.39 Effect of indigenous and injected bacteria ratio on oil recovery. (Credit: from Desouky, S.M., Abdel-Daim, M.M., Sayyouh, M.H., Dahab, A.S., 1996. Modelling and laboratory investigation of microbial enhanced oil recovery. Journal of Petroleum Science and Engineering 15 (2–4), 309–320.)

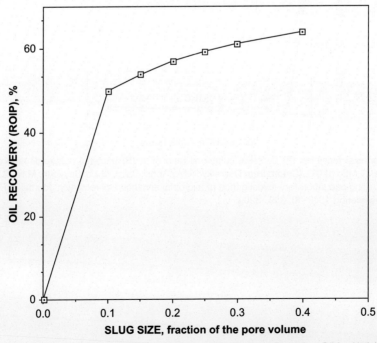

FIG. 4.40 Effect of nutrient slug size on oil recovery. (Credit: from Desouky, S.M., Abdel-Daim, M.M., Sayyouh, M.H., Dahab, A.S., 1996. Modelling and laboratory investigation of microbial enhanced oil recovery. Journal of Petroleum Science and Engineering 15 (2–4), 309–320.)

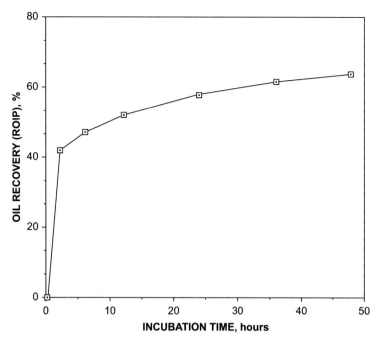

FIG. 4.41 Effect of incubation time on oil recovery. (Credit: from Desouky, S.M., Abdel-Daim, M.M., Sayyouh, M.H., Dahab, A.S., 1996. Modelling and laboratory investigation of microbial enhanced oil recovery. Journal of Petroleum Science and Engineering 15 (2−4), 309−320.)

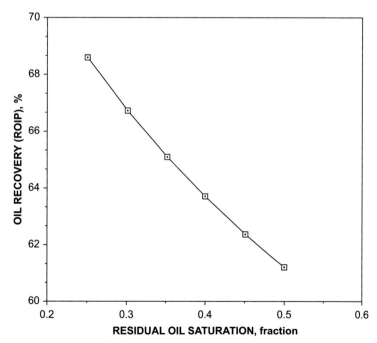

FIG. 4.42 Effect of residual oil saturation on oil recovery. (Credit: from Desouky, S.M., Abdel-Daim, M.M., Sayyouh, M.H., Dahab, A.S., 1996. Modelling and laboratory investigation of microbial enhanced oil recovery. Journal of Petroleum Science and Engineering 15 (2−4), 309−320.)

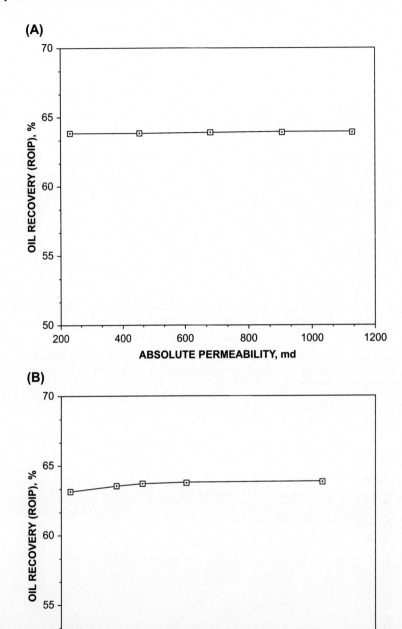

FIG. 4.43 Effects of **(A)** absolute permeability and **(B)** injection rate on oil recovery. (Credit: from Desouky, S.M., Abdel-Daim, M.M., Sayyouh, M.H., Dahab, A.S., 1996. Modelling and laboratory investigation of microbial enhanced oil recovery. Journal of Petroleum Science and Engineering 15 (2–4), 309–320.)

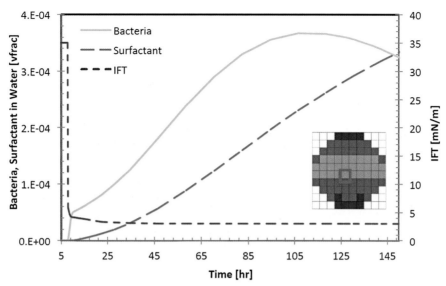

FIG. 4.44 Microbial and surfactant concentrations and IFT change in sand-pack simulation. IFT, interfacial tension. (Credit: from Bültemeier, H., Alkan, H., Amro, M., 2014. A new modeling approach to MEOR calibrated by bacterial growth and metabolite curves. Proceedings of the SPE EOR Conference at Oil and Gas West Asia, Muscat, Oman, 31 March–2 April, 2014.)

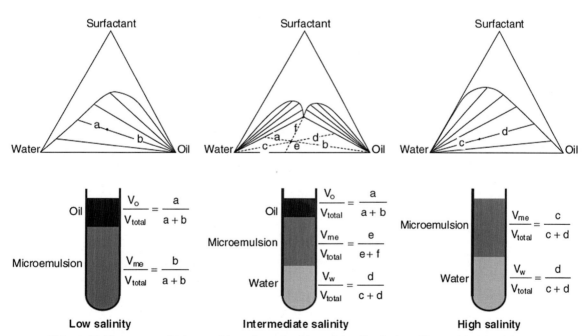

FIG. 4.45 Three types of microemulsions determined by salinity. (Credit: from Sheng, J.J., 2011. Modern Chemical Enhanced Oil Recovery: Theory and Practice. Gulf Professional Publishing, Houston, Texas, USA.)

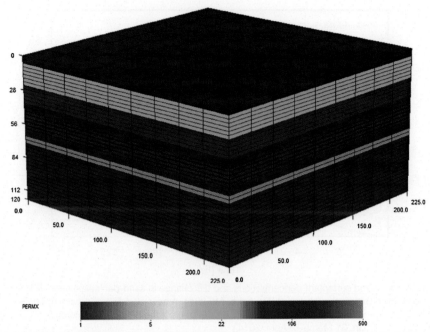

FIG. 4.46 Permeability distribution of reservoir model. (Credit: from Hosseininoosheri, P., Lashgari, H.R., Sepehrnoori, K., 2016. A novel method to model and characterize in-situ bio-surfactant production in microbial enhanced oil recovery. Fuel 183, 501–511.)

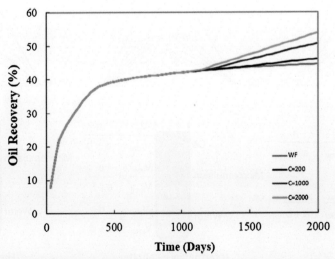

FIG. 4.47 Effect of nutrient concentration on oil recovery. (Credit: from Hosseininoosheri, P., Lashgari, H.R., Sepehrnoori, K., 2016. A novel method to model and characterize in-situ bio-surfactant production in microbial enhanced oil recovery. Fuel 183, 501–511.)

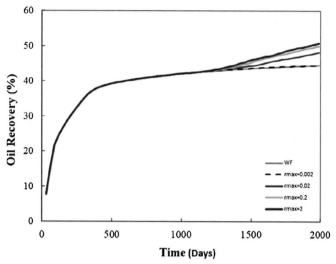

FIG. 4.48 Effect of bacterial growth rate on oil recovery. (Credit: from Hosseininoosheri, P., Lashgari, H.R., Sepehrnoori, K., 2016. A novel method to model and characterize in-situ bio-surfactant production in microbial enhanced oil recovery. Fuel 183, 501−511.)

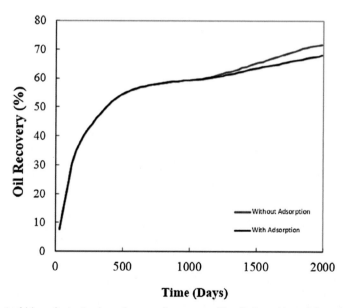

FIG. 4.49 Effect of biosurfactant adsorption on oil recovery. (Credit: from Hosseininoosheri, P., Lashgari, H.R., Sepehrnoori, K., 2016. A novel method to model and characterize in-situ bio-surfactant production in microbial enhanced oil recovery. Fuel 183, 501−511.)

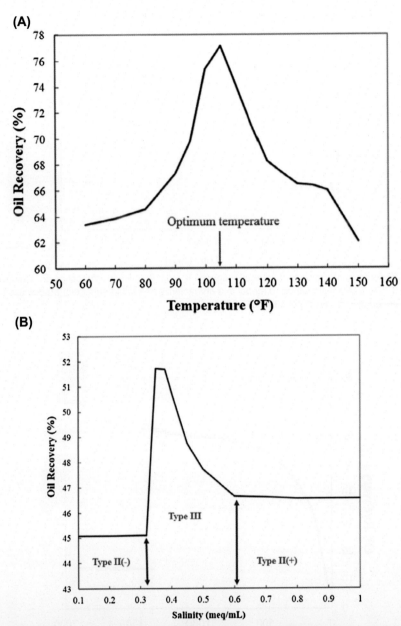

FIG. 4.50 Simulation results of oil recovery factors: **(A)** temperature effect, **(B)** salinity effect. (Credit: from Hosseininoosheri, P., Lashgari, H.R., Sepehrnoori, K., 2016. A novel method to model and characterize in-situ bio-surfactant production in microbial enhanced oil recovery. Fuel 183, 501−511.)

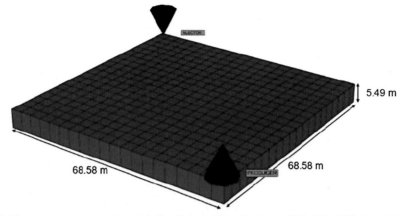

FIG. 4.51 Homogeneous reservoir model. (Credit: from Hong, E., Jeong, M.S., Kim, T.H., Lee, J.H., Cho, J.H., Lee, K.S., 2019a. Development of coupled biokinetic and thermal model to optimize cold-water microbial enhanced oil recovery (MEOR) in homogenous reservoir. Sustainability 11 (6), 1−19.)

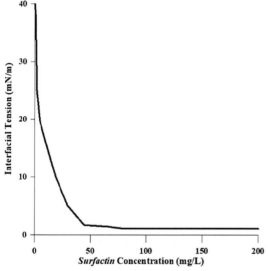

FIG. 4.52 Interfacial tension reduction with biosurfactant concentration. (Credit: from Hong, E., Jeong, M.S., Kim, T.H., Lee, J.H., Cho, J.H., Lee, K.S., 2019a. Development of coupled biokinetic and thermal model to optimize cold-water microbial enhanced oil recovery (MEOR) in homogenous reservoir. Sustainability 11 (6), 1−19 adapted from Amani, H., Sarrafzadeh, M.H., Haghighi, M., Mehrnia, M.R., 2010. Comparative study of biosurfactant producing bacteria in MEOR applications. Journal of Petroleum Science and Engineering 75 (1−2), 209−214.)

process. As a result, the liquid production and water-cut were improved in some wells, but not in the other wells. To further evaluate and understand the well performances, simulation analyses were conducted using the CMG STARS. The target wells were B-1 and B-3. The results from simulations were compared and matched to the real production trends to quantify the various possible treatment effects.

Radial model surrounded with aquifer was proposed to represent the effective reservoir (Fig. 4.64). The drainage radius of about 800−900 ft was needed to match the water-cut at the beginning of the treatment. The grids of near well were refined to verify the near wellbore effects. The reservoir properties of the well vicinity were taken directly from the full-field reservoir model. However, the properties away from the well represented averages for each major sand unit, allowing connection with each unit but not between major units. Such a grid model was believed that it could efficiently describe the dynamic processes near the wellbore, therefore allowing the microbial effects to be simulated. They considered the bacterial growth and death in this simulation. The biochemical generation was assumed to utilize some parts of oil components, and this biochemical could resolve the solid hydrocarbon. Additionally, the adsorption of bacteria and biochemical were included (Fig. 4.65).

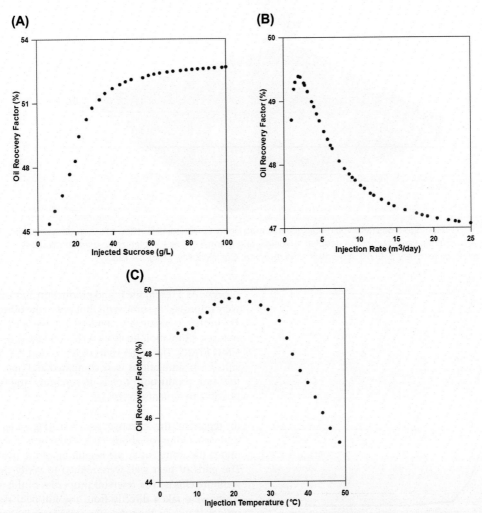

FIG. 4.53 Relationship between oil recovery and injection parameters: **(A)** nutrient concentration, **(B)** injection rate, **(C)** injection temperature. (Credit: from Hong, E., Jeong, M.S., Kim, T.H., Lee, J.H., Cho, J.H., Lee, K.S., 2019a. Development of coupled biokinetic and thermal model to optimize cold-water microbial enhanced oil recovery (MEOR) in homogenous reservoir. Sustainability 11 (6), 1−19.)

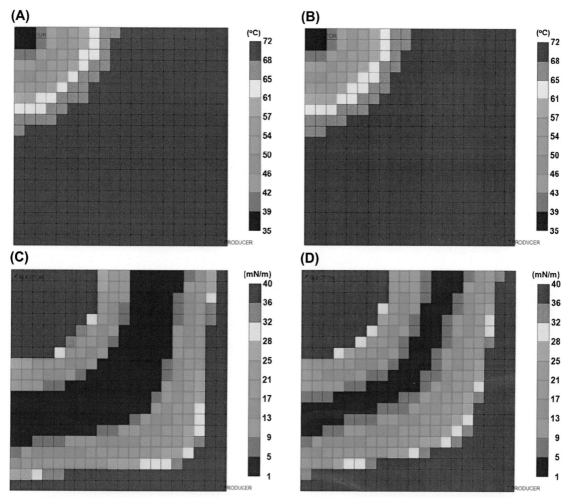

FIG. 4.54 Distributions of temperature and IFT depending on injection rate: **(A)** temperature distribution in 1 m³/day injection rate, **(B)** temperature distribution in 24.8 m³/day injection rate, **(C)** IFT distribution in 1 m³/day injection rate, **(D)** IFT distribution in 24.8 m³/day injection rate. IFT, interfacial tension. (Credit: from Hong, E., Jeong, M.S., Kim, T.H., Lee, J.H., Cho, J.H., Lee, K.S., 2019a. Development of coupled biokinetic and thermal model to optimize cold-water microbial enhanced oil recovery (MEOR) in homogenous reservoir. Sustainability 11 (6), 1—19.)

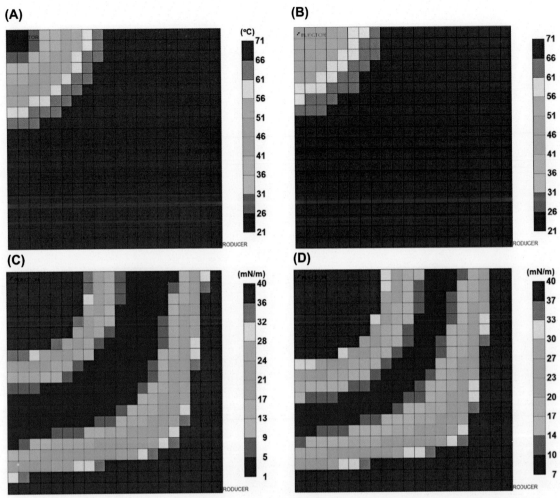

FIG. 4.55 Distributions of temperature and IFT depending on injection temperature: **(A)** temperature distribution in 20°C injection temperature, **(B)** temperature distribution in 45°C injection temperature, **(C)** IFT distribution in 20°C injection temperature, **(D)** IFT distribution in 45°C injection temperature. IFT, interfacial tension. (Credit: from Hong, E., Jeong, M.S., Kim, T.H., Lee, J.H., Cho, J.H., Lee, K.S., 2019a. Development of coupled biokinetic and thermal model to optimize cold-water microbial enhanced oil recovery (MEOR) in homogenous reservoir. Sustainability 11 (6), 1–19.)

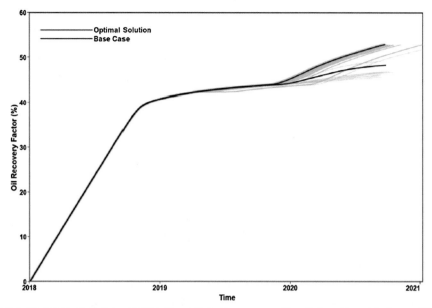

FIG. 4.56 Maximum oil recovery obtained by optimization process. (Credit: from Hong, E., Jeong, M.S., Kim, T.H., Lee, J.H., Cho, J.H., Lee, K.S., 2019a. Development of coupled biokinetic and thermal model to optimize cold-water microbial enhanced oil recovery (MEOR) in homogenous reservoir. Sustainability 11 (6), 1–19.)

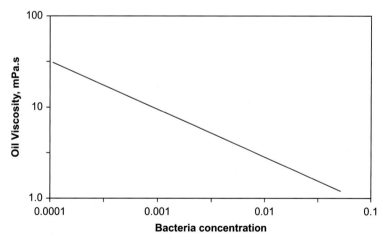

FIG. 4.57 Correlation between oil viscosity and bacteria concentration. (Credit: from Islam, M.R., 1990. Mathematical modeling of microbial enhanced oil recovery. Proceedings of the SPE Annual Technical Conference and Exhibition, New Orleans, LA, USA, 23–26 September, 1990.)

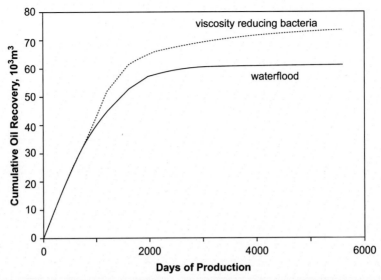

FIG. 4.58 Comparison of oil recovery between waterflooding and viscosity-reducing bacteria process. (Credit: from Islam, M.R., 1990. Mathematical modeling of microbial enhanced oil recovery. Proceedings of the SPE Annual Technical Conference and Exhibition, New Orleans, LA, USA, 23–26 September, 1990.)

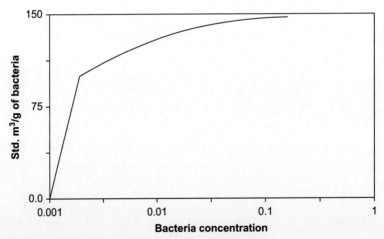

FIG. 4.59 Correlation between CO_2 generation and bacteria concentration. (Credit: from Islam, M.R., 1990. Mathematical modeling of microbial enhanced oil recovery. Proceedings of the SPE Annual Technical Conference and Exhibition, New Orleans, LA, USA, 23–26 September, 1990.)

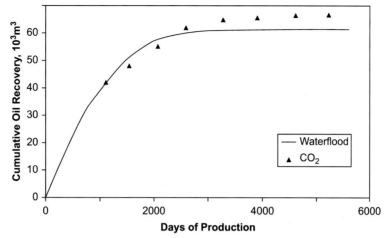

FIG. 4.60 Comparison of oil recovery between waterflooding and CO_2-generating bacteria process. (Credit: from Islam, M.R., 1990. Mathematical modeling of microbial enhanced oil recovery. Proceedings of the SPE Annual Technical Conference and Exhibition, New Orleans, LA, USA, 23–26 September, 1990.)

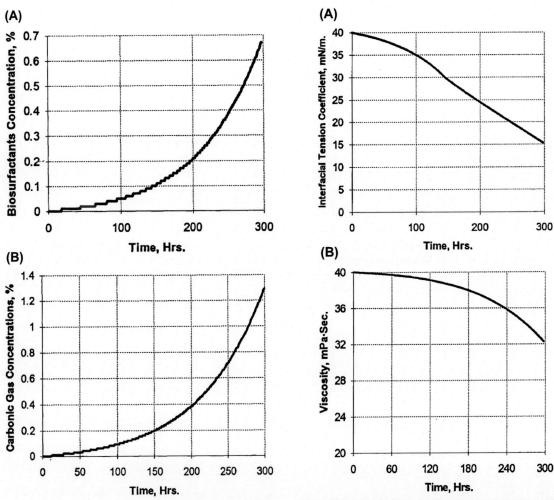

FIG. 4.61 Productions of metabolites: **(A)** biosurfactant, **(B)** carbonic gas. (Credit: from Sitnikov, A.A., Eremin, N.A., 1994. A mathematical model of microbial enhanced oil recovery (MEOR) method for mixed type rock. Proceedings of European Petroleum Conference, London, United Kingdom, 25–27 October, 1994.)

FIG. 4.62 Properties changes: **(A)** interfacial tension, **(B)** oil viscosity. (Credit: from Sitnikov, A.A., Eremin, N.A., 1994. A mathematical model of microbial enhanced oil recovery (MEOR) method for mixed type rock. Proceedings of European Petroleum Conference, London, United Kingdom, 25–27 October, 1994.)

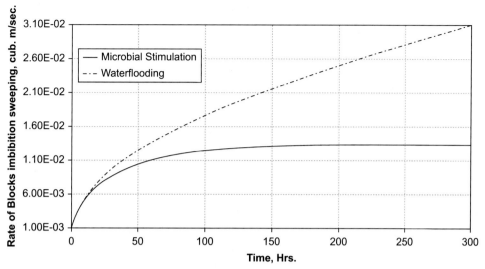

FIG. 4.63 Comparison of sweep efficiency between waterflooding and microbial treatment. (Credit: from Sitnikov, A.A., Eremin, N.A., 1994. A mathematical model of microbial enhanced oil recovery (MEOR) method for mixed type rock. Proceedings of European Petroleum Conference, London, United Kingdom, 25–27 October, 1994.)

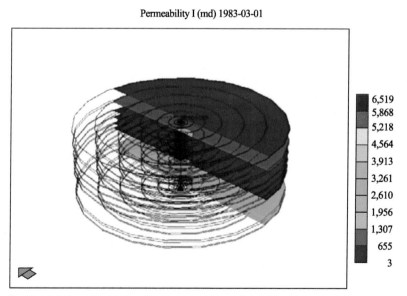

FIG. 4.64 Radial model for microbial treatment extracted form full-field reservoir model. (Credit: from Ibrahim, Z., Omar, M.I., Foo, K.S., Elias, E.J., Othman, M., 2004. Simulation analysis of microbial well treatment of Bokor field, Malaysia. Proceedings of SPE Asia Pacific Oil and Gas Conference and Exhibition, Perth, Australia, 18–20 October, 2004.)

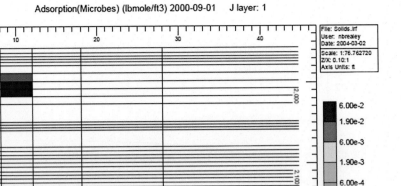

FIG. 4.65 Adsorption of bacteria within 10 ft of well distance. (Credit: from Ibrahim, Z., Omar, M.I., Foo, K.S., Elias, E.J., Othman, M., 2004. Simulation analysis of microbial well treatment of Bokor field, Malaysia. Proceedings of SPE Asia Pacific Oil and Gas Conference and Exhibition, Perth, Australia, 18–20 October, 2004.)

(A)

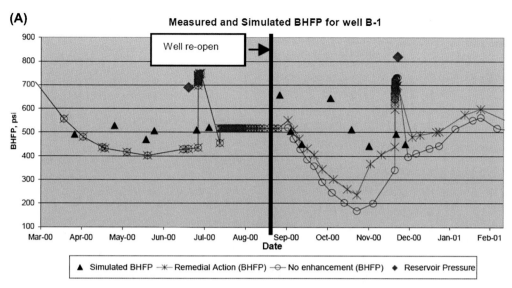

(B)

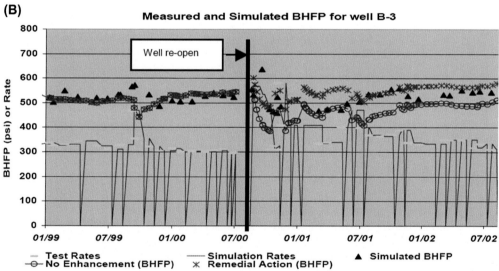

FIG. 4.66 Measured and simulated BHFP for **(A)** well B-1 and **(B)** well B-3. BHFP, bottom hole flowing pressure. (Credit: from Ibrahim, Z., Omar, M.I., Foo, K.S., Elias, E.J., Othman, M., 2004. Simulation analysis of microbial well treatment of Bokor field, Malaysia. Proceedings of SPE Asia Pacific Oil and Gas Conference and Exhibition, Perth, Australia, 18–20 October, 2004.)

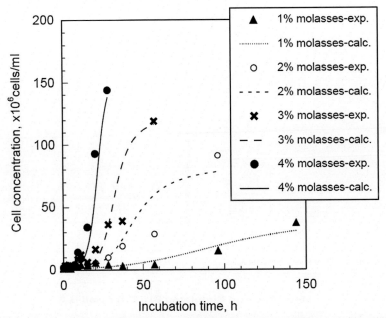

FIG. 4.67 Comparison of experimental and simulated bacterial growth. (Credit: from Sugai, Y., Hong, C., Chida, T., Enomoto, H., 2007. Simulation studies on the mechanisms and performances of MEOR using a polymer producing microorganism Clostridium sp. TU-15A. Proceedings of Asia Pacific Oil and Gas Conference and Exhibition, Jakarta, Indonesia, 30 October–1 November, 2007.)

Fig. 4.66 represented the bottom hole flowing pressure (BHFP) responses before and after microbial remedial actions for well B-1 and B-3. These measured pressures were plotted the simulation results with (remedial action) and without (no enhancement) microbial treatment. The results showed that the case with microbial treatment fitted better with the actual field data. However, in well B-3, the case without microbial treatment showed better fitting results with real field data (Fig. 4.67). Thus, they concluded that B-1 was benefiting from the microbial actions and that B-3 had no clear benefit from the microbial treatment.

Sugai et al. (2007) suggested a numerical model to analyze the process of MEOR using polymer-producing bacteria. The model consisted of two-phase (water and oil) and five components (oil, water, bacteria, nutrient, and polymer). This model included the microbial growth and death, nutrient consumption, polymer generation, and water viscosity increment. On the other hand, they did not consider the absorption of bacteria, nutrient, and metabolite to the rock surface to clarify the effect of recovery enhancement by viscous solution generated by polymer-producing bacteria. They also ignored the diffusion effect because the diffusion was negligible in field scale. The microbial growth followed the Moser equation, and the results were shown in Fig. 4.67. The viscosity results of the culture solution were described in Fig. 4.68.

The numerical simulation was carried out in quarter of a five-spot pattern flooding (Fig. 4.69). The viscosity of oil and reservoir brine were 9.6 and 0.9 cp, respectively. The absolute permeability was 1000 md. The production process consisted of four steps. First, water was injected into the reservoir until water-cut became 90%. Then, the 1-pore volume solution including polymer-producing bacteria and nutrient was injected.

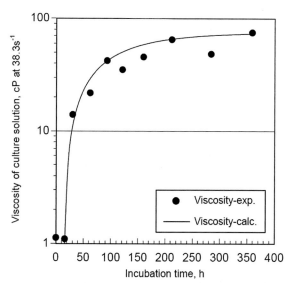

FIG. 4.68 Comparison of experimental and simulated solution viscosity. (Credit: from Sugai, Y., Hong, C., Chida, T., Enomoto, H., 2007. Simulation studies on the mechanisms and performances of MEOR using a polymer producing microorganism Clostridium sp. TU-15A. Proceedings of Asia Pacific Oil and Gas Conference and Exhibition, Jakarta, Indonesia, 30 October–1 November, 2007.)

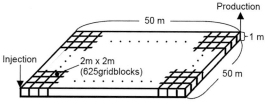

FIG. 4.69 Structure of reservoir model. (Credit: from Sugai, Y., Hong, C., Chida, T., Enomoto, H., 2007. Simulation studies on the mechanisms and performances of MEOR using a polymer producing microorganism Clostridium sp. TU-15A. Proceedings of Asia Pacific Oil and Gas Conference and Exhibition, Jakarta, Indonesia, 30 October–1 November, 2007.)

Next, both injector and producer were shut-in to incubate the bacteria for 10 days. Finally, water was reinjected until the water-cut became 95%. The result was represented in Fig. 4.70. In the MEOR case, oil recovery began to improve after 0.4 pore volume of water injection. This improvement was continued until 0.65 pore volume of water injection. As a result, a 10% increase in oil recovery was achieved in the MEOR process using polymer-producing bacteria.

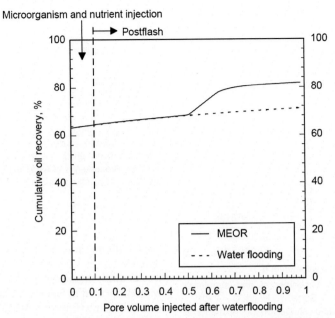

FIG. 4.70 Comparison of cumulative oil recovery between waterflooding and MEOR. Microbial enhanced oil recovery. (Credit: from Sugai, Y., Hong, C., Chida, T., Enomoto, H., 2007. Simulation studies on the mechanisms and performances of MEOR using a polymer producing microorganism Clostridium sp. TU-15A. Proceedings of Asia Pacific Oil and Gas Conference and Exhibition, Jakarta, Indonesia, 30 October–1 November, 2007.)

REFERENCES

Amaefule, J.O., Handy, L.L., 1982. The effect of interfacial tensions on relative oil/water permeabilities of consolidated porous media. Society of Petroleum Engineers Journal 22 (3), 371–381.

Amani, H., Sarrafzadeh, M.H., Haghighi, M., Mehrnia, M.R., 2010. Comparative study of biosurfactant producing bacteria in MEOR applications. Journal of Petroleum Science and Engineering 75 (1–2), 209–214.

Anderson, W.G., 1987. Wettability literature survey part 5: the effects of wettability on relative permeability. Journal of Petroleum Engineers 39 (11), 1453–1468.

Bajpai, R.K., Reuss, M., 1982. Coupling of mixing and microbial kinetics for evaluating the performance of bioreactors. The Canadian Journal of Chemical Engineering 60 (3), 384–392.

Bang, H.W., Caudle, B.H., 1984. Modeling of a micellar/polymer process. Society of Petroleum Engineers Journal 24 (6), 617–627.

Basak, S., Ramaswamy, H.S., Piette, J.P.G., 2002. High pressure destruction kinetics of *Leuconostoc mesenteroides* and *Saccharomyces cerevisiae* in single strength and concentrated orange juice. Innovative Food Science & Emerging Technologies 3 (3), 223–231.

Behesht, M., Roostaazad, R., Farhadpour, F., Pishvaei, M.R., 2008. Model development for MEOR process in conventional non-fractured reservoirs and investigation of physico-chemical parameter effects. Chemical Engineering & Technology 31 (7), 953–963.

Bültemeier, H., Alkan, H., Amro, M., 2014. A new modeling approach to MEOR calibrated by bacterial growth and metabolite curves. In: Proceedings of the SPE EOR Conference at Oil and Gas West Asia, Muscat, Oman, 31 March–2 April, 2014.

Chang, M.M., Bryant, R.S., Chung, T.H., Gao, H.W., Burchfield, T.E., 1991. Modeling and laboratory investigations of microbial transport phenomena in porous media. In: Proceedings of the SPE Annual Technical Conference and Exhibition, Dallas, Texas, USA, 6–9 October, 1991.

CMG, 2018. STARS User Gide. Computer Modelling Group Ltd., Calgary, Alberta, Canada.

Corapcioglu, M.Y., Haridas, A., 1984. Transport and fate of microorganisms in porous media: a theoretical investigation. Journal of Hydrology 72 (1–2), 149–169.

Delshad, M., Asakawa, K., Pope, G.A., Sepehrnoori, K., 2002. Simulations of chemical and microbial enhanced oil recovery methods. In: Proceedings of the SPE/DOE Improved Oil Recovery Symposium, 13–17 April, 2002.

Delshad, M., Delshad, M., Pope, G.A., Lake, L.W., 1987. Two- and three-phase relative permeabilities of micellar fluids. SPE Formation Evaluation 2 (3), 327–337.

Desouky, S.M., Abdel-Daim, M.M., Sayyouh, M.H., Dahab, A.S., 1996. Modelling and laboratory investigation of microbial enhanced oil recovery. Journal of Petroleum Science and Engineering 15 (2–4), 309–320.

Dols, M., Chraibi, W., Remaud-Simeon, M., Lindley, N.D., Monsan, P.F., 1997. Growth and energetics of *Leuconostoc mesenteroides* NRRL B-1299 during metabolism of various sugars and their consequences for dextransucrase production. Applied and Environmental Microbiology 63 (6), 2159–2165.

Foster, W.R., 1973. A low-tension waterflooding process. Journal of Petroleum Technology 25 (2), 205–210.

Fulcher, R.A., Ertekin, T., Stahl, C.D., 1985. Effect of capillary number and its constituents on two-phase relative permeability curves. Journal of Petroleum Technology 37 (2), 249–260.

Green, D.W., Willhite, G.P., 1998. Enhanced Oil Recovery. Society of Petroleum Engineers, Dallas, Texas, USA.

Hand, D.B., 1939. The distribution of a consulate liquid between two immiscible liquids. Journal of Physics and Chemistry 34, 1961–2000.

Healy, R.N., Reed, R.L., 1997a. Immiscible microemulsion flooding. Society of Petroleum Engineers Journal 17 (2), 129–139.

Healy, R.N., Reed, R.L., 1997b. Some physical-chemical aspects of microemulsion flooding: a review. In: Shah, D.O., Schechter, R.S. (Eds.), Improved Oil Recovery by Surfactant and Polymer Flooding. Academic Press, Cambridge, Massachusetts, USA, pp. 383–437.

Henze M., Gujer W., Mino T., Matsuo T., Wentzel, M.C. and Marais G.v.R., Activated Sludge Model No. 2, 1995, IAWQ Scientific and Technical Report No. 3, IAWQ, London.

Hong, E., Jeong, M.S., Kim, T.H., Lee, J.H., Cho, J.H., Lee, K.S., 2019a. Development of coupled biokinetic and thermal model to optimize cold-water microbial enhanced oil recovery (MEOR) in homogenous reservoir. Sustainability 11 (6), 1–19.

Hong, E., Jeong, M.S., Lee, K.S., 2019b. Optimization of non-isothermal selective plugging with a thermally active biopolymer. Journal of Petroleum Science and Engineering 173, 434–446.

Hosseininoosheri, P., Lashgari, H.R., Sepehrnoori, K., 2016. A novel method to model and characterize in-situ biosurfactant production in microbial enhanced oil recovery. Fuel 183, 501–511.

Huh, C., 1979. Interfacial tensions and solubilizing ability of a microemulsion phase that coexists with oil and brine. Journal of Colloid and Interface Science 71 (2), 408–426.

Ibrahim, Z., Omar, M.I., Foo, K.S., Elias, E.J., Othman, M., 2004. Simulation analysis of microbial well treatment of Bokor field, Malaysia. In: Proceedings of SPE Asia Pacific Oil and Gas Conference and Exhibition, Perth, Australia, 18–20 October, 2004.

Islam, M.R., 1990. Mathematical modeling of microbial enhanced oil recovery. In: Proceedings of the SPE Annual Technical Conference and Exhibition, New Orleans, LA, USA, 23–26 September, 1990.

Islam, M.R., Erno, B.P., Davis, D., 1992. Hot gas and waterflood equivalence of in situ combustion. Journal of Canadian Petroleum Technology 31 (8), 44–52.

Islam, M.R., Gianetto, A., 1993. Mathematical modelling and scaling up of microbial enhanced oil recovery. Journal of Canadian Petroleum Technology 32 (4), 30–36.

Jeong, M.S., Lee, J.H., Lee, K.S., 2019a. Critical review on the numerical modeling of in-situ microbial enhanced oil recovery processes. Biochemical Engineering Journal 150, 1–9.

Jeong, M.S., Noh, D., Hong, E., Lee, K.S., Kwon, T., 2019b. Systematic modeling approach to selective plugging using in situ bacterial biopolymer production and its potential for microbial-enhanced oil recovery. Geomicrobiology Journal 36 (5), 468–481.

Lappan, R.E., 1994. Reduction of Porous Media Permeability from in Situ Bacterial Growth and Polysaccharide Production (Ph.D. thesis). University of Michigan, Ann Arbor, Michigan, USA.

Lappan, R.E., Fogler, H.S., 1994. *Leuconostoc mesenteroides* growth kinetics with application to bacterial profile modification. Biotechnology and Bioengineering 43 (9), 865–873.

Leroi, F., Fall, P.A., Pilet, M.F., Chevalier, F., Baron, R., 2012. Influence of temperature, pH and NaCl concentration on the maximal growth rate of *Brochothrix thermosphacta* and a bioprotective bacteria *Lactococcus piscium* CNCM I-4031. Food Microbiology 31 (2), 222–228.

Masalmeh, S.K., 2002. The effect of wettability on saturation functions and impact on carbonate reservoirs in the middle east. In: Proceedings of Abu Dhabi International Petroleum Exhibition and Conference, Abu Dhabi, UAE, 13–16 October, 2002.

Meszaros, G., Chakma, A., Jha, K.N., Islam, M.R., 1990. Scaled model studies and numerical simulation of inert gas injection with horizontal wells. In: Proceedings of the SPE Annual Technical Conference and Exhibition, New Orleans, Louisiana, USA, 23–26 September, 1990.

Mitchell, D.A., von Meien, O.F., Krieger, N., Dalsenter, F.D.H., 2004. A review of recent developments in modeling of microbial growth kinetics and intraparticle phenomena in solid-state fermentation. Biochemical Engineering Journal 17 (1), 15–26.

Moore, T.F., Slobod, R.L., 1955. Displacement of oil by water-effect of wettability, rate, and viscosity on recovery. In: Proceedings of Fall Meeting of the Petroleum Branch of AIME, New Orleans, Louisiana, 2–5 October, 1955.

Nelson, R.C., Pope, G.A., 1978. Phase relationships in chemical flooding. Society of Petroleum Engineers Journal 18 (5), 325–338.

Noh, D.H., Ajo-Franklin, J.B., Kwon, T.H., Muhunthan, B., 2016. P and S wave responses of bacterial biopolymer formation in unconsolidated porous media. Journal of Geophysical Research: Biogeosciences 121 (4), 1158–1177.

Park, C., Marchand, E.A., 2006. Modelling salinity inhibition effects during biodegradation of perchlorate. Journal of Applied Microbiology 101 (1), 222–233.

Park, C., Raines, R.T., 2001. Quantitative analysis of the effect of salt concentration on enzymatic catalysis. Journal of the American Chemical Society 123 (46), 11472–11479.

Patel, J., Borgohain, S., Kumar, M., Rangarajan, V., Somasundaran, P., Sen, R., 2015. Recent developments in microbial enhanced oil recovery. Renewable and Sustainable Energy Reviews 52, 1539–1558.

Ratkowsky, D.A., Olley, J., McMeekin, T.A., Ball, A., 1982. Relationship between temperature and growth rate of bacterial cultures. Journal of Bacteriology 149 (1), 1–5.

Rittmann, B.E., 1982. The effect of shear stress on biofilm loss rate. Biotechnology and Bioengineering 24 (2), 501–506.

Rittmann, B.E., McCarty, P.L., 2001. Environmental Biotechnology: Principles and Applications. McGraw-Hill, New York, USA.

Robertson, E.C., 1988. Thermal Properties of Rocks. US Geological Survey.

Roels, J.A., 1983. Energetics and Kinetics in Biotechnology. Elsevier Science Ltd., Amsterdam, Netherland.

Rosso, L., Lobry, J.R., Bajard, S., Flandrois, J.P., 1995. Convenient model to describe the combined effects of temperature and pH on microbial growth. Applied and Environmental Microbiology 61 (2), 610–616.

Santos, M., Teixeira, J., Rodrigues, A., 2000. Production of dextransucrase, dextran and fructose from sucrose using *Leuconostoc mesenteroides* NRRL B512(f). Biochemical Engineering Journal 4 (3), 177–188.

Sheng, J.J., 2011. Modern Chemical Enhanced Oil Recovery: Theory and Practice. Gulf Professional Publishing, Houston, Texas, USA.

Silfanus, N.J., 1990. Microbial Mechanisms for Enhanced Oil Recovery from High Salinity Core Environments (M.S. thesis). University of Oklahoma, Norman, Oklahoma, USA.

Sitnikov, A.A., Eremin, N.A., 1994. A mathematical model of microbial enhanced oil recovery (MEOR) method for mixed type rock. In: Proceedings of European Petroleum Conference, London, United Kingdom, 25–27 October, 1994.

Steefel, C.I., 2009. CrunchFlow-software for Modeling Multicomponent Reactive Flow and Transport. Lawrence Berkeley National Laboratory, Berkeley, California, USA.

Stewart, T.L., Fogler, H.S., 2001. Biomass plug development and propagation in porous media. Biotechnology and Bioengineering 72 (3), 353–363.

Stewart, T.L., Kim, D.S., 2004. Modeling of biomass-plug development and propagation in porous media. Biochemical Engineering Journal 17 (2), 107–119.

Sugai, Y., Hong, C., Chida, T., Enomoto, H., 2007. Simulation studies on the mechanisms and performances of MEOR using a polymer producing microorganism *Clostridium* sp. TU-15A. In: Proceedings of Asia Pacific Oil and Gas Conference and Exhibition, Jakarta, Indonesia, 30 October–1 November, 2007.

Surasani, V.K., Li, L., Ajo-Franklin, J.B., Hubbard, C., Hubbard, S.S., Wu, Y., 2013. Bioclogging and permeability alteration by *L. mesenteroides* in a sandstone reservoir: a reactive transport modeling study. Energy & Fuels 27 (11), 6538–6551.

Tang, G.Q., Firoozabadi, A., 2002. Relative permeability modification in gas/liquid systems through wettability alteration to intermediate gas wetting. SPE Reservoir Evaluation and Engineering 5 (6), 427–436.

Vilcáez, J., Li, L., Wu, D., Hubbard, S.S., 2013. Reactive transport modeling of induced selective plugging by *Leuconostoc mesenteroides* in carbonate formations. Geomicrobiology Journal 30 (9), 813–828.

Villadsen, J., Nielsen, J., Lidén, G., 2011. Bioreaction Engineering Principles. Springer Science and Business Media, Berlin, Germany.

Wijtes, T., McClure, P.J., Zwietering, M.H., Roberts, T.A., 1993. Modelling bacterial growth of *Listeria monocytogenes* as a function of water activity, pH and temperature. International Journal of Food Microbiology 18 (2), 139–149.

Wolery, T.J., Jackson, K.J., Bourcier, W.L., Bruton, C.J., Viani, B.E., Knauss, K.G., Delany, J.M., 1990. Current status of the EQ3/6 software package for geochemical modeling. In: Melchior, D.C., Bassett, R.L. (Eds.), Chemical Modeling of Aqueous Systems II. American Chemical Society, Washington D.C., USA, pp. 104–116.

Wu, Y., Surasani, V.K., Li, L., Hubbard, S.S., 2014. Geophysical monitoring and reactive transport simulations of bioclogging processes induced by *Leuconostoc mesenteroides*. Geophysics 79 (1), E61–E73.

Xiao, M., Zhang, Z.Z., Wang, J.X., Zhang, G.Q., Luo, Y.J., Song, Z.Z., Zhang, J.Y., 2013. Bacterial community diversity in a low-permeability oil reservoir and its potential for enhancing oil recovery. Bioresource Technology 147, 110–116.

Zhang, X., Knapp, R.M., McInerney, M.J., 1992. A mathematical model for microbially enhanced oil recovery process. In: Proceedings of the SPE/DOE Enhanced Oil Recovery Symposium, Tulsa, Oklahoma, USA, 22–24 April, 1992.

Zwietering, M.H., Wijtzes, T., Rombouts, F.M., Riet, K.V.T., 1993. A decision support system for prediction of microbial spoilage in foods. Journal of Industrial Microbiology 12, 324–329.

CHAPTER 5

Field Applications

5.1 CONSIDERATIONS FOR MEOR IMPLEMENTATION

5.1.1 Injection Strategies

In order to apply the MEOR to the actual field, it is important to determine its application strategy. There are two main types for MEOR applications. The first one is microbial well stimulation or cyclic microbial recovery (CMR) or Huff-and-Puff process. This CMR is widely used all over the world due to its commercial advantages. In CMR, the microbial and nutrient mixtures are introduced into the reservoir during the injection (Huff stage). The injector is then shut-in for incubation period allowing the bacteria to generate the metabolites which help to extract oil. The well is then opened and produces the oil and bioproducts resulting from the incubation (Puff stage) (Fig. 5.1). These processes may be repeated. The main applications are heavier oil reservoirs that suffer from paraffin and asphaltene depositions. The main areas where this method is applied are the United States, Venezuela, China, Indonesia, and to some extent in India (Patel et al., 2015).

The second method is microbial enhanced water flooding or microbial flooding recovery (MFR). The MFR transports the nutrients farther from the injector to the producer. For oil and gas production, the MFR method utilizes a microbial solution injected into the reservoir. In general, the reservoir is preflushed with water before the solution is introduced. After the solution is injected, it is pushed into the reservoir by drive water to form the metabolites necessary for oil extraction. Finally, the extracted oil and microbial products are produced in production wells (Patel et al., 2015) (Fig. 5.2).

5.1.2 Microorganisms

In the practical applications of MEOR to increase oil productivity, specific mechanisms of microorganisms should be applied to different reservoir characteristics. Therefore, it is important to choose the appropriate microorganisms. Consequently, classifications of microbes and their recovery mechanisms are the critical issues that should be taken account before the implementation of MEOR projects.

In Chapter 1, microorganisms can be categorized into two groups of indigenous (in situ) and exogenous (ex situ). In terms of MEOR process, indigenous bacteria have greater potential in that they adapt better to the reservoir environment than exogenous ones. However, the technique also includes the problem of high uncertainty.

In another perspective, bacteria can be classified according to temperature and pressure conditions in the range of their metabolism. This classification is shown in Table 5.1 (Safdel et al., 2017). The need for oxygen for microbial survival can also be a condition of classification. Table 5.2 presents three microbial classifications according to this condition and their products.

Table 5.3 shows the classification by microorganism type or recovery mechanism applied to MEOR processes. The sulfate- and nitrate-reducing bacteria (SRB and NRB) can help increase oil recovery by mechanisms being different from other types of microorganisms. The SRB can reduce the viscosity and interfacial tension by the generation of aliphatic and aromatic hydrocarbons. The NRB can control the souring phenomena to modify the permeability.

The SRB can potentially form a stabilized emulsion system of crude oil and can also reduce surface and interfacial tension by producing metabolites such as aliphatic and aromatic hydrocarbons (Romero-Zerón, 2012; Song et al., 2014). The examples of SRB are *Desulfotomaculum nigrificans*, *Thermodesulfobacterium mobile*, *Archaeoglobus fulgidus*, *Desulfovibrio longus*, and *Desulfobacterium cetonicum* (Song et al., 2014). The activity of these bacteria can lead to the souring phenomena (Youssef et al., 2009).

The NRB are another type of microbial species that control the souring effect by mechanisms such as performing as electron donors, sulfide oxidation, and increasing the redox potential (Youssef et al., 2009). Table 5.4 shows some examples of using the nitrates to control hydrogen sulfide (H_2S) production in the reservoir environment.

5.1.3 Nutrients

Although the application of proper microbes is very important in the successful field implementation of MEOR, supplying the appropriate and sufficient amount of nutrients is also vital for surviving and growth of microbes that finally contributes to increase

Theory and Practice in Microbial Enhanced Oil Recovery. https://doi.org/10.1016/B978-0-12-819983-1.00005-3

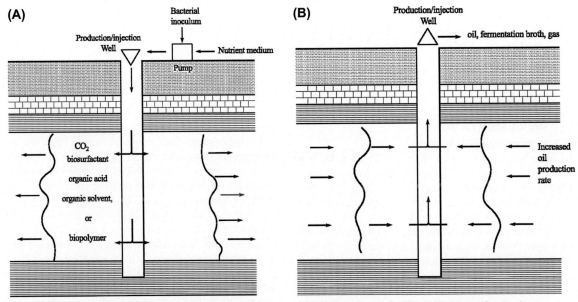

FIG. 5.1 Schemes of cyclic microbial recovery process: **(A)** Huff stage, **(B)** Puff stage. (Credit: from Donaldson, E.C., Chilingarian, G.V., Yen, T.F., 1989. Introduction. In: Donaldson, E.C., Chilingarian, G.V., Yen, T.F. (Eds.), Microbial Enhanced Oil Recovery. In: Chilingarian, G.V. (Ed.), Developments in Petroleum Science, vol. 22. Elsevier, Amsterdam, pp. 1–14.)

in oil recovery. Therefore, introduction and classification of different kinds of nutrients for MEOR applications has a benefit. The microbial growth rate and nutrient concentration are closely related. In other words, the use of a sufficient amount of nutrients with the right composition is critical to produce the desired bioproduct and perform a successful MEOR project. Some bacteria consume the hydrocarbons as their nutrients by degrading long alkyl chains (Sen, 2008; Gudiña et al., 2012). According to MEOR purpose, the nutrients are classified as follows (Safdel et al., 2017):

- Molasses only
- In situ hydrocarbon (crude oil)
- Molasses and nitrogen and phosphorous salts
- Miscellaneous nutrients (Maudgalya et al., 2007)

5.2 CLASSIFICATIONS OF FIELD APPLICATIONS

In the last few decades, a number of field trials have been performed in various countries to evaluate the applicability of MEOR technology. The first reported field trial of MEOR was implemented by the Mobil Research laboratory in Arkansas in 1954. Although it was considered as a marginal success (Yarbrough and Coty, 1983), its

analysis highlighted the complexity of using bacteria in oil recovery process. A large number of bacteria were found in the reservoir, and the necessity of providing proper nutrients and controlling the growth of target bacteria was a critical conclusion. Since then, many field applications have been carried out. Maudgalya et al. (2007) presented the successes and failures in the field applications. They classified 407 field trials which were reported in previous literatures (Hitzman, 1983; Lazar, 1983; Yarbrough and Coty, 1983; Davidson and Russell, 1988; Hitzman, 1988; Oppenheimer and Heibert, 1988; Sheehy, 1990; Bryant et al., 1991, 1994; Jack et al., 1991; Knapp et al., 1991; Lazar, 1991; Lazar et al., 1991, 1992; Wagner, 1991; Matz et al., 1992; Portwood and Hiebert, 1992; Moses et al., 1993; Zhang and Zhang, 1993; Arinbasarov et al., 1995; Portwood, 1995; Wagner et al., 1995; Bryant and Lindsey, 1996; Buciak et al., 1996; Dietrich et al., 1996; Jenneman et al., 1996; Deng et al., 1999; Trebbau et al., 1999; Yusuf et al., 1999; Zhang et al., 1999; Ghazali Abd. Karim et al., 2001; Maure et al., 2001; Brown et al., 2002; Nagase et al., 2002; Hitzman et al., 2004; Strappa et al., 2004). Maudgalya et al. (2007) classified the field trials according to lithology, trial types, recovery mechanisms, microorganisms, nutrients, and reservoir environments.

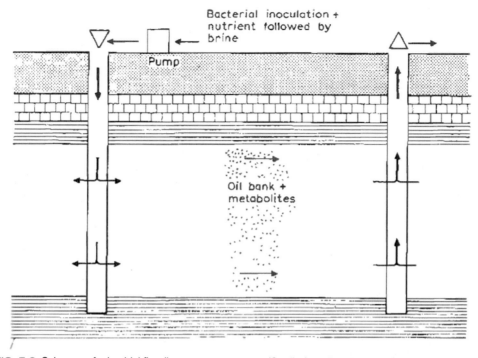

FIG. 5.2 Schemes of microbial flooding recovery process. (Credit: from Donaldson, E.C., Chilingarian, G.V., Yen, T.F., 1989. Introduction. In: Donaldson, E.C., Chilingarian, G.V., Yen, T.F. (Eds.), Microbial Enhanced Oil Recovery. In: Chilingarian, G.V. (Ed.), Developments in Petroleum Science, vol. 22. Elsevier, Amsterdam, pp. 1–14.)

TABLE 5.1
Classification of Microorganisms According to Temperature and Pressure (Safdel et al., 2017).

Classification Parameter	Microorganism Class	Range
Temperature (°C)	Psychrophile	<13
	Mesophile	8–47
	Different classes of thermophiles	42–113
Pressure (MPa)	Piezotolerant	Up to 50
	Piezophiles	Up to 65
	Extreme piezophiles	Up to 100

TABLE 5.2
Classification of Microorganisms According to Respiration (Safdel et al., 2017).

Type of Respiration	Microorganism Family	Products
Aerobic	Corynebacterium	Surfactants
	Psedomonas	Surfactants and polymers
	Xanthomonas	Polymers
	Clostridium	Gases, acids, alcohols, and surfactants
Anaerobic	Desulfovibrio	Gases and acids
	Bacillus	Acids and surfactants
	Leuconostoc	Polymers
Facultative	Arthobacter	Surfactants and alcohols
	Enterobacter	Gases and acids

5.2.1 Lithology and Trial Types

In the MEOR process, the reservoir lithology impacts on the retention of bacterial cells and nutrients. Lithology is largely divided into two categories: sandstone and carbonate, with 314 of 407 projects applied to the

TABLE 5.3
MEOR Classification According to Type of Microorganism or Recovery Process (Kianipey and Donaldson, 1986; Bubela, 1989; Grula et al., 1989; Marsh et al., 1995; Safdel et al., 2017).

TYPE OF MICROORGANISM

Bacillus	Production of gases, alcohols, and biosurfactants
Clostridia	Production of acid and gases
Pseudomonas	Production of biopolymer and biosurfactant along with permeability modification
Sulfate-reducing bacteria (SRB)	Oil biodegradability and viscosity reduction along with production of methane
Nitrate-reducing bacteria (NRB)	Souring control and permeability modification
Others	Oxidation and biodegradability of hydrocarbons along with permeability modification and methane production

RECOVERY PROCESS

Permeability modification	Volumetric sweep efficiency improvement in waterflooding process
Biopolymer, biosurfactants, acids, and alcohol production	Permeability reduction and capillary number enhancement
Oil biodegradation	Production of low viscosity molecules
Gas production	Oil viscosity reduction

TABLE 5.4
The Hydrogen Sulfide Control in Reservoirs (Safdel et al., 2017).

Injection Species	Result	References
Continuous NH_4NO_3 injection	Sulfide levels reduction by 40%–60%	McInerney et al. (1991)
Continuous injection of NO_3^-	H_2S reduction after breakthrough of treated water	Thorstenson et al. (2002) Larsen et al. (2004)
NO_3^- and NO_2 injection	Decrease of sulfide levels dissolved in both production equipment and produced water	Hitzman et al. (2004)
Continuous injection of NO_3^- and PO_4^{2-}	Decrease in sulfide levels; enhancement of nitrate reducer	Telang et al. (1997)

sandstone reservoir and 89 to the carbonate reservoir. The remaining four cases did not provide detailed information. The trial types were classified into three types: well stimulation, waterflooding, and Huff-and-Puff process. In this case, well stimulation meant removing wellbore damage or formation damage, and also included removing paraffin and asphaltene inside and around the wellbore. The results for these are summarized in Table 5.5. Of the 403 projects, except 4 projects for which no detailed information was given, 333 were well stimulation, and the remaining 70 were either waterflooding or Huff-and-Puff process. 45 of the remaining 70 projects were Huff-and-Puff process, and the remaining 25 were waterflooding case. 36 of the Huff-and-Puff tests were classified as successful project in which improvements of oil rates were observed. 20 of the 25 waterflooding cases were designated successful. Trial operators judged the success or failure of their projects. The most dramatic success was the eightfold increase in oil rate, which improved from 9 to 75 bbl/day over 4 months (Arinbasarov et al., 1995). Decline curve analysis predicted a 5% change in oil in place. Of the 70 waterfloodings and Huff-and-Puff, 66 were applied to sandstone reservoirs and 4 were applied to carbonate reservoirs. In particular, the applications of carbonate reservoirs were successful due to matrix acidizing with metabolic acid and decreasing oil viscosity with produced gas (Ghazali Abd. Karim et al., 2001). The *Clostridia* strains were used to generate this acid.

All MEOR cases were applied to the reservoirs with oil saturation of 40%–70%. Though early attempts were mainly applied to depleted reservoirs with very low oil saturation, subsequent tests were performed on reservoirs with higher oil saturation than trapped oil saturation. A number of MEOR applications have been reported in Europe. Most of these were done in

TABLE 5.5
Summary of Results by Application Type (Maudgalya et al., 2007).

Trial Type	SANDSTONE		CARBONATE	
	Success	Failure	Success	Failure
Well stimulation	248	0	84	1
Waterflooding	17	5	3	0
Huff-and-Puff	35	9	1	0

the 1960s, 1970s, and several years thereafter. The United States has the second largest number of applications, which were most active in the 1980s. China has recently been actively implementing MEOR field applications.

5.2.2 Type of Recovery Mechanisms

The main mechanisms of the microbial process include changes in permeability by microorganisms and bioproducts, changes in physical properties by the metabolites, and degradation of high molecular weight hydrocarbons. 70 waterflooding and Huff-and-Puff cases were investigated and the results are shown in Table 5.6. The total number of trials in the table is 80 because more than one main mechanisms have been applied in some cases. In order to avoid confusion and simplify the classification, only the mechanisms that contributed the most to the oil recovery were considered.

The most successful process was the permeability modification. In some successful waterflooding cases, only nutrients were injected to utilize the indigenous bacteria. In three applications, indigenous NRB were utilized. An additional benefit of using them was that the bacteria inhibited oil souring which was induced by sulfide. Even in Huff-and-Puff process, where IFT reduction or oil degradation was the main recovery mechanism, bacteria formed colonies in the highly permeable zone and diverted water into the unswept area. Though few tests are available, sweep efficiency improvement following permeability reduction can be confirmed by production oil analysis or pressure transient analysis.

Oil degradation is the process of breaking high molecular hydrocarbons into smaller molecules to improve the hydrocarbons' mobility and has been successfully applied in both waterflooding and Huff-and-Puff process. This can be verified by chromatographs which represented the changes in the composition of produced oil samples. The viscosity reduction can be measured with viscometers.

TABLE 5.6
Summary of Results by Recovery Mechanisms (Maudgalya et al., 2007).

Type of Recovery Mechanism	No. of Trials	Success	Failure
Permeability modification	10	7	3
CO_2 production	10	9	1
Generation of biopolymer, biosurfactant, alcohols, and acid	26	20	6
Oil degradation	34	29	5

Production of CO_2 and/or methane to displace oil due to increasing gas saturation or decreasing oil viscosity was generally successful. CO_2 was generated during bacteria activities or chemical reactions between carbonate rocks and metabolic acids. In early tests using SRB, there was a problem that caused oil souring. Therefore, later trials used *Clostridia* species to produce the gases.

The wettability alteration due to biosurfactant production has been used as a major mechanism in a few waterflooding and Huff-and-Puff processes. Since these biosurfactants, alcohols, and biopolymers have not been detected in the product stream, there were also questions about their practical applications (Maudgalya et al., 2007). Based on preliminary laboratory core analysis and subsequent increase in oil rate, their effects on recovery enhancement were analyzed.

According to Moses et al. (1993), three of four carbonate reservoirs have been successful in matrix acidizing by generated acid and increasing oil recovery by generated gases to reduce the viscosity. In the fourth field application, too low matrix permeability was analyzed as a failure factor.

5.2.3 Microorganisms

Field applications can also be classified by the type of bacteria used in MEOR process (Table 5.7). Microbial adaptability and robustness in the reservoir conditions are very important in that they affect the formation of the desired bioproduct and oil recovery. Either indigenous or injected bacteria can be applied. When the indigenous bacteria are used, only nutrient injection is required. The microorganisms used in the MEOR method and the main mechanisms are as follows:

- *Bacillus*: production of biosurfactant, alcohols, and gases (Marsh et al., 1995; Kianipey and Donaldson, 1986)
- *Clostridia*: production of acids and gases (methane by methanogenesis) (Marsh et al., 1995; Grula et al., 1989; Kianipey and Donaldson, 1986)
- *Pseudomonas*: production of biosurfactant and biopolymer for the permeability modification (Kianipey and Donaldson, 1986)
- SRB: oil degradation, oil viscosity reduction, and methane production by methanogenesis (Bubela, 1989)

- NRB: permeability modification and souring control (Bubela, 1989)
- Others: oxidation and degradation of hydrocarbons, permeability modification, and methane production to lower gas viscosity (Bubela, 1989)

The *Bacillus* and *Clostridia* were the most commonly utilized microorganisms. The *Clostridia* spores were utilized more frequently. Most experiments applied a combination of bacteria to take advantage of their different abilities. Most applications using two or more microbes attempted to exploit their different abilities. SRB were used in early time, but were avoided later. In situ NRB successfully modified permeability and had a benefit of reducing oil souring.

Most of successful trials used anaerobic microorganisms. Though a few early field applications used aerobic microorganisms, the danger of air injection and the uncertainty of the effects prevented their use. The effects of biosurfactants, biopolymers, and alcohols generated by bacteria such as *Bacillus* were also inconclusive because they have not been observed in product streams during composition analysis.

Though the oil-degrading bacteria have been reported in most applications, the type of bacteria actually applied was not clear. In addition, there was no relationship between the application results and the species of bacteria used. Microbial behavior is usually inconsistent and does not guarantee the success of the next attempt even if particular bacteria has shown successful results.

5.2.4 Nutrients

Since nutrients are the biggest part of the MEOR cost, it is important to use the appropriate type and amount. Molasses is the most commonly used carbon source because it is easy to use as a slurry. Other nutrients include nitrates and phosphorous salts provided by fertilizer such as ammonium phosphate, superphosphate, ammonium nitrate, and sodium nitrate (Bubela, 1989; Maudgalya et al., 2007). Table 5.8 shows the field applications according to nutrients. The results indicated that the phosphorous and nitrogen fertilizers were present as inorganic components due to availability and cost.

5.2.5 Reservoir Environments

As the reservoir environments have a great influence on microbial growth and metabolic activities, screening is very important for MEOR application. Detailed screening results are presented in Chapter 1. This section describes the field application results for reservoir permeability, temperature, and salinity (Table 5.9).

TABLE 5.7
Status of Microorganisms Used in the Field Application (Maudgalya et al., 2007).

Type of Bacteria	No. of Trials	Waterflooding	Huff-and-Puff
Bacillus species (biosurfactant production)	11	10	1
Clostridium species (gas and acid production)	37	13	24
Pseudomonas species (biosurfactant and biopolymer production)	14	7	7
Nitrate-reducing bacteria (permeability modification)	3	2	1
Sulfate-reducing bacteria (oil degradation)	15	2	13
Unknown and proprietary bacteria	39	11	28

Most of MEOR processes were applied to reservoirs with permeability ranging from 70 to 1,000 md, of which more than three-quarters were successful. Only one application was reported in carbonate reservoir with 1–10 md permeability (Maudgalya et al., 2007). Though acidizing was also performed in this reservoir, it failed. A few applications were carried out in 10–75 md reservoirs. Injected bacteria were sometimes observed at several times higher than the injection concentration in neighboring wells, which was the basis for identifying their motility and proliferation. The probability of MEOR success according to salinity was high when it was less than 100,000 ppm. In the early applications, salinity conditions were not presented. All MEOR methods were applied to reservoirs below 200°F. This temperature is considered as the maximum temperature at which the effect of bacteria used in MEOR appears. Although there is no direct correlation between reservoir depth and MEOR efficiency, deep

reservoirs are generally not suitable for MEOR due to high salinity, low permeability, and high temperature environment.

5.3 WORLD MEOR APPLICATIONS

Since the mid-1950s, many MEOR field applications have been applied in the United States, Eastern Europe, and the Soviet Union (Hitzman, 1983). Then, the oil crisis of 1970s became a trigger in many countries to raise interest in MEOR. Table 5.10 provides information on MEOR implemented in different countries (Lazar et al., 2007; Patel et al., 2015; Safdel et al., 2017). The abbreviations for type of MEOR application in the table means cyclic microbial recovery (CMR), microbial flooding recovery (MFR), and microbial selective plugging recovery (MSPR). Safdel et al. (2017) graphically illustrated the world's applications of MEOR and its results (Figs. 5.3–5.6).

In United States, the first MEOR field application was implemented in the Lisbon field, Union County, Arkansas (Hitzman, 1983, 1988; Grula and Russell, 1985; Zajic and Smith, 1987; Bryant and Douglas, 1988; Jenneman et al., 1993; Nelson and Schneider, 1993). 2% beet molasses solution was continuously injected for 6 months with 4,000 gallons of *C. acetobutylicum* suspension. Breakthrough of metabolic products was observed at 80–90 days after injection. Produced water samples contained source bacteria, while *C. acetobutylicum* was not detected. No bacteria and fermentation products were detected in the produced oil samples. Yarbrough and Coty (1983)

TABLE 5.8
Classification of Field Applications According to Nutrients (Maudgalya et al., 2007).

Type of Nutrient	No. of Applications
Molasses and N and P fertilizer	27
Only molasses	23
In situ hydrocarbon	17
Others	7

TABLE 5.9
Field Applications According to Reservoir Conditions (Maudgalya et al., 2007).

Property	Lithology	Value	No. of Success	No. of Failure
Permeability (md)	Sandstone	1–10	0	0
		10–75	6	1
		75–1,000	41	12
		1,000–10,000	1	1
	Carbonate	1–10	2	0
		10–75	1	0
		75–1,000	1	0
		1,000–10,000	0	0
Temperature (°F)	—	50–200	48	18
		>200	0	0
Salinity (ppm)	—	<1,000	0	0
		1,000–100,000	13	1
		>100,000	6	6

TABLE 5.10
MEOR Applications Around World (Lazar et al., 2007; Patel et al., 2015).

Country	Type of MEOR Application	Microbial System	Nutrients	Oil Production	References
United States	CMR, MFR, MSPR	- Pure or mixed cultures of *Bacillus*, *Clostridium*, *Pseudomonas*, gram-negative rods - Mixed cultures of hydrocarbon-degrading bacteria - Mixed cultures of marine source bacteria - Spore suspension of *Clostridium* - Indigenous stratal microflora - Slime-forming bacteria - Ultramicrobacteria	- Molasses 2%−4% - Molasses and ammonium nitrate addition - Free corn syrup + mineral salts - Maltodextrine and organic phosphate esters (OPE) - Salt solution - Sucrose 10% + Peptone 1% + NaCl 0.5% −30% - Brine supplemented with nitrogen and phosphorous sources and nitrate - Biodegradable paraffinic fractions + mineral salts - Naturally contain inorganic and organic materials + N, P sources	Improvement	Hitzman (1983, 1988), Yarbrough and Coty (1983), Grula and Russell (1985), Zajic and Smith (1987), Bryant and Douglas (1988), Bryant et al. (1990), Jenneman et al. (1993), and Nelson and Schneider (1993)
Canada	MSPR	- Pure culture of *Leuconostoc mesenteroides*	- Dry sucrose + sugar beet molasses dissolved in water	Not yet reported	Jack and Stehmeier (1988), Jack (1991), and Cusack et al. (1992)
Russia	MFR, MSPR	- Pure cultures of Clostridium tyrobutiricum - Bacteria-mixed cultures - Indigenous microflora of water injection and water formation - Activated sludge bacteria - Naturally occurring microbiota of industrial (food) wastes	- Molasses 2%−6% with nitrogen and phosphorous salt addition - Water injection with nitrogen and phosphorous salt and air addition - Wastewaters with addition of biostimulators and chemical additives - Industrial wastes with salts addition - Dry milk 0.04%	Improvement	Senyukov et al. (1970), Ivanov et al. (1993), Wagner et al. (1993), Svarovskaya (1995), and Nazina (2004)

Former Czechoslovakia	CMR, MFR	- Hydrocarbon-oxidizing bacteria (predominant *Pseudomonas* sp.) - Sulfate-reducing bacteria	- Molasses	Improvement	Dostaleck and Spurny (1958) and Updegraff (1990)
Poland	MFR	- Mixed bacteria cultures (*Arthrobacter, Clostridium, Mycobacterium, Pseudomonas, Peptococcus*)	- Molasses 2%	Improvement	Karaskiewicz (1974)
Romania	MFR	- Adapted mixed enrichment cultures (predominant: *Clostridium, Bacillus, Pseudomonas,* and other gram-negative rods)	- Molasses 2%−4%	Improvement	Lazar (1998) and Lazar et al., (1999)
Hungary	MFR	- Mixed sewage-sludge bacteria cultures (predomi-nant: *Clostridium, Pseudomonas, Desulfovibrio*)	- Molasses 2%−4% with addition of sugar and nitrogen and phosphorous sources	Improvement	Yaranyi (1968) and Dienes and Yaranyi (1973)
Bulgaria	CMR	- Indigenous oil-oxidizing bacteria from water injection and water formation	- Water containing air + ammonium and phosphate ions - Molasses 2%	Improvement	Groudeva et al. (1993)
Former East Germany	MFR	- Mixed cultures of thermophilic: *Bacillus* and *Clostridium* - Indigenous brine microflora	- Molasses 2%−4% with addition of nitrogen and phosphorous sources	Improvement	Wagner et al. (1993)
England	MSPR	- Naturally occurring anaerobic strain, high generator of acids - Special starved bacteria, good producers of exopolymers	- Soluble carbohydrate sources - Suitable growth media (type E and G)	Not all reported	Moses et al. (1993)
Norway	MFR	- Nitrate-reducing bacteria naturally occurring in North Sea water	- Nitrate and 1% carbohydrates addition to injected sea water	Not yet reported	Awan et al. (2008) and Rassenfoss (2011)
Netherlands	MSPR	- Slime-forming bacteria (*Beta-coccus dextranicus*)	- Sucrose-molasses 10%	Improvement	Hitzman (1988);

Continued

TABLE 5.10
MEOR Applications Around World (Lazar et al., 2007; Patel et al., 2015).—cont'd

Country	Type of MEOR Application	Microbial System	Nutrients	Oil Production	References
Australia	MFR	- Ultramicrobacteria with surface-active properties	- Formulate suitable base media	Improvement	Sheehy (1991)
China	CMR, MFR, MSPR	- Mixed enriced bacterial cultures of *Bacillus, Pseudomonas, Eurobacterium, Fusobacterium, Bacteroides* - Slime-forming bacteria: *Xanthomonas campestris, Brevibacterium viscogenes, Corynebacterium gumiform* - Microbial products as biopolymers, biosurfactants	- Molasses 4%−6% - Molasses 5% + - Residue sugar 4% + - Crude oil 5% - Xanthan 3% in waterflooding	Improvement	Wang (1991), Wang et al. (1993), and He et al. (2000)
Malaysia	CMR	—	—	Not all reported	Ghazali Abd. Krim et al. (2001)
Saudi Arabia	CMR, MFR, MSPR	- Adequate bacterial inoculum according to requirements of each technology	- Adequate nutrients for each technology	Not yet reported	Sim et al. (1997) and Al-Sulaimani et al. (2011)
Trinidad-Tobago	CMR	- Fac. Anaerobic bacteria high producers of gases	- Molasses 2%−4%	Not yet reported	Maharaj et al. (1993)
Venezuela	CMR	- Adapted mixed enrichment cultures	- Molasses	Not yet reported	Bastardo et al. (1993)
Argentina	CMR	—	—	Improvement	Maure et al. (1999) and Strappa et al. (2004)

reported an MEOR application using molasses to generate volatile acids and CO_2. The injection of *Clostridium* was confirmed to significantly increase the oil production. The actual oil production rate from a well was 3.5 times higher than previous estimates and the overall oil production increased by approximately 250%. This rapid increase in production was analyzed by four mechanisms: surfactant production, CO_2 production, acid production by carbonate dissolution with CO_2, and oil viscosity reduction by CO_2 dissolution. In addition, permeability modification may have occurred by selective plugging. Application of MEOR in Oklahoma's North Burbank Unit (NBU) reservoir resulted in a 33% reduction in effective permeability (Jenneman et al., 1996; Sen, 2008). After injecting the nutrients (maltodextrin, MD) used in the MEOR, the

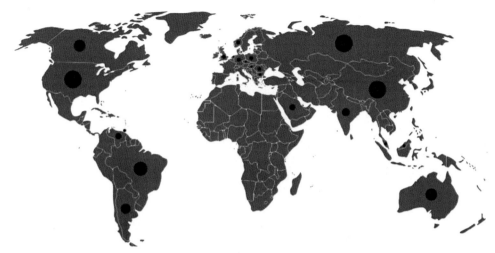

FIG. 5.3 Map of global MEOR application status. (Credit: from Safdel, M., Anbaz, M.A., Daryasafar, A., Jamialahmadi, M., 2017. Microbial enhanced oil recovery, a critical review on worldwide implemented field trials in different countries. Renewable and Sustainable Energy Reviews 74, 159–172.)

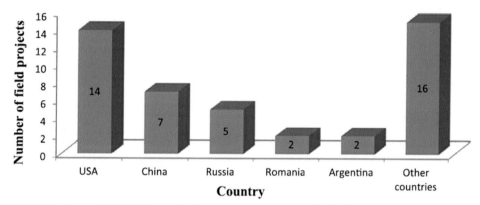

FIG. 5.4 Number of MEOR field applications by each country. (Credit: from Safdel, M., Anbaz, M.A., Daryasafar, A., Jamialahmadi, M., 2017. Microbial enhanced oil recovery, a critical review on worldwide implemented field trials in different countries. Renewable and Sustainable Energy Reviews 74, 159–172.)

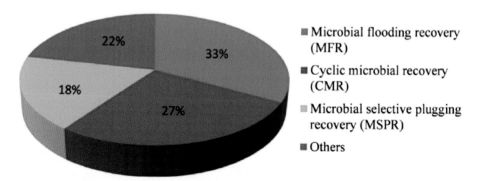

FIG. 5.5 Classification by MEOR application type. (Credit: from Safdel, M., Anbaz, M.A., Daryasafar, A., Jamialahmadi, M., 2017. Microbial enhanced oil recovery, a critical review on worldwide implemented field trials in different countries. Renewable and Sustainable Energy Reviews 74, 159–172.)

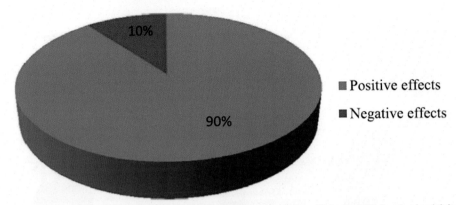

FIG. 5.6 Effects of MEOR applications. (Credit: from Safdel, M., Anbaz, M.A., Daryasafar, A., Jamialahmadi, M., 2017. Microbial enhanced oil recovery, a critical review on worldwide implemented field trials in different countries. Renewable and Sustainable Energy Reviews 74, 159–172.)

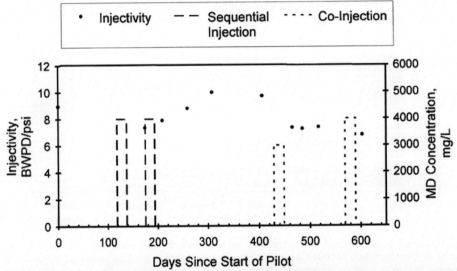

FIG. 5.7 Injectivity from pressure falloff test at injector 16W21 in NBU reservoir. (Credit: from Jenneman, G.E., Moffitt, P.D., Young, G.R., 1996. Application of a microbial selective-plugging process at the North Burbank Unit: prepilot tests. SPE Production and Facilities 11 (1), 11–17.)

effective permeability of water decreased from 300 to 200 md, and the resulting injectivity was shown in Fig. 5.7. The microbial permeability profile modification technique was implemented in North Blowhorn Creek Oil Unit (NBCU) located in Alabama (Brown et al., 2002). It is expected that additional oil recovery ($4-6 \times 10^5$ bbl) would extend the economic life of the oil field by 60–137 months. Fig. 5.8 illustrated a positive effect of selective plugging on oil recovery, and Fig. 5.9 showed that plugging slowed down the oil production decline of NBCU. In addition, the oil recovery of the limestone reservoir in Bebee field,

Oklahoma, was increased by in situ stimulating *Bacillus* strains to generate biosurfactants (Youssef et al., 2007).

In Canada, researchers performed MEOR field applications using selective plugging by *Leuconostoc*. Cusak et al. (1992) proposed a new concept of selective plugging based on utilizing ultramicrobacteria generated by selective starvation. Another new concept of microbial plugging was based on using biomineralization, which is the formation of calcite cements for sand consolidation and fracture closure in carbonate formation (Jack et al., 1991; Jack, 1993).

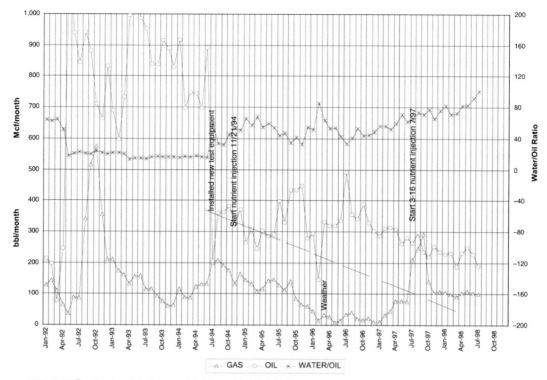

FIG. 5.8 Production data from well 2-13 no. 1 in NBCU. (Credit: from Brown, L.R., Vadie, A.A., Stephens, J.O., 2002. Slowing production decline and extending the economic life of an oil field: new MEOR technology. SPE Reservoir Evaluation and Engineering 5 (1), 33—41.)

In the Soviet Union, MEOR technology began in the 1960s. 54 m^3 of mixed microbial cultures were grown in 4% molasses solution and injected into Sernovodek oil field (Kuznetsov et al., 1963). The well was closed for 6 months for microbial metabolic activity in the reservoir. The well head pressure was set to 1.5 atm. When the well resumed production, oil rate was increased for a while and then dropped again (Senyukov et al., 1970; Ivanov et al., 1993). Some MEOR methods were also applied to Tataryia oil fields in Russia (Lazar, 1998).

MEOR field applications have also been studied in Eastern European countries such as Czechoslovakia, Poland, Romania, and Hungary. Dostalek and Spurny (1958) in Czechoslovakia reported soil *Clostridium* isolation, which can grow in petroleum, or carbohydrate, or yeast extract, and can generate large amounts of gas. The laboratory experiment indicated that produced gas was the critical factor for displacing oil. In another field trial in 1954, mixed bacteria of *Desulfovibrio* and *Pseudomonas* were injected with molasses. In three of the seven applications, oil recovery increased.

Oil production in a certain well improved by 12% —36%, and overall production increased 7% during 6-month application. No increase in oil recovery was observed without nutrient injection.

In Romania, a number of field trials were conducted from 1971 to 1982. Many successes in both the microbial flooding and Huff-and-Puff processes have been reported in Romanian oil fields. Adapted mixed enrichment cultures and molasses were injected into reservoirs after pilot test to verify the feasibility (Lazar, 1998). 20% molasses medium was injected with bacteria isolated from formation water. Flooding was carried out for 6—8 months to generate metabolites, and two of seven reservoirs showed increased oil production. The causes of failures were analyzed in four ways: low permeability rock, lack of strata continuity from the injector to producer, movement of unconsolidated sand particle, high temperature, and high salinity of formation water.

In Hungary, a mixed bacterial cultured in a medium containing molasses, potassium nitrate, sodium phosphate, and sucrose were injected into reservoir (Yaranyi,

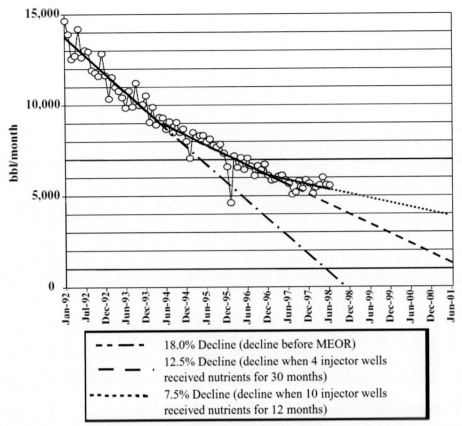

FIG. 5.9 Changes in oil production decline with MEOR application in NBCU. (Credit: from Brown, L.R., Vadie, A.A., Stephens, J.O., 2002. Slowing production decline and extending the economic life of an oil field: new MEOR technology. SPE Reservoir Evaluation and Engineering 5 (1), 33–41.)

1968). After 8 months of observation, oil production increased in reservoirs with a permeability range of 600–700 md. However, there was no increase in oil production in the 10–70 md permeability reservoir.

In Germany, *Clostridia* species and molasses have been injected into the carbonate reservoir to improve successful oil recovery (Wagner et al., 1993). In 1958, the Netherlands researchers applied selective plugging with *Betacoccus dextranicus*, which was effective in improving oil recovery and water-oil ratio. The selective plugging phenomenon was utilized in two applications (Hitzman, 1988). In addition to *B. dextranicus*, *Bacillus polymyxa*, and *Clostridium gelatinosum* were also used with molasses. As a result, an increase in oil recovery was found in one case, and an improvement in water-oil ratio was found in the other case. Only one case of MEOR in Norway has been reported for offshore oil fields in Norne (Awan et al., 2008; Rassenfoss, 2011).

In Australia, a new concept of MEOR technique using ultramicrobacteria was developed. These ultramicrobacteria were produced by stimulating indigenous microbial flora with manipulated nutrients. This microbial system has been applied to Alton oil field in Queensland, Australia, with increased oil recovery (Sheehy, 1990, 1991). Fig. 5.10 showed the oil production increased by 40% after microbial treatment (the test in the figure meant the result after MEOR and the control meant the result before MEOR).

China is a leading country in the MEOR technology field applications (Lazar et al., 2007). The biopolymers produced by *L. Mesenteroides* and *P. aeruginosa* strains were applied to Chinese oil fields. *Brevibacterium viscogenes*, *Corynebacterium gumiform*, and *Xanthomonas campestris* were also applied, which use hydrocarbons to produce biopolymers. Wang (1991) represented the results of MEOR applications in China oil field (Lazar et

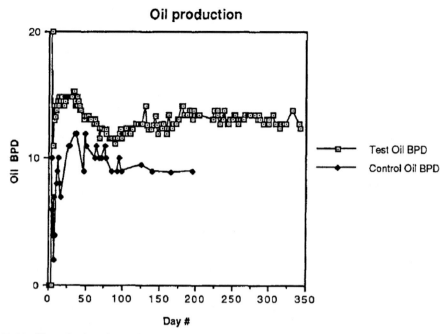

FIG. 5.10 Oil production change by microbial treatment. (Credit: from Sheehy, A.J., 1990. Field studies of microbial EOR. Proceedings of SPE/DOE Enhanced Oil Recovery Symposium, Tulsa, Oklahoma, USA, 22–25 April, 1990.)

al., 2007). The MEOR field trials in China operated by China National Petroleum Company (CNPC) were implemented using the microbial waterflooding. Oil production increased significantly from the implementation (Sun, 2014; Guo et al., 2015). Moreover, Huff-and-Puff processes have been applied to hundreds of single wells in Shengli oil field, resulting in satisfactory oil production and water-cut reduction (Lazar et al., 2007; Sun, 2014; Guo et al., 2015). Successful MEOR applications have been reported in Daqing, Dagang Kongdian, Xinjiang Liuzhongqu, Fuyu in Jilin, and Huabe Baolige oil fields (Nagase et al., 2001; Wang et al., 2005; Guo et al., 2015). The MEOR was applied in Fuyu oil field in Jilin (Nagase et al., 2002). This sandstone reservoir has a permeability of 240 md, and the oil viscosity is 4 cp. The *Enterobacter cloacae* were used with molasses. Oil production doubled with MEOR during 6 months (Fig. 5.11). Verification of MEOR applicability to high temperature Liaohe oil field was performed (Nazina et al., 2000; He et al., 2003). As a result, they found that the Liaohe oil field has a high potential for improving the oil recovery because of the presence of diverse thermophilic microbial community (Nazina

et al., 2000). In Kongdian reservoirs of Dagang oil field, a study of MEOR applicability has also been carried out. Since anaerobic thermophilic fermentative, SRB and methanogenic bacteria were found in the reservoir, the probability of success of the technology was highly evaluated (Nazina et al., 2007). In Daqing oil field, *Bacillus*, *Pseudomonas*, *Phodococcus*, *Dietzia*, *Clavibacter*, and other bacteria were found in formation water (Nazina et al., 2003). After the MEOR process was applied for 35–40 days in 1997, both increased oil production and reduced water-cut were observed in 17 of 25 wells. Another pilot test utilized metabolic products of *Pseudomonas aeruginosa* to increase oil recovery (Li et al., 2002).

There was also an example of applying MEOR to Malaysia oil field. Three wells in Bokor offshore field were selected for microbial treatment. The treatment contributed to water-cut reduction and oil recovery improvement (Fig. 5.12). Moreover, skin factor and permeability were reduced in two wells (Ghazali Abd. Karim et al., 2001; Sabut et al., 2003; Ibrahim et al., 2004).

In India, MEOR applications in Indian oil field have been conducted by Oil and Natural Gas Corporation

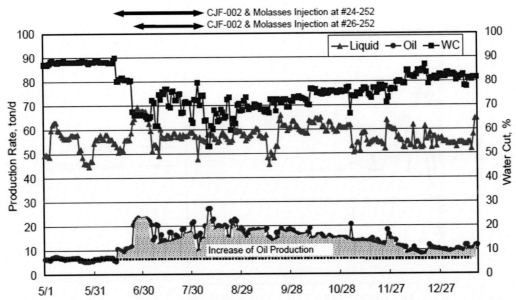

FIG. 5.11 Oil production behavior of whole test area in Fuyu oil field. (Credit: from Nagase, K., Zhang, S.T., Asami, H., Yazawa, N., Fujiwara, K., Enomoto, H., Hong, C.X., Liang, C.X., 2002. A successful field test of microbial EOR process in Fuyu oilfield, China. Proceedings of SPE/DOE Improved Oil Recovery Symposium, Tulsa, Oklahoma, USA, 13–17 April, 2002.)

(ONGC), The Energy and Resources Institute (TERI), and Institute of Reservoir Studies (IRS). Some trials were implemented using the Huff-and-Puff process and indigenously developed MEOR method with anaerobic extremophiles which were isolated from the candidate reservoirs (Patel et al., 2015). By applying this technique, oil production was tripled in 9 of 12 wells and water-cuts were greatly reduced (Patel et al., 2015).

There was a report of MEOR applicability in the Arab reservoir conditions, covering more than 300 formations in seven Arab countries (Saudi Arabia, Egypt, Kuwait, Qatar, UAE, Iraq, and Syria) (Sim et al., 1997). The MEOR was expected to recover up to 30% of the residual oil in the Arab reservoir environments. Experiments on the applicability to the Omani oil field were also conducted by Sultan Qaboos University in the Sultanate of Oman (Al-Sulaimani et al., 2011). In another experiment (Al-Hattali et al., 2012), *Bacillus licheniformis* were used with nutrients including various nitrogen sources, yeast extract, peptone, and urea. As a result, the effect of selective plugging on oil recovery was confirmed in a fractured reservoir.

The MEOR was applied using Huff-and-Puff process in the Piedras Coloradas field in Argentina (Maure et al., 1999). Microbial treatment was performed on two wells of Barrancas formation and four wells of Blanco

formation. The permeability of Barrancas formation was 120 md and the temperature was 170°F. Blanco formation showed 5–10 md permeability and 180°F temperature conditions. The wells were shut-in for 72 hours after microbial injection. The oil viscosity decreased through a series of processes, resulting in increased oil production and reduced water-cut (Fig. 5.13). Another MEOR application in Argentina was by microbial flooding in the Papagayos reservoir in Vizcacheras field (Strappa et al., 2004). The reservoir had a permeability of 1 Darcy, a pressure of 1,400 psi, and a temperature of 198°F. The reservoir was waterflooded before MEOR application and showed a high water-cut of 96%. Following microbial injection, oil recovery increased in nine production wells (Fig. 5.14).

The Trinidad and Tobago Oil Company Limited (TRINTOC) operates approximately 1,300 oil wells, 75% of which produce less than 15 barrels per day (Patel et al., 2015). Since Trinidad and Tobago is a major country of sugar production, MEOR using sugar has been proposed as an alternative oil production technology (Maharaj et al., 1993). In Peru, the MEOR pilot tests were carried out in Providencia and Lobitos oil fields. The oil recovery increased in 36.5% (Fig. 5.15) and 46.5% (Fig. 5.16), respectively (Maure et al., 2005).

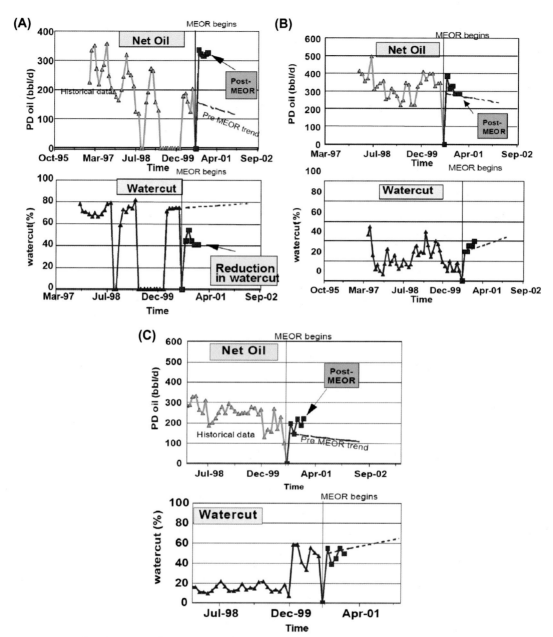

FIG. 5.12 Improvements of oil recovery and water-cut after MEOR in Bokor field: **(A)** well B-1, **(B)** well B-2, and **(C)** well B-3. (Credit: from Ghazali Abd. Karim, M., Hj Salim, M.A., Md. Zain, Z., Talib, N.N., 2001. Microbial enhanced oil recovery (MEOR) technology in Bokor field, Sarawak. Proceedings of SPE Asia Pacific Improved Oil Recovery Conference, Kuala Lumpur, Malaysia, 6–9 October, 2001.)

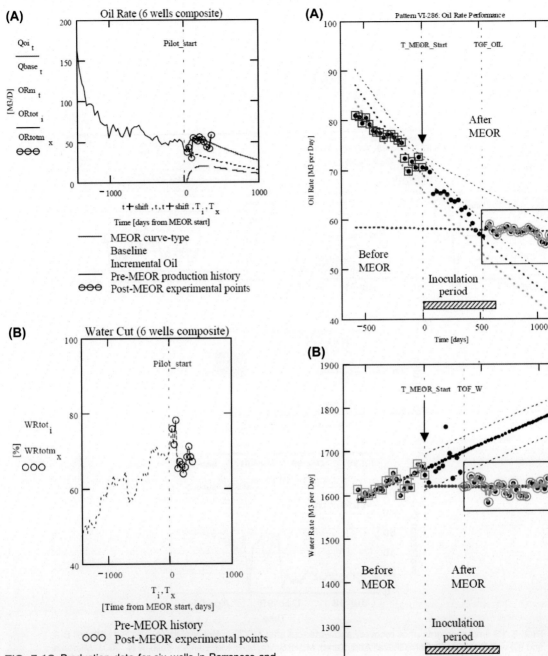

FIG. 5.13 Production data for six wells in Barrancas and Blanco formations: **(A)** oil rate, **(B)** water-cut. (Credit: from Maure, M.A., Dietrich, F.L., Diaz, V.A., Argañaraz, H., 1999. Microbial enhanced oil recovery pilot test in Piedras Coloradas field, Argentina. Proceedings of Latin American and Caribbean Petroleum Engineering Conference, Caracas, Venezuela, 21–23 April, 1999.)

FIG. 5.14 Production data of VI-286 pattern in Vizcacheras field: **(A)** oil rate, **(B)** water-cut. (Credit: from Strappa, L.A., De Lucia, J.P., Maure, M.A., Lopez Llopiz, M.L., 2004. A novel and successful MEOR pilot project in a strong water-drive reservoir Vizcacheras field, Argentina. Proceedings of SPE/DOE Symposium on Improved Oil Recovery, Tulsa, Oklahoma, USA, 17–21 April, 2004.)

(A)

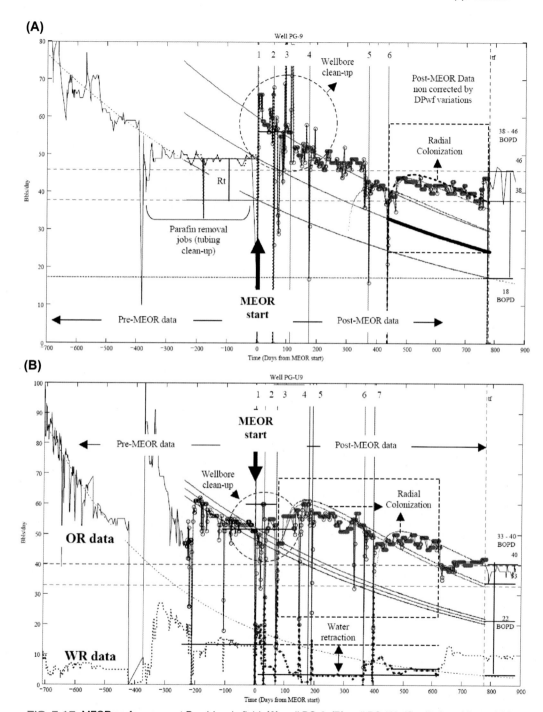

FIG. 5.15 MEOR performance at Providencia field: **(A)** well PG-9, **(B)** well PG-U9. (Credit: from Maure, M.A., Saldana, A.A., Juarez, A.R., 2005. Biotechnology application to EOR in Talara off-shore oil fields, Northwest Peru. Proceedings of SPE Latin American and Caribbean Petroleum Engineering Conference, Rio de Janeiro, Brazil, 20–23 June, 2005.)

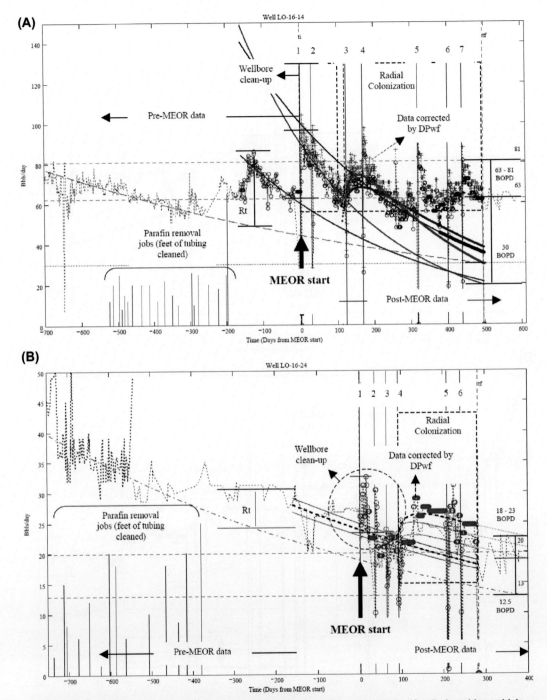

FIG. 5.16 MEOR performance at Lobitos field: **(A)** well LO16-14, **(B)** well LO16-24. (Credit: from Maure, M.A., Saldana, A.A., Juarez, A.R., 2005. Biotechnology application to EOR in Talara off-shore oil fields, Northwest Peru. Proceedings of SPE Latin American and Caribbean Petroleum Engineering Conference, Rio de Janeiro, Brazil, 20–23 June, 2005.)

REFERENCES

Al-Hattali, R., Al-Sulaimani, H., Al-Wahaibi, Y., Al-Bahry, S., Elshafie, A., Al-Bemani, A., Joshi, S., 2012. Improving sweep efficiency in fractured carbonate reservoirs by microbial biomass. In: Proceedings of SPE EOR Conference at Oil and Gas West Asia, Muscat, Oman, 16—18 April, 2012.

Al-Sulaimani, H., Joshi, S., Al-Wahaibi, Y., Al-Bahry, S., Elshafie, A., Al-Bemani, A., 2011. Microbial biotechnology for enhancing oil recovery: current developments and future prospects. Journal of Biotechnology, Bioinformatics and Bioengineering 1 (2), 147—158.

Arinbasarov, M.U., Murygina, V.P., Mats, A.A., 1995. Chemical and biological monitoring of MIOR on the pilot area of Vyngapour oil field, West Siberia, Russia. In: Proceedings of the Fifth International Conference on Microbial Enhanced Oil Recovery and Related Biotechnology for Solving Environmental Problems, Plano, Texas, USA, 11 September, 1995.

Awan, A.R., Teigland, R., Kleppe, J., 2008. A survey of North Sea enhanced-oil-recovery projects initiated during the years 1975 to 2005. SPE Reservoir Evaluation and Engineering 11 (3), 497—512.

Bastardo, H., Vierma, L., Estevez, A., 1993. Microbial characteristics and metabolic activity of bacteria from Venezuelan oil wells. In: Premuzic, E.T., Woodhead, A. (Eds.), Microbial Enhancement of Oil Recovery—Recent Advances. In: Chilingarian, G.V. (Ed.), Developments in Petroleum Science, vol. 39. Elsevier, Amsterdam, pp. 307—318.

Brown, L.R., Vadie, A.A., Stephens, J.O., 2002. Slowing production decline and extending the economic life of an oil field: new MEOR technology. SPE Reservoir Evaluation and Engineering 5 (1), 33—41.

Bryant, R.S., Burchfield, T.E., Dennis, D.M., Hitzman, D.O., 1990. Microbial-enhanced waterflooding: Mink Unit project. SPE Reservoir Engineering 5 (1), 9—13.

Bryant, R.S., Burchfield, T.E., Dennis, D.M., Hitzman, D.O., Porter, R.E., 1991. Microbial enhanced waterflooding: a pilot study. In: Donaldson, E.C. (Ed.), Microbial Enhancement of Oil Recovery—Recent Advances. In: Chilingarian, G.V. (Ed.), Developments in Petroleum Science, vol. 31. Elsevier, Amsterdam, pp. 399—419.

Bryant, R.S., Douglas, J., 1988. Evaluation of microbial systems in porous media for EOR. SPE Reservoir Engineering 3 (2), 489—495.

Bryant, R.S., Lindsey, R.P., 1996. World-wide applications of microbial technology for improving oil recovery. In: Proceedings of SPE/DOE Improved Oil Recovery Symposium, Tulsa, Oklahoma, USA, 21—24 April, 1996.

Bryant, R.S., Stepp, A.K., Bertus, K.M., Burchfield, T.E., Dennis, M., 1994. Microbial enhanced waterflooding field tests. In: Proceedings of SPE/DOE Improved Oil Recovery Symposium, Tulsa, Oklahoma, USA, 17—20 April, 1994.

Bubela, B., 1989. Geobiology and microbiologically enhanced oil recovery. In: Donaldson, E.C., Chilingarian, G.V., Yen, T.F. (Eds.), Microbial Enhanced Oil Recovery. In: Chilingarian, G.V. (Ed.), Developments in Petroleum Science, vol. 22. Elsevier, Amsterdam, pp. 75—97.

Buciak, J., Vazquez, A., Frydman, R., Mediavilla, J., Bryant, R., 1996. Enhanced oil recovery by means of microorganisms: pilot test. SPE Advanced Technology Series 4 (1), 144—149.

Cusack, F.M., Singh, S., Novosad, J., Chmilar, M., Blenkinsopp, S.A., Costerton, J.W., 1992. The use of ultra-microbacteria for selective plugging in oil recovery by waterflooding. In: Proceedings of International Meeting on Petroleum Engineering, Beijing, China, 24—27 March, 1992.

Davidson, W.S., Russell, H.H., 1988. A MEOR test pilot in the Loco field. In: Proceedings of Symposium on Applications of Microorganisms to Petroleum Technology, Bartlesville, Oklahoma, USA, 12—13 August, 1987.

Deng, D., Li, C., Ju, Q., Wu, P., Dietrich, F.L., Zhou, Z.H., 1999. Systematic extensive laboratory studies of microbial EOR mechanisms and microbial EOR application results in Changqing oilfield. In: Proceedings of SPE Asia Pacific Oil and Gas Conference and Exhibition, Jakarta, Indonesia, 20—22 April, 1999.

Dienes, M., Yaranyi, I., 1973. Increase of oil recovery by introducing anaerobic bacteria into the formation Demjen field. Koolaj ex Foldgaz 106 (7), 205—208.

Dietrich, F.L., Brown, F.G., Zhou, Z.H., Maure, M.A., 1996. Microbial EOR technology advancement: case studies of successful projects. In: Proceedings of SPE Annual Technical Conference and Exhibition, Denver, Colorado, USA, 6—9 October, 1996.

Donaldson, E.C., Chilingarian, G.V., Yen, T.F., 1989. Introduction. In: Donaldson, E.C., Chilingarian, G.V., Yen, T.F. (Eds.), Microbial Enhanced Oil Recovery. In: Chilingarian, G.V. (Ed.), Developments in Petroleum Science, vol. 22. Elsevier, Amsterdam, pp. 1—14.

Dostalek, M., Spurny, M., 1958. Bacterial release of oil. A preliminary trial in an oil deposit. Folia Biologica 4, 166—172.

Ghazali Abd. Karim, M., Hj Salim, M.A., Md. Zain, Z., Talib, N.N., 2001. Microbial enhanced oil recovery (MEOR) technology in Bokor field, Sarawak. In: Proceedings of SPE Asia Pacific Improved Oil Recovery Conference, Kuala Lumpur, Malaysia, 6—9 October, 2001.

Groudeva, V.I., Ivanova, I.A., Groudev, S.N., Uzunov, G.G., 1993. Enhanced oil recovery by stimulating the activity of the indigenous microflora of oil reservoirs. In: Proceedings of Biohydrometallurgical Technologies, Warrendale, Pennsylvania, USA, 1993.

Grula, E.A., Russell, H.H., Bryant, D., Kenaga, M., 1989. Oil displacement by anaerobic and facultatively anaerobic bacteria. In: Donaldson, E.C., Chilingarian, G.V., Yen, T.F. (Eds.), Microbial Enhanced Oil Recovery. In: Chilingarian, G.V. (Ed.), Developments in Petroleum Science, vol. 22. Elsevier, Amsterdam, pp. 113—123.

Grula, M.M., Russell, H.H., 1985. Isolation and Screening of Anaerobic *Clostridia* for Characteristics Useful in Enhanced Oil Recovery (Technical Report). Oklahoma State University, Stillwater, OK.

Gudiña, E.J., Pereira, J.F.B., Rodrigues, L.R., Coutinho, J.A.P., Texeira, J.A., 2012. Isolation and study of microorganisms from oil samples for application in microbial enhanced oil recovery. International Biodeterioration & Biodegradation 68, 56–64.

Guo, H., Li, Y., Yiran, Z., Wang, F., Wang, Y., Yu, Z., Haicheng, S., Yuanyuan, G., Chuyi, J., Xian, G., 2015. Progress of microbial enhanced oil recovery in China. In: Proceedings of SPE Asia Pacific Enhanced Oil Recovery Conference, Kuala Lumpur, Malaysia, 11–13 August, 2015.

He, Z., Mei, B., Wang, W., Sheng, J., Zhu, S., Wang, L., Yen, T.F., 2003. A pilot test using microbial paraffin-removal technology in Liahe oilfield. Petroleum Science and Technology 21 (1–2), 201–210.

He, Z., She, Y., Xiang, T., Xue, F., Mei, B., Li, Y., Ju, B., Mei, H., Yen, T.F., 2000. MEOR pilot sees encouraging results in Chinese oil field. Oil & Gas Journal 98 (4), 46–52.

Hitzman, D.O., 1983. Petroleum microbiology and the history of its role in enhanced oil recovery. In: Proceedings of the 1982 International Conference on Microbial Enhancement of Oil Recovery, Afton, Oklahoma, USA, 16–21 May, 1982.

Hitzman, D.O., 1988. Review of microbial enhanced oil recovery field tests. In: Proceedings of Symposium on Applications of Microorganisms to Petroleum Technology, Bartlesville, Oklahoma, USA, 12–13 August, 1987.

Hitzman, D.O., Dennis, M., Hitzman, D.C., 2004. Recent successes: MEOR using synergistic H_2S prevention and increased oil recovery systems. In: Proceedings of SPE/DOE Symposium on Improved Oil Recovery, Tulsa, Oklahoma, USA, 17–21 April, 2004.

Ibrahim, Z., Omar, M.I., Foo, K.S., Elias, E.J., Othman, M., 2004. Simulation analysis of microbial well treatment of Bokor field, Malaysia. In: Proceedings of SPE Asia Pacific Oil and Gas Conference and Exhibition, Perth, Australia, 18–20 October, 2004.

Ivanov, M.V., Belyaev, S.S., Borzenkov, I.A., Glumov, I.F., Ibatullin, R.R., 1993. Additional oil production during field trials in Russia. In: Premuzic, E.T., Woodhead, A. (Eds.), Microbial Enhancement of Oil Recovery—Recent Advances. In: Chilingarian, G.V. (Ed.), Developments in Petroleum Science, vol. 39. Elsevier, Amsterdam, pp. 373–381.

Jack, T.R., 1991. Microbial enhancement of oil recovery. Current Opinion in Biotechnology 2 (3), 444–449.

Jack, T.R., 1993. M.O.R.E. to M.E.O.R.: an overview of microbially enhanced oil recovery. In: Premuzic, E.T., Woodhead, A. (Eds.), Microbial Enhancement of Oil Recoverye—Recent Advances. In: Chilingarian, G.V. (Ed.), Developments in Petroleum Science, vol. 39. Elsevier, Amsterdam, pp. 7–16.

Jack, T.R., Stehmeier, L.G., 1988. Selective plugging in watered out oil wells. In: Proceedings of the Symposium on Applications of Microorganisms to Petroleum Technology, Bartlesville, Oklahoma, USA, 12–13 August, 1987.

Jack, T.R., Stehmeier, L.G., Islam, M.R., Ferris, F.G., 1991. Microbial selective plugging to control water channeling. In: Donaldson, E.C. (Ed.), Microbial Enhancement of Oil Recovery—Recent Advances. In: Chilingarian, G.V. (Ed.), Developments in Petroleum Science, vol. 31. Elsevier, Amsterdam, pp. 433–440.

Jenneman, G.E., Clark, J.B., Moffitt, P.D., 1993. A nutrient control process for microbially enhanced oil recovery applications. In: Premuzic, E.T., Woodhead, A. (Eds.), Microbial Enhancement of Oil Recovery—Recent Advances. In: Chilingarian, G.V. (Ed.), Developments in Petroleum Science, vol. 39. Elsevier, Amsterdam, pp. 319–333.

Jenneman, G.E., Moffitt, P.D., Young, G.R., 1996. Application of a microbial selective-plugging process at the North Burbank Unit: prepilot tests. SPE Production and Facilities 11 (1), 11–17.

Karaskiewicz, I., 1974. The Application of Microbiological Method for Secondary Oil Recovery from the Carpathian Crude Oil Reservoir. Widawnistwo "SLASK" Katowice, pp. 1–67.

Kianipey, S.A., Donaldson, E.C., 1986. Mechanisms of oil displacement by microorganisms. In: Proceedings of SPE Annual Technical Conference and Exhibition, New Orleans, Louisiana, USA, 5–8 October, 1986.

Knapp, R.M., McInerney, M.J., Menzie, D.E., 1991. Microbial Field Pilot Study (Technical Report). Oklahoma University, Norman, Oklahoma, OK.

Kuznetsov, S.I., Broneer, P.T., Ivanov, M.V., lyalikova, N.N., Oppenheimer, C.H., 1963. Introduction to Geological Microbiology. McGraw-Hill, New York, NY.

Larsen, J., Rod, M.H., Zwolle, S., 2004. Prevention of reservoir souring in the Halfdan field by nitrate injection. In: Proceedings of CORROSION 2004, New Orleans, Louisiana, USA, 28 March–1 April, 2004.

Lazar, I., 1983. Microbial enhancement of oil recovery in Romania. In: Proceedings of the 1982 International Conference on Microbial Enhancement of Oil Recovery, Afton, Oklahoma, USA, 16–21 May, 1982.

Lazar, I., 1991. MEOR field trials carried out over the world during the last 35 years. In: Donaldson, E.C. (Ed.), Microbial Enhancement of Oil Recovery—Recent Advances. In: Chilingarian, G.V. (Ed.), Developments in Petroleum Science, vol. 31. Elsevier, Amsterdam, pp. 485–530.

Lazar, I., 1998. International MEOR application for marginal wells. Pakistan Journal of Hydrocarbon Research 10 (1), 11–30.

Lazar, I., Dobrota, S., Stefanescu, M., Sandulescu, L., Constantinescu, P., Morosanu, C., Botea, N., Iliescu, O., 1991. Preliminary results of some recent MEOR field trials in Romania. In: Donaldson, E.C. (Ed.), Microbial Enhancement of Oil Recovery—Recent Advances. In: Chilingarian, G.V. (Ed.), Developments in Petroleum Science, vol. 31. Elsevier, Amsterdam, pp. 365–385.

Lazar, I., Petrisor, I.G., Yen, T.F., 2007. Microbial enhanced oil recovery (MEOR). Petroleum Science and Technology 25 (11), 1353–1366.

Lazar, I., Stefanescu, M.M., Dobrota, S.C., 1992. MEOR, the suitable bacterial inoculum according to the kind of technology used: results from Romania's last 20 years' experience. In: Proceedings of SPE/DOE Enhanced Oil Recovery Symposium, Tulsa, Oklahoma, USA, 22–24 April, 1992.

Lazar, I., Voicu, A., Nicolescu, C., Mucenica, D., Dobrota, S., Petrisor, I.G., Stefanescu, M., Sandulescu, L., 1999. The use of naturally occurring selectively isolated bacteria for inhibiting paraffin deposition. Journal of Petroleum Science and Engineering 22 (1–3), 161–169.

Li, Q., Kang, C., Wang, H., Liu, C., Zhang, C., 2002. Application of microbial enhanced oil recovery technique to Daqing oilfield. Biochemical Engineering Journal 11 (2–3), 197–199.

Maharaj, U., May, M., Imbert, M.P., 1993. The application of microbial enhanced oil recovery to Trinidadian oil wells. In: Premuzic, E.T., Woodhead, A. (Eds.), Microbial Enhancement of Oil Recovery–Recent Advances. In: Chilingarian, G.V. (Ed.), Developments in Petroleum Science, vol. 39. Elsevier, Amsterdam, pp. 245–263.

Marsh, T.L., Zhang, X., Knapp, R.M., McInerney, M.J., Sharma, P.K., Jackson, B.E., 1995. Mechanisms of microbial oil recovery by *Clostridium acetobutylicum* and *Bacillus* strain JF-2. In: Proceedings of International Conference on Microbial Enhanced Oil Recovery and Related Biotechnology for Solving Environment Problems, Dallas, Texas, USA, 11–14 September, 1995.

Matz, A.A., Borisov, A.Y., Mamedov, Y.G., Ibatulin, R.R., 1992. Commercial (pilot) test of microbial enhanced oil recovery methods. In: Proceedings of SPE/DOE Enhanced Oil Recovery Symposium, Tulsa, Oklahoma, USA, 22–24 April, 1992.

Maure, M.A., Dietrich, F.L., Diaz, V.A., Argañaraz, H., 1999. Microbial enhanced oil recovery pilot test in Piedras Coloradas field, Argentina. In: Proceedings of Latin American and Caribbean Petroleum Engineering Conference, Caracas, Venezuela, 21–23 April, 1999.

Maure, M.A., Dietrich, F.L., Gomez, U., Vallesi, J., Irusta, M., 2001. Waterflooding optimization using biotechnology: 2-year field test, La Ventana field, Argentina. In: Proceedings of SPE Latin American and Caribbean Petroleum Engineering Conference, Buenos Aires, Argentina, 25–28 March, 2001.

Maure, M.A., Saldana, A.A., Juarez, A.R., 2005. Biotechnology application to EOR in Talara off-shore oil fields, Northwest Peru. In: Proceedings of SPE Latin American and Caribbean Petroleum Engineering Conference, Rio de Janeiro, Brazil, 20–23 June, 2005.

Maudgalya, S., Knapp, R.M., McInerney, M., 2007. Microbially enhanced oil recovery technologies: a review of the past, present and future. In: Proceedings of Production and Operations Symposium, Oklahoma City, Oklahoma, USA, 31 March–3 April, 2007.

McInerney, M.J., Sublette, K.L., Montgomery, A.D., 1991. Microbial control of the production of sulfide. In: Donaldson, E.C. (Ed.), Microbial Enhancement of Oil Recovery–Recent Advances. In: Chilingarian, G.V. (Ed.), Developments in Petroleum Science, vol. 31. Elsevier, Amsterdam, pp. 441–449.

Moses, V., Brown, M.J., Burton, C.C., Gralla, D.S., Cornelius, C., 1993. Microbial hydraulic acid fracturing. In: Premuzic, E.T., Woodhead, A. (Eds.), Microbial Enhancement of Oil Recovery–Recent Advances. In: Chilingarian, G.V. (Ed.), Developments in Petroleum Science, vol. 39. Elsevier, Amsterdam, pp. 207–229.

Nagase, K., Zhang, S.T., Asami, H., Yazawa, N., Fujiwara, K., Enomoto, H., Hong, C.X., Liang, C.X., 2001. Improvement of sweep efficiency by microbial EOR process in Fuyu oilfield, China. In: Proceedings of SPE Asia Pacific Oil and Gas Conference and Exhibition, Jakarta, Indonesia, 17–19 April, 2001.

Nagase, K., Zhang, S.T., Asami, H., Yazawa, N., Fujiwara, K., Enomoto, H., Hong, C.X., Liang, C.X., 2002. A successful field test of microbial EOR process in Fuyu oilfield, China. In: Proceedings of SPE/DOE Improved Oil Recovery Symposium, Tulsa, Oklahoma, USA, 13–17 April, 2002.

Nazina, T.N., 2004. Analysis of microbial community from water-flooded oil field by chromatography-mass spectrometry. Mikrobiologiya 63, 876–882.

Nazina, T.N., Grigor'yan, A.A., Shestakova, N.M., Babich, T.L., Ivoilov, V.S., Feng, Q., Ni, F., Wang, J., She, Y., Xiang, T., Luo, Z., Belyaev, S.S., Ivanov, M.V., 2007. Microbiological investigations of high-temperature horizons of the Kondian petroleum reservoir in connection with field trial of a biotechnology for enhancement of oil recovery. Microbiology 76 (3), 287–296.

Nazina, T.N., Sokolova, D.S., Grior'yan, A.A., Xue, Y.F., Belyaev, S.S., Ivanov, M.V., 2003. Production of oil-releasing compounds by microorganisms from the Daqing oil field, China. Microbiology 72 (2), 173–178.

Nazina, T.N., Xue, Y.F., Wang, X.Y., Belyaev, S.S., Ivanov, M.V., 2000. Microorganisms of the high-temperature Liaohe oil field of China and their potential for MEOR. Resource and Environmental Biotechnology 3, 149–160.

Nelson, L., Schneider, D.R., 1993. Six years of paraffin control and enhanced oil recovery with the microbial product, Para-Bac^{TM}. In: Premuzic, E.T., Woodhead, A. (Eds.), Microbial Enhancement of Oil Recovery–Recent Advances. In: Chilingarian, G.V. (Ed.), Developments in Petroleum Science, vol. 39. Elsevier, Amsterdam, pp. 355–362.

Oppenheimer, C.H., Heibert, F.K., 1988. Microbial enhanced oil production field tests in Texas. In: Proceedings of Symposium on Applications of Microorganisms to Petroleum Technology, Bartlesville, Oklahoma, USA, 12–13 August, 1987.

Patel, J., Borgohain, S., Kumar, M., Rangarajan, V., Somasundaran, P., Sen, R., 2015. Recent developments in microbial enhanced oil recovery. Renewable and Sustainable Energy Reviews 52, 1539–1558.

Portwood, J.T., 1995. A commercial microbial enhanced oil recovery technology: evaluation of 322 projects. In: Proceedings of SPE Production Operations Symposium, Oklahoma City, Oklahoma, USA, 2–4 April, 1995.

Portwood, J.T., Hiebert, F.K., 1992. Mixed culture microbial enhanced waterflood: tertiary MEOR case study. In: Proceedings of SPE Annual Technical Conference and Exhibition, Washington, DC, USA, 4–7 October, 1992.

Rassenfoss, S., 2011. Form bacteria to barrels: microbiology having an impact on oil fields. Journal of Petroleum Engineers 63 (11), 32–38.

Romero-Zerón, L., 2012. Introduction to Enhanced Oil Recovery (EOR) Processes and Bioremediation of Oil-Contaminated Sites. Books on Demand.

Sabut, B., Salim, M.A.H., Hamid, A.S.A., Khor, S.F., 2003. Further evaluation of microbial treatment technology for improved oil production in Boor field, Sarawak. In: Proceedings of SPE International Improved Oil Recovery

Conference in Asia Pacific, Kuala Lumpur, Malaysia, 20–21 October, 2003.

Safdel, M., Anbaz, M.A., Daryasafar, A., Jamialahmadi, M., 2017. Microbial enhanced oil recovery, a critical review on worldwide implemented field trials in different countries. Renewable and Sustainable Energy Reviews 74, 159–172.

Sen, R., 2008. Biotechnology in petroleum recovery: the microbial EOR. Progress in Energy and Combustion Science 34 (6), 714–724.

Senyukov, W.M., Yulbarisov, E.M., Taldykina, N.N., Shishenina, E.P., 1970. Microbial method of treating a petroleum deposit containing highly mineralized stratal waters. Mikrobiologiya 39, 705–710.

Sheehy, A.J., 1990. Field studies of microbial EOR. In: Proceedings of SPE/DOE Enhanced Oil Recovery Symposium, Tulsa, Oklahoma, USA, 22–25 April, 1990.

Sheehy, A.J., 1991. Microbial physiology and enhanced oil recovery. In: Donaldson, E.C. (Ed.), Microbial Enhancement of Oil Recovery–Recent Advances. In: Chilingarian, G.V. (Ed.), Developments in Petroleum Science, vol. 31. Elsevier, Amsterdam, pp. 37–44.

Sim, L., Ward, O.P., Li, Z.Y., 1997. Production and characterization of a biosurfactant isolated from *Pseudomonas aeruginosa* UW-1. Journal of Industrial Microbiology & Biotechnology 19 (4), 232–238.

Song, W., Ma, D., Zhu, Y., Wei, X., Wu, J., Li, S., Chen, X., Zan, C., 2014. The role of sulphate-reducing bacteria in oil recovery. International Journal of Current Microbiology and Applied Sciences 3 (7), 385–398.

Strappa, L.A., De Lucia, J.P., Maure, M.A., Lopez Llopiz, M.L., 2004. A novel and successful MEOR pilot project in a strong water-drive reservoir Vizcacheras field, Argentina. In: Proceedings of SPE/DOE Symposium on Improved Oil Recovery, Tulsa, Oklahoma, USA, 17–21 April, 2004.

Sun, S., 2014. Field practice and analysis of MEOR in Shengli oilfield. Journal of Oil and Gas Technology 36 (2), 149–152.

Svarovskaya, 1995. Physicochemical principles of models for determination the optimal composition of raw materials for petrochemical processes. Khimicheskaya Promyshlennost 3, 182–184.

Telang, A.J., Ebert, S., Foght, J.M., Westlake, D.W.S., Jenneman, G.E., Gevertz, D., Voordouw, G., 1997. Effect of nitrate injection on the microbial community in an oil field as monitored by reverse sample genome probing. Applied and Environmental Microbiology 63 (5), 1785–1793.

Thorstenson, T., Sunde, E., Bodtker, G., Lillebo, B.L., Torsvik, T., Beeder, J., 2002. Biocide replacement by nitrate in sea water injection systems. In: Proceedings of CORROSION 2002, Denver, Colorado, USA, 7–11 April, 2002.

Trebbau, G.L., Nunez, G.J., Caira, R.L., Molina, N.Y., Entzeroth, L.C., Schneider, D.R., 1999. Microbial stimulation of Lake Maracaibo oil wells. In: Proceedings of SPE Annual Technical Conference and Exhibition, Houston, Texas, USA, 3–6 October, 1999.

Updegraff, D.M., 1990. Early research on microbial enhanced oil recovery. Developments in Industrial Microbiology 31, 135–142.

Wagner, M., 1991. Microbial enhancement of oil recovery from carbonate reservoirs with complex formation characteristics. In: Donaldson, E.C. (Ed.), Microbial Enhancement of Oil Recovery–Recent Advances. In: Chilingarian, G.V. (Ed.), Developments in Petroleum Science, vol. 31. Elsevier, Amsterdam, pp. 387–398.

Wagner, M., Lungerhausen, D., Murtada, H., Rosenthal, G., 1995. Development and application of a new biotechnology of the molasses in-situ method; detailed evaluation for selected wells in the Romashkino carbonate reservoir. In: Proceedings of the Fifth International Conference on Microbial Enhanced Oil Recovery and Related Biotechnology for Solving Environmental Problems, Plano, Texas, USA, 11 September, 1995.

Wagner, M., Lungerhansen, D., Nowak, U., Ziran, B., 1993. Microbially improved oil recovery from carbonate. Biohydrometalurg Technology 2, 695–710.

Wang, X.Y., 1991. Advances in research, production and application of biopolymers used for EOR in China. In: Donaldson, E.C. (Ed.), Microbial Enhancement of Oil Recovery–Recent Advances. In: Chilingarian, G.V. (Ed.), Developments in Petroleum Science, vol. 31. Elsevier, Amsterdam, pp. 467–481.

Wang, X., Chen, Z., Li, X., Lu, Y., Zhang, S., Li, X., Ni, J., Duan, G., 2005. Microbial flooding in Guan 69 block, Dagang oilfield. Petroleum Exploration and Development 32 (2), 107–109.

Wang, X.Y., Yue, Y.F., Xie, S.H., 1993. Characteristics of enriched cultures and their application to MEOR field tests. In: Premuzic, E.T., Woodhead, A. (Eds.), Microbial Enhancement of Oil Recovery–Recent Advances. In: Chilingarian, G.V. (Ed.), Developments in Petroleum Science, vol. 39. Elsevier, Amsterdam, pp. 335–348.

Yaranyi, I., 1968. Bezamolo a nagylengyel tezegeben elvegzett koolaj mikrobiologiai kiserletkrol. M All Faldany Intezet Evi Jelentese A Evval 423–426.

Yarbrough, H.F., Coty, V.F., 1983. Microbially enhanced oil recovery from the upper Cretaceous Nacatoch formation, Union County, Arkansas. In: Proceedings of the 1982 International Conference on Microbial Enhancement of Oil Recovery, Afton, Oklahoma, USA, 16–21 May, 1982.

Youssef, N., Elshahed, M.S., McInerney, M.J., 2009. Microbial processes in oil fields: culprits, problems, and opportunities. In: Laskin, A.I., Sariaslani, S., Gadd, G.M. (Eds.), Advances in Applied Microbiology, vol. 66. Elsevier, Amsterdam, pp. 141–251.

Youssef, N., Simpson, D.R., Duncan, K.E., McInerney, M.J., Folmsbee, M., Fincherm, T., Knapp, R.M., 2007. In situ biosurfactant production by *Bacillus* strains injected into a limestone petroleum reservoir. Applied and Environmental Microbiology 73 (4), 1239–1247.

Yusuf, A., Kadarwati, S., Nurkamelia, Sumaryana, 1999. Field test of the indigenous microbes for oil recovery, Ledok field, Central Java. In: Proceedings of SPE Asia Pacific Improved

Oil Recovery Conference, Kuala Lumpur, Malaysia, 25–26 October, 1999.

Zajic, J.E., Smith, S.W., 1987. Oil separation relating to hydrophobicity and microbes. In: Kosaric, N.C.C., Cairns, W.L., Gray, N.C.C. (Eds.), Biosurfactants and Biotechnology. Routledge, New York, NY, pp. 121–142.

Zhang, C.Y., Zhang, J.C., 1993. A pilot test of EOR by in-situ microorganism fermentation in the Daqing oilfield. In: Premuzic, E.T., Woodhead, A. (Eds.), Microbial Enhancement of Oil Recovery—Recent Advances. In: Chilingarian, G.V. (Ed.), Developments in Petroleum Science, vol. 39. Elsevier, Amsterdam, pp. 231–244.

Zhang, Y., Xu, Z., Ji, P., Hou, W., Dietrich, F., 1999. Microbial EOR laboratory studies and application results in Daqing oilfield. In: Proceedings of SPE Asia Pacific Oil and Gas Conference and Exhibition, Jakarta, Indonesia, 20–22 April, 1999.

Index

A

Acids, 4t–5t
Advancing contact angle, 68–70
Aerobes, 28
Agrobacterium biobar, 39–41
Alcaligenes faecalis, 39–41
Alginate, 46–47
  chemical and physicochemical
    characteristics, 46
  microbial production, 46
Anaerobes, 28
Anaerobic hydrocarbon
  biodegradation, 28
Aureobasidium pullulans, 41–42

B

Bacillus cereus
  ATCC 10987 and ATCC 14579
    strains, 54
  biofilm production, 53–54
  biooxidation reaction, 54
  biosurfactant production, 54
  extracellular polymeric substance
    (EPS), 53–54
Bacillus licheniformis
  biomass production, 52–53
  core flooding study, 53
  EPS production, 52
  laboratory cultivation tests, 52–53
Bacillus stearothermophilus, 16–17
Bacillus subtilis
  surfactin production
    concentration measurement,
      74–75, 76f–77f
    interfacial property modifications,
      75, 78t
    interfacial tension and contact
      angle, 73, 73f–74f
    interfacial tension (IFT) reduction,
      75
    low bond axisymmetric drop shape
      analysis (LBADSA) method, 73
    oil droplets, 74f–75f
    sessile drop method, 73–74
    stability, 75–79
    survivability, 72–73
Batch model simulation results, 129,
  132f
Bingham yield stress model, 94,
  125
Bioacids and solvents, 2, 9
Bioaugmentation, 38–39

Bioclogging, *Leuconostoc mesenteroides*
  column experiments
    biopolymer accumulation, 96–97
    ceramic cores, 94
    dextran estimation method, 97–98
    growth medium, 94–95, 95t–96t
    low permeability formations, 95
    permeability, 94
    permeability-porosity relation-
      ships, 98–99
    permeability reduction, 95–96,
      101f
    phases, 94–95
    setup, 95–96, 97f
    sucrose-fed cores, 95
    sucrose-rich media, 98, 99f
  pore-scale observations
    capillary grain-coating model
      (CGCM), 103
    dextran and cultured cells, 91–92,
      92f
    growth medium, 91, 91t
    Kozeny grain models, 102–103
    micromodel, 93
    nutrient injection, 94
    parallel capillary tube model,
      100–102
    permeability reduction and
      biopolymer saturation,
      103–104, 104f
    shear stress, 94
Biofilms
  advantages, 47
  architecture, 49–50, 50f
  biofilm-bacterial species, 51–54
    Bacillus cereus, 53–54
    Bacillus licheniformis, 52–53
    Pseudomonas aeruginosa, 51–52
    Shewanella oneidensis MR-1, 54
  cell-to-cell communication, 48
  cyclic dimeric guanosine
    monophosphate, 48–49, 50f
  definition, 47–48
  factors influencing, 49–50, 51t
  formation, 48–49, 49f
  genetic exchange, 48
  microbial nutrient niche, 48
  permeability control, 48
  selective plugging mechanism, 49f
  self-defense system, 47–48
  structure, 50–51
  wettability modification, 48

Biogases, 2, 9
Biogenic gases
  CH_4-producing bacteria, 55–56
  CO_2-producing bacteria, 55
  H_2-producing bacteria, 55
  N_2-producing bacteria, 55
Biological growth and metabolism
  kinetics
  environmental factors
    Arrhenius equation, 111
    association and dissociation
      constants, 112–113
    cardinal pH model, 112
    cardinal temperature model, 112
    decimal reduction time and
      pressure, 113–114
    enzyme activity, 112
    overall reaction, 112–113
    protein denaturation, 111
    salinity effect model, 113
    salinity inhibition constant, 113
    square root method, 111–112
    temperature, 111
  growth rate
    Contois model, 111
    empirical kinetics, 109
    endogenous decay, 109
    exponential form, 110
    fast acceleration and slow
      deceleration model, 110
    linear form, 110
    logistic form, 110
    microbial process model, 109
    Monod model, 111
    Moser and Tessier equations, 111
    overall decay rate and oxidation
      decay rate, 109
    oxidation rate, 109
    rate-limiting substrate, 109
    specific growth rate, 109, 110f
    substrate utilization, 109
  metabolites generation model, 114
Biomass, 2, 4t–5t
Biomass-associated products (BAP),
  114
Biopolymers, 4t–5t
  alginate, 46–47
    chemical and physicochemical
      characteristics, 46
    microbial production, 46
  bioaugmentation, 38–39
  biostimulation, 38–39

Note: Page numbers followed by "f" indicate figures and "t" indicate tables.

Biopolymers (*Continued*)
  condensation reactions, 39, 40f
  curdlan, 39–41
    applications, 40–41
    chemical and physicochemical
      characteristics, 39–40
    microbial production, 39
    structure, 41f
  definition, 39
  dextran, 42–43
    application strategies, 43
    chemical and physicochemical
      characteristics, 42–43, 43f
    microbial production, 42
  extracellular polymeric substances
    (EPS), 47
  levan, 45–46
    applications, 46
    chemical and physicochemical
      characteristics, 45–46
    chemical structure, 46f
    microbial production, 45
  microbial exopolysaccharides (EPSs),
    39
  phosphorylated intermediates, 39,
    41f
  pullulan, 41–42
    applications, 42
    carbon sources, 41–42
    chemical and physiochemical
      characteristics, 42
    microbial production, 41–42
    nitrogen sources, 41–42
  scleroglucan, 43–44
    applications, 44
    chemical and physicochemical
      characteristics, 43–44
    microbial production, 43
    structure, 44f
  selective plugging strategy, 38–39
  xanthan gum, 44–45
    applications, 45
    chemical and physicochemical
      characteristics, 44–45
    microbial production, 44
    structure, 45f
Biostimulation, 38–39
Biosurfactants, 4t–5t
  advantages, 32
  biodegradability, 32
  vs. chemical surfactants, 32–33, 34t
  contact angle modification, 33f
  definition, 32–33
  emulsification capability, 33
  hydrophobic part, 32
  interfacial tension alteration, 33f
  low toxicity, 32
  lyophilic part, 32
  producibility, 32
  production, 33–38, 35t
    Acinetobacter spp, 38
    Arthrobacter spp, 38
    Bacillus species, 36–37, 36f
    Pseudomonas species, 37–38

Biosurfactants (*Continued*)
  Rhodococcus spp, 38
  structural characteristics, 32
  surface tension alteration
    cell concentrations, 9, 10f
    critical micelle concentration
      (CMC), 7–8
    crude biosurfactants, 7–8
    oil recovery, 8–9
    oil-washing experiments, 8f
  water-soluble surfactant, 32

C
Candida tropicalis MTCC230, 82–83
Capillary grain-coating model
  (CGCM), 103
Capillary number, 115–116
Capillary pressure, 70–71
Capillary rise method, 69t
Carbohydrates, 39
Cardinal temperature model, 112
Carman-Kozeny equation, 115
Catenary pores, 85, 85f
Chemical enhanced oil recovery
  (CEOR) approach, 1
Closed pores, 85, 85f
Clostridia, 28
Contact angle hysteresis, 68–70
Contois model, 111
Critical micelle concentration (CMC),
  71, 72f
Cul-de-sac pores, 85, 85f
Curdlan, 6, 39–41
  applications, 40–41
  chemical and physicochemical
    characteristics, 39–40
  microbial production, 39
  structure, 41f
Cyclic microbial recovery (CMR), 169,
  170f

D
Darcy's law, 86, 100–102
Darcy velocity, 116
Denitrification, 55
Denitrifying bacteria, 55
Dextran, 42–43
  application strategies, 43
  chemical and physicochemical
    characteristics, 42–43, 43f
  microbial production, 42
Dex-transucrase, 42
Du Noüy Ring method, 69t

E
Emulsifiers, 2, 4t–5t
Enterobacter cloacae, 81–82
Exopolysaccharides, 6
Exopolysaccharides (EPSs), 39
Ex situ microbial enhanced oil
  recovery (MEOR) process, 1
Extracellular DNA (eDNA), 53–54
Extracellular polymeric substance
  (EPS)

Extracellular polymeric substance (EPS)
  (*Continued*)
  hydrophobicity, 51
  selective plugging strategy, 47
  structure, 50–51

F
Facultative aerobes, 28
Field applications
  Arab reservoir conditions, 184
  Argentina, 184, 186f
  Australia, 182–183, 183f
  Canada, 181
  China, 183, 184f
  classifications, 170–175
  Czechoslovakia, 181
  Eastern European countries, 181
  Germany, 182
  global application, 176t–178t, 179f
  Hungary, 182
  India, 184
  injection strategies
    cyclic microbial recovery (CMR),
      169, 170f
    microbial flooding recovery (MFR),
      169, 171f
  lithology and trial types, 171–173
  Lobitos field, 188f
  Malaysia oil field, 183–184, 185f
  microorganisms, 169, 171t–172t,
    174
  nutrients, 169–170, 174, 175t
  Providencia field, 187f
  recovery mechanisms, 173, 173t
  reservoir environments, 174–175,
    175t
  Romania, 181–182
  Soviet Union, 181
  United States, 180, 181f–182f
  Vizcacheras field, 186f
Fluid saturation, 86
Fracture porosity, 86

G
Gases, 4t–5t
Geothermal temperature, 28
Gigascopic heterogeneity, 90t
Grain-coating capillary model
  (GCCM), 101–102
Grain-coating Kozeny model
  (GCKM), 103

H
Halophiles, 27–28
Hand's rule, 138
Heterogeneous reservoirs
  petrophysical properties, 88–89, 90t
  physical indices, 88–89
  scales, 89f, 90t
  selective plugging
    bioaugmentation, 89
    biomasses, 89
    biopolymers, 89
    biostimulation, 89

Heterogeneous reservoirs (*Continued*)
    clogging mechanism, 89, 91f
    screening criteria, 89–91
    temperature, 89–91
    thief zones, 89
    water flooding, 89
Homogeneous reservoir model, 153f
Huff-and-Puff process, 172–173
Hydrocarbon metabolism, 4t–5t
Hyperthermophiles, 27–28

I
In situ microbial enhanced oil
    recovery (MEOR) process, 1, 2f
Interfacial tension (IFT)
    biosurfactant production
      Candida tropicalis MTCC230,
        82–83
      economics improvement, 84t–85t
      Enterobacter cloacae, 81–82
      Fusarium sp, 82
      rhamnolipid, *Pseudomonas* spp,
        79–81
      Rhodococcus spp, 81, 82f
      surfactin, *Bacillus* spp, 72–79
    capillary pressure, 70–71
    dimensions, 67
    experimental system, 73f
    interfacial force, 67
    liquid-liquid interface, 67, 67f
    measurement method, 69t, 70f
    reduction, 71
    surface wettability
      adhesive force, 67–68
      cohesive force, 67–68
      contact angle hysteresis, 68–70
      multiphase pore fluids, 68
      reservoir rocks, 68
    surfactant efficiency and effectiveness,
      71–72

K
Kaolinites, 11
Kozeny-Carman model, 87–88
Kozeny grain models, 102–103, 102f

L
Langmuir isotherm equation, 115
Levan, 45–46
    applications, 46
    chemical and physicochemical
      characteristics, 45–46
    chemical structure, 46f
    microbial production, 45
Lipopeptides, 9
Lithology
    bacteria cells and nutrient transport,
      12
    cell wall destructions, 12, 14f
    clay swelling, 11–12
    kaolinites, 11
    montmorillonites, 11
    sedimentary rocks, 11
    silicates and carbonates, 11–12

Lithology and trial types, 171–173
Low bond axisymmetric drop shape
    analysis (LBADSA) method, 73

M
Macroscopic heterogeneity, 90t
Maximum bubble pressure method,
    69t
Megascopic heterogeneity, 90t
Mesophiles, 27
Mesoscopic heterogeneity, 90t
Methane-producing bacteria, 16
Methanobacterium
    thermoautotrophicum, 16
Methanogens, 28, 30
Michaelis-Menten kinetics, 114
Microaerophilic aerobes, 28
Microbial enhanced oil recovery
    (MEOR)
    bioacids and solvents, 2, 9
    biogases, 2, 9
    biomass, 2
    biopolymers, 2
    biosurfactants, 2
    degradation/cleanup, 9–11, 11f
      alkane compositions, 11, 13f
      crude oil solubility, 11, 12f
      light oils, 9–11
      oil-degrading bacteria, 11
    emulsifiers, 2
    ex situ method, 1
    field applications. *See* Field
      applications
    interfacial tension (IFT) and
      wettability. *See* Interfacial tension
      (IFT)
    metabolites, 2, 4t–5t
    microbiology and microbial
      products. *See* Microbiology and
      microbial products
    oil emulsification, 3f
    oil viscosity reduction, solvent-and
      acid-producing microorganisms
      Bacillus subtilis, 83
      Clostridium acetobutylicum, 83
      permeability, 83
      pressure support, 83–85
      Thermoanaerobacter ethanolicus,
        83
    reservoir conditions, 1
    reservoir rocks
      permeability, 86
      porosity, 85–86
      porosity-permeability relations,
        86–88, 87f–88f
    selective plugging, 6f
      Bacillus licheniformis BNP29, 3–6
      biomass plugging, 3
      biopolymers, 6
      criteria, 3–6
      curdlan, 6
      emulsions, 6–7
      heterogeneous reservoirs, 88–91
      indigenous microbes, 3

Microbial enhanced oil recovery
    (MEOR) (*Continued*)
      Leuconostoc mesenteroides. See Bio-
        clogging, Leuconostoc
        mesenteroides
      porous media, 3–6
      reservoir zone permeability, 2–3
      xanthan gum, 6
    in situ method, 1, 2f
    subsurface environment. *See* Subsur-
      face environment
    sulfate-reducing bacteria, 1
    surface tension alteration
      biosurfactants, 7–8
      cell concentrations, 9, 10f
      critical micelle concentration
        (CMC), 7–8
      oil-washing experiments, 8f
    water flooding, 2
    wettability alteration, 7
Microbial flooding recovery (MFR),
    169, 171f
Microbiology and microbial products
    acids, 57
    biofilms
      advantages, 47
      architecture, 49–50, 50f
      biofilm-bacterial species, 51–54
      cell-to-cell communication, 48
      cyclic dimeric guanosine
        monophosphate, 48–49, 50f
      definition, 47–48
      factors influencing, 49–50, 51t
      formation, 48–49, 49f
      genetic exchange, 48
      microbial nutrient niche, 48
      permeability control, 48
      selective plugging mechanism, 49f
      self-defense system, 47–48
      structure, 50–51
      wettability modification, 48
    biopolymers
      alginate, 46–47
      dextran, 42–43
      extracellular polymeric substances
        (EPS), 47
      levan, 45–46
      scleroglucan, 43–44
      xanthan gum, 44–45
    biosurfactants
      advantages, 32
      biodegradability, 32
      vs. chemical surfactants, 32–33,
        34t
      contact angle modification, 33f
      definition, 32–33
      emulsification capability, 33
      hydrophobic part, 32
      interfacial tension alteration, 33f
      low toxicity, 32
      lyophilic part, 32
      producibility, 32
      production, 33–38, 35t
      structural characteristics, 32

Microbiology and microbial products (*Continued*)
    water-soluble surfactant, 32
  microbial ecology, deep subsurface
    microorganisms metabolite, 27
    oxygen, 28
    pH, 27
    salinity, 27–28
    temperature, 27
  reservoir environments
    bacteria and archaea, 29t–32t
    indigenous microbial communities, 28–30
    living microorganism metabolites, 28
    methanogens, 30
    pH, 30–31, 31t
    salinity, 31–32, 32t
    temperature, 28–30, 29t–30t
  solvents, 56–57
Microscopic heterogeneity, 90t
Monod degradation equations, 114
Monod model, 111
Montmorillonites, 11
Moser and Tessier equations, 111

N
Nitrate-reducing bacteria (NRB), 169
Nitrate-reducing bacteria (SRB), 169
Nonwetting fluid, 68
Normalized capillary desaturation curve, 116f
Numerical simulation
  CO_2 generation and bacteria concentration, 144, 158f
  interfacial tension reduction
    absolute permeability and injection rate, 131–134, 148f
    bacteria adsorption, 153, 162f
    bacterial growth, 137–138, 151f
    biosurfactant adsorption, 151f
    biosurfactant-based MEOR simulations, 138
    biosurfactant concentration, 153f
    bottom hole flowing pressure (BHFP) responses, 163f, 164
    cold-water MEOR, 140
    cumulative oil recovery, 164–165, 166f
    experimental and simulated bacterial growth, 164f
    five-spot pattern flooding, 164–165
    homogeneous reservoir model, 153f
    incubation time, 131–134, 147f
    indigenous and injected bacteria ratio, 131–134, 146f
    Langmuir type isotherm model, 138–140
    metabolite productions, 160f
    microbial and surfactant concentrations, 149f
    numerical model, 134–137

Numerical simulation (*Continued*)
    nutrient concentration, 140–142
    nutrient slug size, 131–134, 146f
    oil recovery and injection parameters, 154f
    one-dimensional model, 131
    optimization process, 140–142, 157f
    permeability distribution, 150f
    pore volume, 138
    properties changes, 160f
    radial model, 161f
    relative permeability curves, 131
    reservoir model, 131, 143f, 165f
    residual oil saturation, 147f
    salinity, 138, 149f
    simulation results, 152f
    surfactant phase behavior model, 138
    sweep efficiency, 161f
    temperature effect, 140, 155f–156f
    three-dimensional model, 138–140
    viscosity results, 164, 165f
    water flooding, oil recovery, 131, 143f
  mathematical model, 144
  mechanisms, 142–144
  oil viscosity and bacteria concentration, 144, 157f–158f
  permeability alteration
    aqueous and solid phases, 128
    bacteria growth, 126
    bacteria injection, 118f–119f
    bacterial plugging, 117–118
    bacteria transport, 117
    batch model simulation results, 129, 132f, 139f
    biofilm evolution and removal model, 124–125, 126f
    biokinetics, 123
    biomass distribution, 123–124, 125f
    chemotaxis effect, 119, 120f
    clogging and declogging rates, 119
    experimental results, 121, 127f
    flow equations, 117
    fractional flow, 128–129, 131f
    hydrodynamic alteration, 128–129, 130f
    injection strategy and oil recovery, 141f
    King Island gas field, 127–128, 129f
    mathematical models, 119–122
    mechanistic modeling approach, 129
    microbial models, 122–123
    microbial profiles, 119, 120f
    network model, 125
    oil recovery factors, 129–130, 134f–135f
    one-dimensional analysis, 117–118
    permeability reduction factor (PRF), 123

Numerical simulation (*Continued*)
    plugging types, 122, 123f
    porosity-permeability correlation, 121–122
    reactive transport models, 125–128
    realistic model application, 142f
    recovery process, 130–131
    sand-pack simulation results, 129, 133f
    selective plugging, 126–127, 128f
    sensitivity analysis and optimization, 129–130, 137f
    shear stress, 125
    synthetic reservoir, 129–130, 134f
    temperature effect, microbial biokinetics, 130, 138f
    three-dimensional case, 118
    two-dimensional model, 130–131, 140f
    water diversion effect, 129–130, 136f
    water-oil ratios, 119, 121f
  radial model, 153
Nutrients, 169–170

O
Obligate aerobes, 28
Obligate anaerobes, 28
Oil-brine interfacial tension (IFT), 71
Oil emulsification, 2, 3f
Oil viscosity reduction, solvent-and acid-producing microorganisms
  Bacillus subtilis, 83
  Clostridium acetobutylicum, 83
  permeability, 83
  pressure support, 83–85
  Thermoanaerobacter ethanolicus, 83
Oil-washing experiments, 8f

P
Parallel capillary tube model, 100–102
Pendant drop method, 69t
Permeability reduction factor (PRF), 123, 124f
pH
  microbial ecology, deep subsurface, 27
  reservoir environments, 30–31, 31t
Polypeptides, 39
Polysaccharide-type biopolymers, 39
Pore-filling capillary model (PFCM), 102
Pore-filling Kozeny model (PFKM), 103–104
Pseudomonas aeruginosa
  biofilm production, 51–52
  elastic modulus and yield stress, 51–52
  PAO1 and PANO67 strains, 51
  rhamnolipid production, 51, 52f
    AP02-1 strain, 81
    stability, 79–80

Pseudomonas aeruginosa (*Continued*)
   surface tension and emulsification
      index measurements, 79–80,
      80f
Pseudomonas stutzeri, rhamnolipid
   production, 81
Psychrophiles, 27
Pullulan, 41–42
   applications, 42
   carbon sources, 41–42
   chemical and physiochemical
      characteristics, 42
   microbial production, 41–42
   nitrogen sources, 41–42

R
Receding contact angle, 68–70
Reservoir rocks
   permeability
      definition, 85
      dimension, 86
      flow rate, 86
   porosity
      definition, 85
      fluid saturation, 86
      morphological pore types, 85, 85f
      pore types, 85, 85f
      porous media, 85
      primary pores, 86, 86t
      secondary porosity, 86
   porosity and permeability relation,
      86–88
Residual saturation, 116
Rhamnolipid, 9
   Pseudomonas spp, 79–81
   pseudomonas stutzeri, 81

S
Salinity
   microbial ecology, deep subsurface,
      27–28
   reservoir environments, 31–32, 32t
Salinity effect model, 113
Scleroglucan, 43–44
   applications, 44
   chemical and physicochemical
      characteristics, 43–44
   microbial production, 43
   structure, 44f
Secondary porosity, 86
Sedimentary rocks, 11
Selective plugging, 6f
   Bacillus licheniformis BNP29, 3–6
   biofilms, 49f
   biomass plugging, 3
   biopolymers, 6, 38–39
   criteria, 3–6
   curdlan, 6
   emulsions, 6–7

Selective plugging (*Continued*)
   extracellular polymeric substance
      (EPS), 47
   heterogeneous reservoirs, 88–91
      bioaugmentation, 89
      biomasses, 89
      biopolymers, 89
      biostimulation, 89
      clogging mechanism, 89, 91f
      screening criteria, 89–91
      temperature, 89–91
      thief zones, 89
      water flooding, 89
   indigenous microbes, 3
   Leuconostoc mesenteroides. *See* Bio-
      clogging, Leuconostoc
      mesenteroides
   permeability alteration, 126–127,
      128f
   porous media, 3–6
   reservoir zone permeability, 2–3
   xanthan gum, 6
Sessile drop method, 73–74
Shewanella oneidensis MR-1, 54
Simulation. *See also* Numerical
   simulation
   interfacial tension reduction
      capillary number, 115–116
      Darcy velocity, 116
      interstitial velocity, 116
      residual saturation, 116
      solubilization parameter, 115
   permeability alteration, 115
   relative permeability curve
      end-point relative permeability,
         117
      relative permeability function, 116
      residual saturation, 116–117
      two-phase flow system, 116–117
      water and oil, 117f
Sloughing, 94
Soluble microbial products (SMPs),
   114
Solution-induced porosity, 86
Solvents, 2, 4t–5t
Spinning drop method, 69t
Substrate-utilization-associated
   products (UAP), 114
Subsurface environment
   porosity and permeability
      frequency diameter, 12–14
      microbial growth and metabolism,
         14
      pore geometry factor, 12–14
   pressure
      Bacillus stearothermophilus, 16–17
      bio-physical conditions, 16–17
      ocean floor, 16
      Streptococcus faecalis variant, 16–17

Subsurface environment (*Continued*)
   reservoir depth, 14
   reservoir screening criteria, 21t
      permeability and pore size, 20
      pH, 20
      pressure, 20
      salinity, 20
      temperature, 19–20
   salinity/pH
      Bacillus species, 18
      bacterial growth, 18–19
      Clostridium acetobutylicum
         metabolism, 19
      halophiles, 18
      heavy metals, 18–19
      Leuconostoc Mesenteroides
         metabolites, 19f
      moderate halophiles, 18
      NaCl concentration, polymer
         production, 17–18, 18f
   temperature
      Galapagos vents, 16
      geothermal gradients, 15
      hydrothermal vents, 16
      methane-producing bacteria, 16
      thermal springs and fumaroles,
         15–16
      thermophiles, 16
      thermophilic bacteria, 15–16
Sulfate-reducing bacteria (SRB), 169
Surface tension, 67

T
Thermoanaerobacter ethanolicus, 83
Thermophiles, 27
Thermoproteus tenax, 16
Tortuosity, 87–88, 102–103
Trinidad and Tobago Oil Company
   Limited (TRINTOC), 184
Two-dimensional hypothetical
   reservoir model, 140f

W
Wettability alteration, 7, 7f
Wetting fluid, 68
Wilhelmy plate method, 69t

X
Xanthan gum, 6, 44–45
   applications, 45
   chemical and physicochemical
      characteristics, 44–45
   microbial production, 44
   structure, 45f
Xanthomonas campestris, 44

Z
Zymomonas mobilis, 45